Human Factors in Air Traffic Control

Human Factors in
Air Traffic Control

V. David Hopkin

Human Factors Consultant
Fleet, Hants

Taylor & Francis
Publishers since 1798

UK   Taylor & Francis Ltd, 4 John St, London WC1N 2ET
USA   Taylor & Francis Inc., 1900 Frost Road, Suite 101, Bristol, PA 19007

British Library Cataloguing in Publication Data
A catalogue record for this book is available from the British Library

ISBN 0 85066 823 9 (cased)
        0 7484 0357 4 (paperback)

Library of Congress Cataloging in Publication Data are available

Cover design by Amanda Barragry

Set by Santype International Ltd, Salisbury, Wilts

Printed in Great Britain by Burgess Science Press, Basingstoke, on paper which has a specified pH value on final paper manufacture of not less than 7.5 and is therefore 'acid free'.

To Betty

Contents

List of figures

† These figures are in the colour plate section between pages 238 and 239.

Preface

This text is an attempt to describe every application of human factors as a discipline to air traffic control. As such it cannot succeed fully, but the reasons why such an aim is intrinsically over-ambitious are instructive. The boundaries of the subject matter of human factors are vague but expanding, and the legitimacy of some human factors applications near or beyond those boundaries is disputed. Each reader will deplore the omission of topics that should have been included or the inclusion of topics that should have been omitted, for there is no complete consensus on what the subject matter should cover. In fact, much of the structure of the subject matter is quite old and recognizable from previous texts (Hopkin, 1970; 1982a), and some themes can be traced to Fitts (1951a). In the interim, hardly any topics have been dropped but new ones are being added continually. I have tried to mention them, and a few pending ones.

Any text has to partition its contents under headings, and this text does so as a matter of expediency. Yet the partitioning of the subject matter of human factors in this way is inherently foreign to its manner of application, and is misleading if it seems to imply that real human factors problems arise and remain confined within simple headings. Practical experience teaches that the human factors problems that occur in real air traffic control are never as neat.

The text has a single author. This may have some advantages, such as a consistent voice, a unitary approach, and quite a complete structure. It has the disadvantage that my own biases imbue everything and are unredressed, so that some will discern a lack of balance. I am aware that my bias towards captious comments is insufficiently balanced by mellower comments where praise is due, but human factors, as a discipline that purports to understand human beings better than most, cannot boast about its record of self-promotion. In fact, much pioneering human factors work has been done on air traffic control systems and has been successful though not well publicized.

A word is in order about the cited references. Because the text has a wide coverage, it cannot deal with any particular topic in great detail without exaggerating its relative significance. I have tried to prefer references from obtainable sources, and to provide references that will allow the reader to pursue topics further than I do. Many of the cited references contain extensive bibliographies. Throughout the world, there must by now have been thousands of studies related to human factors aspects of air traffic control. A few of the best-known centres of research, listed on pp. 443–5, have each issued numerous reports of their work which has often included major human factors contributions, but these reports were never intended to be of interest beyond the system to which they were applied and most of them are now archived. The few mentioned in this text are either recent, cover topics not otherwise dealt with, or are of interest for another reason such as methodology. For many years, I participated in the programme of the Air Traffic Control Evaluation Unit in the United Kingdom, and I am a joint author of about 50 reports issued by that Unit or through the United Kingdom Civil Aviation Authority. None of them is cited directly in this text, but this practical experience has influenced my views.

The concept of "human factors" can cause grammatical difficulties. I have treated it as a singular noun to denote a discipline, and also adjectivally. Except for a few of the first references to it as a discipline, I have not used initial capitals. In accordance with current conventions, I have eschewed sexist language, except in some historical references to man–machine systems and the allocation of functions to man, where I have retained the original wording.

The practice of human factors in air traffic control is inherently a team activity in an interdisciplinary work environment. I therefore owe a great debt to many colleagues over the years, far too many to name. They have contributed to any merits of this text, but all its deficiencies are mine alone. The kindness of colleagues is not fittingly repaid by attributing to them views they may not share.

I have received much help from my family, particularly from my wife to whom the book is dedicated, from my daughters-in-law Anne and Samantha, and from my eldest son Anthony who compiled and drew Figures 16 and 17. My thanks also to the UK Civil Aviation Authority for Figures 2, 5, 7, 8, 10, 11, 13, 26, 27 and 28, and to the UK Ministry of Defence for Figures 1, 3, 4, 6, 9, 12, 14, 15 and 18–25 inclusive.

V. David Hopkin
Church Crookham
Hampshire
UK

1

Historical introduction

1.1 Human factors

The origins of Human Factors as a profession with its own domain of expertise are not all self-evident. Various influences have fostered its emergence as
an independent discipline and contributed to its consolidation, but the gifts of
hindsight are needed to retrace them. It was not apparent beforehand that
these influences would coalesce into the single discipline called Human Factors,
any more than it is apparent now what the long-term future of Human Factors
as a discipline will be. In retrospect, the following themes deserve some
mention as formative influences on Human Factors.

Human Factors can be subsumed under Applied Psychology, particularly if
Applied Psychology is interpreted, according to the distinction drawn by
Hearnshaw (1987), as the extension of psychological methods to real problems
rather than the extrapolation of laboratory findings to real-life settings. Early
texts on Industrial Psychology (e.g. Myers, 1929) heralded many topics that
have become familiar human factors themes, although their connotations have
changed. Industrial Psychology evolved into Occupational Psychology as its
influence spread to non-industrial jobs, and a broader range of factors was
accepted as relevant. Sometimes Industrial Psychology and Organizational
Psychology are equated (Reber, 1985), although the latter now extends beyond
industrial and non-industrial jobs to embrace a diversity of social structures
(Bradley and Hendrick, 1994). Ergonomics, evolving contemporarily with
Human Factors, also recognized the mutual interactions of humans and
equipment within the workspace, but usually viewed them from a more interdisciplinary perspective (Murrell, 1965).

A different kind of influence on Human Factors was the treatment of
humans and machines as comparable components of a human–machine
system. The principles of scientific management enunciated by Taylor (1911)
and the study of human movement patterns at work (Gilbreth, 1919) which
developed into time and motion study (Barnes, 1958) exemplified this
approach, with its quest to improve productivity and motivation and its development of techniques for job analysis and design. Similar measures were

applied to describe humans and machines. The machines might be modified to increase the pace, smoothness and efficiency of human movements, but not solely to satisfy human aspirations. However, some individual human characteristics, associated for example with skill, fatigue or errors, were represented in Craik's (1947) pioneering theory of the human as an engineering system, which was in tune with the beginnings of information theory, cybernetics and computer technology (Turing, 1950). The classification and allocation of functions as more suitable for humans or machines, which was originally formulated for application in air traffic control (Fitts, 1951a), has endured far longer than its originators advocated, for they foresaw that technological advances could invalidate rigid function allocation. They did not recognize such inherent limitations of the approach as its competitiveness (Jordan, 1968), its exclusion of some kinds of human–machine relationship (Hopkin, 1982a), and its over-emphasis on those human functions with a feasible machine equivalent.

A further influence came from the performance of the human tasks that resulted from technological innovations in equipment, especially within large systems. The main impetus originated during the Second World War, particularly with radar displays and vigilance tasks (Mackworth, 1950), when human rather than equipment limitations seemed to determine what was achievable. The extensive early psychological research on equipment design sponsored by the United States Armed Forces was initially labelled Engineering Psychology or Human Engineering (McCormick, 1957), then was called Human Factors Engineering, and eventually became designated as Human Factors, partly because of the eponymous journal. Originally, the main purpose of Human Factors was to aid design: the Tufts College handbook of Human Engineering Data (1949) and Woodson's (1954) text were for designers to use, and Sinaiko's (1961) collection of papers emphasized design. Fitts' chapter about equipment design in Stevens' definitive *Handbook of Experimental Psychology* (Fitts, 1951b) testifies alike to the already established role within experimental psychology for Human Factors, and to the confusion over what it should be called. Perhaps it was applied experimental psychology (Chapanis *et al.*, 1949). Human Factors became concerned extensively with the development and evaluation of the human contributions to the functioning of large systems, expressed in systems terms (Meister and Rabideau, 1965). Many studies of large systems as functioning entities were conducted, including studies of air traffic control systems, yet it often proved difficult to trace whether the findings produced by such major efforts and lavish resources had any significant lasting influence at all on the system design or functioning (Parsons, 1972), and studies of that magnitude became less common.

A more human-centred influence on Human Factors concerned the attributes of work that could satisfy human needs and aspirations. Job satisfaction and job enrichment, the quality of working life, and work as a social activity were studied, and theories propounded (Herzberg, 1966; Maslow, 1976). Morale, degree of autonomy, status, responsibilities, self-esteem, and the esteem of colleagues seemed relevant. The significance of team roles, ethos, and professional norms and standards was sometimes conceded grudgingly, when more traditional and orthodox influences had failed to account adequately for what actually occurred. While the motives for acknowledging their significance were occasionally humanitarian, they were more commonly associated with the unacceptable costs of high staff turnover rates, strike-prone workforces and poorly motivated staff. Users' attitudes, as well as technical efficiency, could be crucial to the successful implementation of changes, although their willing adoption of new equipment did not guarantee that they would actually use it in accordance with the designers' intentions.

Another influence was the study of individual differences, generally for the purpose of curtailing them as a source of unwanted variance. Selection procedures sought to choose from a wider population a more homogeneous group possessing identified measurable attributes. Training procedures then instilled common knowledge, skills and practices so that all individuals performed tasks similarly. The allocation of people to their initial jobs, and subsequent career development procedures, acknowledged that some residual individual differences remained, and attempted to utilize them by matching individuals with job requirements, thereby reducing further the individual differences between people doing the same job.

Operators often had little control over the task demands imposed on them by the system. The effects of systems on the operators who worked within them constituted another seminal influence on Human Factors. These effects, whether couched in such system concepts as efficiency, safety and system integrity, or in such human concepts as stress, boredom, health and well-being, were mediated by general differences between categories of people, as in their age or their experience, and by differences between individuals, as in their capabilities, adaptability or tolerance.

A final kind of influence on Human Factors dealt with organizational settings and contexts. Broad examples include conditions of employment, management practices, legislative requirements and organizational structures. More specific aspects are work–rest cycles, policies on supervision and assistance, the provision of supporting services, opportunities for mutual help, and physical and environmental features of the workspace.

Gradually Human Factors as a discipline has evolved to embrace all the above influences, to permit them to interact, and to absorb further influences as they arise. Recent additional influences have come from computer sciences and artificial intelligence, from cognitive theories, from changed social expectations about work and its rewards, from cultural ergonomics, and from repercussions on legal responsibility or financial accountability.

1.2 Air traffic control

When commercial aviation began, aircrew lacked ground-based navigation aids. To avoid becoming lost, they relied on direct observation, their own deductions and simple instruments (Williams, 1990; Kendal, 1990). Only in the immediate vicinity of airports was there any assistance, as the experiences of pioneer aviators make clear (Saint-Exupéry, 1939). The Second World War, with its requirements for air operations in poor visibility and at night, led to rudimentary air traffic control based on procedural principles similar to some of those still followed beyond radar coverage. The safe, orderly and expeditious flow of air traffic became established as the objective of air traffic control. Safety was paramount. Aircraft receiving an air traffic control service would remain safely separated from each other at all times, and would not collide with the ground or with other obstacles. Their pilots would never become lost, but would be guided to their destinations and warned while airborne of hazards such as severe weather.

Towards the end of the Second World War, the Convention on International Civil Aviation, subsequently dubbed the Chicago Convention, led to the formulation of agreed standard practices as a prerequisite for international air traffic control, and to the founding of the International Civil Aviation Organization (ICAO) as the regulatory body for the specification and implementation of common air traffic control procedures and practices and for compliance with them. The need for international agreements and collaboration was thus acknowledged from the early days of air traffic control. It predates most technological advances and the vast expansion of air traffic and air users. A judicious balance between national sovereignty and the international regulation of air traffic was struck, taking account of national differences in geography, traffic demands, and the political and financial priorities accorded to air traffic control. At the end of the Second World War, aeronautical, technical and navigational advances had made long flights feasible and reliable, aircraft construction had become a major industry, airfields that could handle many large aircraft had been built, and plenty of experienced pilots were keen to continue flying.

The greatest initial demand for commercial flights was between main centres of population, and the most direct routes were naturally preferred. Many of these routes were eventually marked by ground-based beacons emitting signals that could be sensed by aircraft flying along the route. Separate and approximately parallel routes might be allocated to aircraft flying in opposite directions. Alternatively, minimum permissible lateral separations were applied between opposite direction traffic within the same route, airway or air corridor. Aircraft in level flight were separated also by height, with different flight levels for different types of aircraft and for traffic flying in different directions. Minimum longitudinal separations between consecutive aircraft at the same flight level on the same route could be expressed as distances or times, with some provision for faster aircraft to overtake slower ones, subject to an adequate lateral or height separation between them. Safe separation between aircraft involves the three spatial dimensions and time. The actual magnitudes of the separations required for safety are not universal but depend on the nature and quality of the navigational information.

When the advent of radar improved the quality of that information, the minimum lateral and longitudinal separations between aircraft could safely be reduced, so that the controller could handle more air traffic within a given airspace while complying with the new standards. Beyond radar coverage, for example in oceanic regions, very large separations between routes and between consecutive aircraft at the same height on the same route still had to be maintained. However, the information about consecutive aircraft on final approach to the same runway at an airport might be of such high quality that the minimum separation between them was no longer determined by it but by regulations that required a following aircraft to be sufficiently separated the aircraft ahead of it not to be put at hazard by wake vortices and t lence.

As the number of aircraft and the demand for air traffic control increased, so did the workload of the controller. The concept of the sec applied to the region of airspace, defined by geographical and height aries, within which one controller or one team of controllers was re for providing the air traffic control service. For a time, further increases demand could be accommodated by reducing the size of sectors. However, the handover of responsibility for the control of each aircraft as it left one sector and entered the next imposed communications workload on both the controllers and on the pilot. At some point, further partitioning of sectors becomes counterproductive as a response to increased traffic demands because of the extra coordination and liaison tasks introduced by sectorization. Other solutions have to be devised that enable the controller to maintain safe separations between aircraft and provide an efficient air traffic control service without becoming overburdened.

Several developments have seemed helpful. One is the improvement in navigational information, expected to continue with the future provision of data derived from satellites. Another is the automation of tasks, particularly those associated with the routine gathering, assimilation, collation, updating and discarding of information. A further development is the provision of computer assistance for decision making, for problem solving and for prediction, often by presenting ready-made solutions for the controller to accept, reject or modify, instead of requiring the controller to fulfil these functions unaided. A current trend is the evolution from tactical to strategic air traffic control, associated with planning in advance to prevent problems from arising rather than resolving problems that have already arisen, and with organizing air traffic into flows rather than dealing with aircraft singly.

The extent of direct human involvement in air traffic control is a further issue. Technological and computer advances, particularly in software, continually extend the feasible options. Fewer technological limitations now impose inappropriate roles on humans simply because no machine can fulfil them. Many air traffic control functions could in principle be wholly automatic or wholly manual, or could employ any intermediate stage of automated assistance. A broader range of questions about suitable future roles for controllers

and pilots can be posed because the technological means to implement alternative policies on preferred human and machine roles are becoming available. Technology has also revived the issue of how far air traffic control should remain exclusively ground-based, when sensors on board aircraft can detect other nearby aircraft and direct how to avoid them. Such a tactical aid is not easily reconciled with the evolution of air traffic control towards the longer-term planning of traffic flows, which presumes that single aircraft will not manoeuvre unexpectedly at short notice.

Developments in navigation, in communications, in computer technology and software, in system planning and strategic control techniques, in artificial intelligence, in human–machine interface design, and in many other technical fields combine to offer a wealth of options for future air traffic control. Few of these developments are exclusively for air traffic control. For most of them, air traffic control is one of many possible applications. Some technological innovations may have no air traffic control applications. Others may seem feasible yet bring no major benefits. A few may lead to spectacular improvements. Almost all will require considerable adaptation before they can benefit air traffic control fully.

Air traffic control is so complex that any innovation can be expected to introduce both advantages and disadvantages, so that its adoption must depend on a favourable balance between them. It is therefore vital to identify all of them before any changes are made. Safety takes precedence. Any lapses make headlines, a reflection alike of their potentially catastrophic consequences and of their current rarity and resultant newsworthiness. Air traffic control must evolve to cope with future traffic, but its safety record must be maintained and preferably enhanced.

1.3 Human factors in air traffic control

The emergence of Human Factors as a discipline cannot be traced to any single event, but the application of Human Factors to air traffic control can be. Although a few small studies related to air traffic control preceded it (Hopkin, 1970), the report by Fitts and his colleagues (1951a), which formulated a long-term programme of human factors research on air traffic control, became a classic reference, and was widely cited for many years even by those with no interest in air traffic control. Fitts deliberately proposed research capable of yielding findings that could be applied broadly and would not be invalidated by relatively trivial system changes. As a result, his report has been applied beyond air traffic control to many large human–machine systems, and his principles and recommendations, including the assignment of roles to men and to machines that gained the sobriquet "Fitts' list", have been treated as universals, though not by Fitts himself who was much more circumspect. With some minor changes of terminology but not of substance, almost all the research objectives in Fitts' report are still valid, a testament to its prescience and perspicacity in discerning key issues and to the universality of many human factors problems within large human–machine systems.

Fitts advocated a systems approach, but it was not as narrow as the approach actually adopted in air traffic control (Parsons, 1972) and elsewhere (Meister and Rabideau, 1965), although the need to consider broader influences was recognized and voiced (Sinaiko and Buckley, 1957). In all these early human factors studies on air traffic control, the main emphases were on the performance of tasks and the selection of controllers. The limitations and underlying assumptions of their mechanistic approach were sometimes recognized, but even criticisms of the approach as "Procrustean" (Taylor and Garvey, 1959) achieved few modifications of it. Nevertheless much valuable and durable work on such topics as workspaces and task designs was accomplished. A series of formal inquiries into the origins of widespread disquietude and grievances among air traffic controllers, particularly in the United States in the 1970s, exposed a gulf between the real human factors issues deemed by controllers to be important and the issues actually addressed during systems design or in contemporary research. Human factors studies of air traffic control paid insufficient heed to preferences or individual differences, tended to presume that every question must have a single optimum answer (Hopkin, 1970), and lacked subtlety in their treatment and classification of controllers' skills (Older and Cameron, 1972).

Over 20 years ago, the human factors literature on air traffic control had already become diffuse and rather inaccessible (Hopkin, 1970), and no comprehensive bibliography of it existed. Despite modern storage and retrieval systems, this remains true and for the same reasons. Much relevant work is never published or widely disseminated or else appears in obscure sources. While the results of simplistic laboratory studies of dubious relevance to real life were being applied, the findings from major air traffic control systems experiments that employed prodigal facilities and resources languished, even after Parsons' (1972) description and critique of them.

Human factors studies of air traffic control form two main categories. Some studies belong to continuous work programmes extending over many years, which utilize dedicated air traffic control simulation facilities and in-house resources or employ contractors under the auspices of national or international agencies. Other studies apply the known relevant expertise of a contractor or academic department to a particular air traffic control problem for a short time, without becoming involved in wider air traffic control issues. This dichotomy has characterized the human factors studies on air traffic control in most nations, including the United States, which potentially has the largest resources for such studies. The influence of psychology on these air traffic control studies has been on their measures and methodologies and their reliance on experimental data, rather than on the application of psychological theories and constructs to the studies themselves or to their interpretation. The conclusion of a retrospective appraisal of the psychological influences exemplified by these studies might be that they constituted a necessary but not a sufficient approach to answer human factors questions in air traffic control.

Many of the simpler problems that are most amenable to orthodox experimentation and measurement can be solved at once by applying to them the substantial body of valid data already available. Examples include human factors recommendations on physical environments, the design of workspaces, task analysis procedures, the legibility and coding of displays, and data input devices. Many of the more difficult problems that are less amenable to orthodox methods remain substantially unresolved. Examples include the roles and influence of human needs and aspirations, the causes and consequences of boredom, the optimum mutual adaptability between human and machine, and criteria to define the maximum attainable human task performance. Human Factors in air traffic control is evolving slowly from a mechanistic to a more socio-technical approach. Increased acknowledgement of the importance of human factors influences on the efficiency and safety of air traffic control may partly explain the current modest expansion in several nations of the human factors resources devoted to it. The associated debates on how any additional resources should be deployed reveal a need for better means to cost and validate the efficacy of human factors contributions to air traffic control, so that limited resources can be applied most effectively and never wasted on trying to improve the unimprovable.

Compared with the human factors resources devoted to aircrew and cockpits, those for air traffic control have always been meagre. Some kinds of study conducted in aircraft cockpits, for example on team roles and interactions (Foushee and Helmreich, 1988), have been rare in air traffic control, though they are needed. Procedures to convert the human factors recommendations from accident and incident investigations into appropriate design modifications are still being extended to air traffic control. However, air traffic control may have benefited from the experience of others in adopting a cautious approach to automated aids. Comparatively recently, a place has been found for Human Factors in air traffic control within reviews of air traffic control (Pozesky, 1989), in human factors texts on aviation (Wiener and Nagel, 1988), in compendia on aviation psychology (Jensen, 1989), in human factors handbooks (Salvendy, 1987), and in aviation human factors digests (ICAO, 1993).

To resolve the human factors problems in air traffic control most effectively and economically it is prudent to discover whether each problem is specific to air traffic control, reappears in some broader context such as aviation, or is endemic in other large human–machine systems and organizational structures such as military systems (Gal and Mangelsdorff, 1991). The degree of generality determines the ways in which each problem can be studied and the range of applicable evidence. Air traffic control resources may have to be used for any problems that do not recur elsewhere. Partitioning the subject-matter of Human Factors is an unavoidable expedient in a text of this kind, but belies the interactive nature of Human Factors and is potentially misleading because real-life problems are never confined within neat headings. Failure to recognize all the influences on them may produce inadequate solutions and encourage their formulation in restrictive terms that exclude crucial aspects of them.

Many organizations have at some time sponsored or reported human factors work on air traffic control. A surprisingly broad range of disciplines and professional bodies have published accounts of such work. Many official bodies concerned with air traffic control research have devoted resources to its human factors aspects. These include governmental and international agencies, professional air traffic control groups and associations, equipment manufacturers, training and teaching establishments, industrial organizations, academic departments and professional institutions. In several West European nations and in the United States, there are dedicated air traffic control research facilities that issue in-house reports of their continuous research programmes which include human factors aspects of air traffic control. Some of them are listed on pp. 443–5. Human Factors in relation to air traffic control has received particular attention in NATO activities, through a series of publications related to it (Baker, 1962; Hopkin, 1970, 1982a), a major text on air traffic control that includes human factors chapters (Benoit, 1975), a series of air traffic control conferences with papers on Human Factors (Benoit, 1973, 1980, 1983, 1986; Benoit and Israel, 1976) and specific air traffic control meetings that have addressed human factors issues comprehensively (e.g. Wise *et al.*, 1991, 1993).

The international nature of air traffic control, the universal demands for handling more traffic, the worldwide marketing of technological advances, and the quest for effective uses of automation have combined to produce similar human factors problems in air traffic control in many countries, with a consequent need to coordinate human factors efforts internationally to avoid duplication. Recent increases in collaboration, in joint studies and in the exchange of ideas should encourage the more fruitful deployment of scarce resources.

When air traffic control equipment is highly reliable, a majority of the errors and failures that do occur become attributable to human inadequacy and fallibility. The attribution may be direct when an individual controller acts inappropriately, or indirect when human planners and system designers have inadvertently built into the system intrinsic sources of human error originating from interactions among factors. Greater understanding of the causes of human errors and failures has expanded the role of Human Factors in their diagnosis and prevention.

Future air traffic control systems will incorporate more technology, computing, automated assistance, and strategic methods, yet human air traffic controllers will remain for the foreseeable future. They will continue to need selection and training, and opportunities to use and hence maintain their knowledge, experience and skills. They will identify with their profession, close ranks if it is threatened, and seek some challenge and interest in their work. They should take pride in what they achieve, have high morale, and encourage

others to join their profession because it offers good jobs, conditions and prospects. They must have confidence in the safety and efficiency of their
equipment, information sources, procedures and instructions. Recent plans
have begun to recognize the full range of human factors issues in air traffic
control (FAA, 1990; Pitts *et al.*, 1993).

In one sense the objective of human factors contributions to air traffic
control is the same as that of air traffic control itself, namely the safe, orderly
and expeditious flow of air traffic. A secondary but essential human factors
objective is to provide air traffic controllers with satisfying and worthwhile
jobs, and tasks that are well matched with human skills and abilities and
performed by a competent and well-motivated workforce. Failure to achieve
this secondary objective would undermine the achievement of the main one.

2

The air traffic control system

2.1 Principles of air traffic control

If an air traffic control system is to achieve its traditional objective of ensuring the safe, orderly and expeditious flow of air traffic, the controller must have access to the following kinds of information about all current and pending air traffic, and be able to fulfil the following broad objectives:

1. The identity of every aircraft must be known positively, so that none is mistaken for another, and instructions to the pilot of one aircraft are not implemented by the pilot of another.
2. The controller must know the performance and manoeuvring capabilities of each aircraft type, such as its maximum flight level and rate of climb, and all air traffic control instructions must conform with these capabilities because the principles for controlling aircraft as traffic must accord with their characteristics as vehicles.
3. The route and current position of each aircraft, the flight level, speed and heading of each aircraft that is in a constant state, and the changes of state of aircraft that are turning, climbing, descending, accelerating or decelerating, must all be known.
4. There must be a means of communication between the air traffic control system and each aircraft, usually including speech between pilots and controllers.
5. There must be sufficient evidence about the position of each aircraft in relation to others for the controller to ensure that every aircraft always remains safely separated from all others.
6. There must be standardized methods, procedures, instructions, and message formats and contents, with rules about when and where air traffic control regulations apply.
7. The depiction of the information about each aircraft must allow it to be related to the corresponding information about other aircraft under air traffic control at the same time.
8. It must be possible to hand over the air traffic control responsibility for an

aircraft safely from one controller to another, in a way that is unambiguous to both controllers and to the pilot.

9. An official record of air traffic control executive actions and their consequences is essential for each flight, as proof of its occurrence and to facilitate retrospective enquiries about it.

In addition to manuals used to teach air traffic control professionally, there are courses to coach applicants for professional training as controllers (e.g. Turner, 1990), and introductions to air traffic control as a career (e.g. Luffsey, 1990). The published texts on air traffic control are for a variety of intended readerships. The best-known early text was by Gilbert (1973). Adair (1985) provides an illustrated and comprehensive but succinct text on British air traffic control for the general reader, which is also suitable as an introduction to air traffic control for human factors specialists. The more recent text by Graves (1992) is less comprehensive, giving broad principles briefly and more detail about the air traffic control at specific airports, and may be more suitable for the traveller than for the student. Raylor (1993) provides a recent simple introduction. The text by Duke (1986) for aviation enthusiasts includes basic principles for the air traffic control user. Buck (1984) puts more emphasis on the vocabulary of air traffic control. The guide by Brenlove (1987), written from the pilot and controller viewpoints, emphasizes United States rather than European systems, and general aviation rather than commercial air traffic. Nolan (1990) covers air traffic control in the United States comprehensively by presenting at college level similar material to that of the air traffic control manuals, and explains well why air traffic control functions as it does. Field's earlier text (1980) provides an introduction to air traffic control, and his later more comprehensive one (Field, 1985) presents an international perspective but draws most examples from United Kingdom or European airspace. From time to time, many non-specialist magazines publish illustrated articles about air traffic control which are well prepared and informative but often over-dramatic.

To people watching an air display, aircraft seem to be moving very fast though most are flying well below their normal cruising speed. The passenger in an aircraft cruising at high altitude has little impression of its speed, except near cloud. Such an aircraft requires a long distance to complete quite small changes of state. The only way that an aircraft can halt in relation to other aircraft traffic is by circling. Any such manoeuvre occupies extensive airspace because the minimum safe separation standards between aircraft must be maintained throughout the manoeuvre. An air traffic control system can be full to capacity when to the layman there seems to be few aircraft in the sky and long intervals between them.

The safe separation of aircraft involves all three spatial dimensions and time. Air routes between main destinations were originally as short and direct as possible, taking advantage of familiar landmarks but avoiding hazards such as high land. Routes became designated by ground-based beacons to aid navigation, and elaborate route networks evolved to handle heavy traffic. Where airspace is limited, all aircraft under air traffic control may have to follow designated routes, the traffic-handling capacity of which is increased by combining accurate navigational data with computer assistance and by constricting freedom of manoeuvre in order to reduce the separation standards between aircraft. Numerous aircraft can fly most efficiently from one place to another as traffic if they are in unidirectional flows, with only small speed differences between aircraft at the same height within each flow, so that any potential infringement of the margins separating aircraft develops gradually and can be corrected. If there is plenty of airspace, many aircraft may not follow fixed routes in traffic flows but each may fly the most direct route, and aircraft are kept safely apart either by tactical instructions during their flights or by preplanning their flight levels and times of arrival at the positions where their flight paths will intersect or amalgamate. A future extension of the latter principle envisages that each aircraft would fly the shortest route between its departure and destination airports, but requires complex computer assistance to maintain safe separations between aircraft and close adherence to planned computed flight paths.

The planning of air traffic flows in response to traffic demands can be represented as a hierarchy of control loops, each with its characteristic time span (Ratcliffe, 1975). The highest loops, with time spans expressed in years, refer to constraints on operational air traffic control that were introduced during the planning or procurement of the air traffic control system. Loops with time spans expressed in days or hours relate to the practical planning, management and organization of air traffic flows in the context of estimated traffic demands. Actual operational air traffic control includes forward planning loops with time spans typically between 15 and 30 minutes, and shorter tactical intervention loops. The shortest loops, with time spans expressed in seconds, are uncommon but exemplified by emergencies requiring immediate tactical intervention to avoid the ground or another aircraft in close proximity, detected by the controller, by the pilot visually or by an airborne collision avoidance system or a ground proximity warning device. Such tactical collision avoidance may provide an ultimate safeguard, but its actual use should be rare because it denotes the failure of all the higher loops and represents an unpredictable manoeuvre within a flow. Air traffic control is evolving towards broad strategic planning based on the capacity limits of large regions of airspace, and toward the pre-planning of traffic flows within each region to minimize the need for tactical interventions. Plans for the evolution of air traffic control have to be quite long term (Hunt and Zellweger, 1987; Perry and Adam, 1991; Stonor, 1991), and based on forecast traffic demands (McAlindon, 1991).

The role of air traffic control depends on phase of flight. The primary distinction is between controlled airspace with an air traffic control service that is

usually mandatory, and uncontrolled airspace where any service is usually
advisory. In most countries, uncontrolled airspace is generally below normal
cruising flight levels and away from airports. For commercial aircraft, different
but coordinated air traffic control services are commonly provided for the following phases of flight in many countries:

1. manoeuvres within an airport and its immediate vicinity, including take-off,
    landing and ground movements;
2. flights within the terminal manoeuvring area around an airport, including
    the organization of aircraft approaching from different directions into one
    or more streams for landing, and the initial routeing of departing aircraft
    according to their destinations;
3. flights cruising en route between the terminal areas of different airports,
    including the initial transition between the terminal area and the cruising
    state by climbing or descending;
4. flights cruising beyond radar coverage, typically over oceans.

The conduct of air traffic control utilizes many kinds of information, including the following:

1. General rules and conventions
    Examples are: international agreements; rules of the air; legal rights, obli-
    gations and responsibilities; conventions for describing types of aircraft
    and types of air traffic; divisions of airspace; classifications of air traffic
    control services provided; and policies on planned system capacities and
    safety standards.
2. Specific universal practices
    Examples are: international standardized conventions for representing air
    traffic control data, for the format, sequence, content and language of air
    traffic control messages, for the applicable minimum separation standards
    between aircraft, and for the training of controllers.
3. Characteristics of aircraft types and users
    Air traffic control instructions must be compatible with aircraft capabil-
    ities in terms of their manoeuvrability, endurance, speeds, on-board
    equipment, and tolerance of extreme conditions such as bad weather, and
    in terms of the interacting effects between aircraft such as wake vortices.
    Different classes of aircraft are subject to different air traffic control regu-
    lations and priorities in accordance with their controllability, manoeuvra-
    bility and functions. Air traffic control as a service must try to meet the
    particular requirements of all its users.
4. Geographical factors
    Geographical factors influence the demands on air traffic control and con-
    strain the ways in which the demands can be satisfied. The terrain influ-
    ences the location of airports, the orientation of runways, and approach
    and take-off paths. Air traffic control must accept numerous related con-

straints, such as topographical influences on its route structures and the handover of responsibilities at political boundaries.

5. *Air traffic control functions*

An air traffic control centre serves a geographical region; an air traffic control tower or visual control room serves an airport. Their functions are distinct though not independent. The air traffic control service offered reflects the densities and mix of traffic, and depends on the facilities and equipment provided.

6. *Air traffic control policy*

National policies on air traffic control are influenced by its perceived relative importance in terms of resourcing and funding, by projected traffic demands, and by requirements to conform with international standards. Ecological factors such as noise abatement and fuel conservation also affect policies. Local attitudes are often ambivalent towards new or expanded airports, which can bring jobs and improved ground transportation along with some disruption and extra burdens on local services. Many factors affect decisions about airport construction, expansion and capacity. Occasionally, air traffic control factors crucial to safety, such as the orientation of new runways in relation to existing ones, have an over-riding influence on policy, but usually air traffic control must accommodate many external constraints which preclude optimum structuring of the airspace according to air traffic control criteria. Specific national policies on air traffic control cover such matters as recruitment, staffing levels, rostering, work–rest cycles, conditions of employment, and the professional status of air traffic control.

7. *System specifications*

Within the constraints of policies, funding, geography, forecast demands, national and international rules and conventions, and ecological and other factors, air traffic control systems are planned and designed. Plans include route structures, aids, ground-based facilities, communications, and design concepts for workspaces, procedures and instructions.

8. *Navigation aids*

Air traffic control is wholly dependent on the effective integration of the different kinds of navigation information available to it from such sources as non-directional beacons, direction measuring equipment, Doppler, area navigation, inertial navigation systems, radar, satellites, and data automatically transponded between air and ground.

9. *Computational and communications facilities*

Air traffic control depends on how the information that has been gathered is used for purposes of computing, planning, prediction, problem solving and decision making. Information is acquired through communications between pilots and controllers and through facilities at centres and towers. It can be transmitted between machines, between humans, and between humans and machines. The means employed to gather, transmit, modify and use information determine the air traffic control service that can be offered.

10. *Local information sources*

    These include all the information at an air traffic control location that influences the nature or timing of the procedures and practices there. Examples would be the length of the runways, the visibility of taxiing aircraft from the tower, and the range of the local radars.

11. *Facilities within the workspace*

    These facilities determine what the controller can do. They encompass the workspace design and the human–machine interface specification. They include the choice and portrayal of the presented information, the means to store and manipulate data, the communications facilities, all human-initiated activities, and the forms of computer assistance for the controller. They largely determine how the air traffic control system can and does function at a specific locality.

12. *Short-term and temporary information*

    Examples are temporary restrictions and regulations, serviceability states of equipment and of installations such as runways, and information from relevant but external sources such as flight service stations, flight information services and weather stations.

13. *Dynamic information*

    This refers to all information in every form specific to the actual or pending air traffic. It is the context for every specific action by the controller, though each action is mediated by all the preceding kinds of information.

14. *Team information*

    This covers the allocation and supervision of work and responsibilities, the relations of the activities of each controller to the activities of the others, and the norms and standards of achievement.

15. *Professional expertise*

    The largest source of information available to the air traffic controller is his or her own professional knowledge and skills acquired through training and experience, by which all other information is interpreted and made meaningful. This source encompasses all judgements about safety and emergencies, the degree of trust in equipment and procedures, and the controller's understanding of the consequences of actions. Without it, all the other kinds of information could be available but there would be no air traffic control.

In some respects, air traffic control is a subsystem within a larger passenger-handling system, preceded by check-in, emigration and security procedures and followed by customs, immigration and baggage-handling procedures. Optimizing the efficiency of air traffic control usually seems a self-evident objective, unless marginal benefits for air traffic control incur disproportionate penalties for other aspects of passenger handling (Hopkin, 1991a).

2.2 Types and classifications of air traffic control

Various categories of user compete for the finite available airspace. They may share the same airspace, permanently or temporarily, if their requirements can be reconciled; otherwise they have to be kept apart. Some categories, not mutually exclusive, are subsonic and supersonic aircraft; military and civil aircraft; fixed-wing aircraft and helicopters; commercial and general aviation aircraft; and powered and unpowered aircraft.

Aircraft flights have many purposes. Supersonic aircraft fly long distances at high speed and great height, but have to be integrated with other traffic near airports. Originally, different kinds of military aircraft fulfilled different roles, but recently the trend has been towards aircraft with multi-role capabilities. Commercial airliners carry passengers and/or freight. General aviation is the broad concept applied to most powered flight that is not military or commercial: it encompasses business flying, flying for pleasure, training flights, air taxis, leisure and tourist flights, emergency flights, rescue flights, patrol or policing flights, agricultural flights (crop spraying), flights for sports such as parachuting, display or demonstration flights, research and development flights, and flights for a multiplicity of other purposes such as surveillance, photography, news coverage, television and advertising.

Air traffic control seeks to provide a safe and efficient service to all who request it and to all who fly where air traffic control is mandatory. Constraints and mutual interference attributable to different user requirements must be minimized. Powered aircraft have to avoid unsteerable air balloons and unpowered gliders. Agricultural aircraft must not be crop spraying in a military training area while nap-of-the-earth flying is in progress. While contrary objectives should all be achieved efficiently if possible with the minimum of disruption to others, this must not be at the cost of air safety. Air traffic control tries to be impartial, within the overriding requirements of safety and of aircraft capability (not willingness) to comply with air traffic control instructions.

Most introductions to air traffic control refer primarily to civil commercial aircraft that are well equipped with navigation aids. They are flown regularly on designated routes by experienced pilots familiar with air traffic control in general and with the local air traffic control procedures at the main airports on the routes. Congestion and traffic handling capacity are the main air traffic control problems. Whereas the civil controller is usually responsible for the orderly flow of all controlled air traffic within a particular region of airspace, the military controller often retains the control responsibility for a small number of designated aircraft wherever they go and whatever they do. As a result, each military aircraft tends to generate much more air traffic control work and the military controller can handle a smaller maximum number of aircraft. Military and civil air traffic control may have different objectives. The military air traffic controller may provide safe passage and separation for a military aircraft transitting between its airfield and an exercise region and then keep other traffic away from that region during the exercise. Another military air traffic control objective may be to arrange for aircraft to converge and fly in formation, which is almost the opposite of the paramount civil air traffic control objective to maintain safe separations between aircraft at all times. Military air traffic control may rely heavily on radar, whereas civil air traffic control uses radar but may be conducted without it in light traffic, and must be so conducted beyond radar coverage (Bradbury, 1991a).

Air traffic control can be categorized according to the type of control exercised, its legal status, and whether it is mandatory or advisory. Some kinds of flight require more freedom of manoeuvre and more individual air traffic control treatment than others. An important factor is whether an aircraft is part of a traffic flow in which all aircraft must comply with the same air traffic control instructions and constraints.

Fixed-wing and rotary-wing aircraft can have different air traffic control requirements because the capability of rotary aircraft to hover makes them much more flexible in some respects. They may have to respond at short notice to emergencies with less preplanning of routes, and may have shorter flight endurance. The choice of rotary-wing aircraft may be determined by a requirement to land in a confined space, such as on a ship, a helicopter landing pad, a building, or an oil rig. Rotary-wing aircraft traffic on designated routes can become quite dense, as on regular flights to an oceanic oilfield.

Chartered aircraft often carry large numbers of passengers on standard airways with navigation aids, and can then be treated in air traffic control terms as commercial traffic, whereas executive aircraft may not follow standard routes but may have to be treated as general aviation traffic, even though their performance characteristics and on-board equipment may be equivalent to those of commercial aircraft.

The concept of general aviation covers a great variety of flights. At one extreme are aircraft equipped to full commercial standards with highly experienced pilots; at the other extreme are minimally equipped aircraft flown by minimally experienced pilots. Their presence may be known to air traffic control, but not their intentions. The controller of general aviation traffic may provide an advisory rather than a mandatory service, but the pilot must not misconstrue the nature of the service offered. Many human factors problems for the controller may be associated with general aviation traffic because of the numerous possible routes and airfields, the large performance and equipment differences between aircraft, and the gross differences between pilots in their experience of flying and knowledge of air traffic control.

2.3 *International, national and regional differences*

Some differences are inevitable for a variety of reasons, but the main trend is to reduce them (Lindgren, 1991; Marten, 1993; Walker, 1993). Factors tending to increase differences include demands for air traffic services, policies for integrating military and civil air traffic, the amount of general aviation traffic, geographical and meteorological influences, air traffic control equipment and facilities installed, and national policies on air traffic control resources. Factors tending to reduce differences include internationally agreed air traffic control rules and procedures, standardization of airborne and ground-based equipment and facilities, international manufacturing and marketing of air traffic control systems, commonality in air traffic control training, and international pressures for greater uniformity in controllers' working conditions and working hours in the interests of aviation safety.

Inevitably, maximum awareness of air traffic control and its requirements occurs where the demand for air traffic control exceeds the available traffic-handling capacities. The main potential benefits of technological advances, such as satellite-derived information, data links, computer assistance, and innovations in displays, controls and communications, are in regions of congested airspace. Improvements in efficiency and recoupment of the costs of technological advances may be difficult to demonstrate if traffic is so sparse that most flights can follow preferred paths unimpeded. Although the human factors problems studied most in air traffic control have been associated with high traffic levels and technological advances, other human factors problems arise with these low traffic densities. The forms of computer assistance proposed hitherto are intended to help controllers who are too busy rather than controllers who have insufficient work.

The practical options for resolving competing demands for airspace depend on geographical factors. A country with a large land mass could in principle reserve extensive designated regions for military air traffic whereas a small country may not have this option. The relative importance of civil and military air traffic control also reflects policies on military air strength. Nations adopt different practices to separate or integrate civil and military air traffic, and to allocate airspace to each and coordinate them. If there are separate systems for civil and military air traffic control, coordination between them may rely on tactical procedures applied to each flight; at the other extreme, a single fully integrated air traffic control system may handle all civil and military traffic. Many nations adopt practices between these extremes, with practical coordination of military and civil traffic but some residual competitiveness over airspace. Military and civil air traffic controllers may respect each others' procedures and skills, while acknowledging that they are not fully interchangeable. However, others outside air traffic control may suggest a sharing of military and civil facilities and equipment for political reasons or in order to cut costs. The feasibility and safety of this depend partly on previous policies and on the relationship between military and civil air traffic control. This relationship can become strained if the replacement of civil by military controllers is adopted to resolve an industrial dispute, or if the control over air traffic becomes a symbol of power in times of political turmoil.

Common international and national practices apply most readily to en route traffic, where many aircraft are in level flight at constant speeds and control responsibility is handed on through several sectors, and to oceanic traffic where the controllers of different nations must employ compatible procedures to hand over responsibility. Characteristics of an airport or a terminal area that can curtail commonality include the number and orientation of airways that converge on it; the structure of the terminal area and the airport layout, especially the number and orientation of runways; the available navigation aids; the terrain; populated areas near the airport; noise level regulations; airport facilities; the separation standards in force; the amount, distribution and mix of air traffic; and the flying experience of typical airport users. Nevertheless, much commonality remains. The main human factors problems associated with air traffic control towers are similar for most towers, although the solutions vary with local conditions such as the customary viewing directions from the tower in relation to the sun, the nature and density of the traffic, policies on night-time flights, and the visibility of taxiways and aprons from the tower. All solutions must comply with broader regulations, standards and guidelines.

2.4 Standardization

International bodies, especially the International Civil Aviation Organization (ICAO), define recommended rules, regulations and practices, and thus promote standardization in air traffic control (Bradbury, 1991b). The number of aircraft manufacturers in the world is limited, and so is the number of companies that manufacture and market air traffic control equipment. Many nations have no capability to manufacture their own air traffic control systems, and therefore manufacturers tend to become an influence towards more standardization. Manufacturing competitors must meet international requirements, and provide air traffic control systems in different countries that are mutually compatible with the equipment on board aircraft flying between them. The systems must all be used in the same way, which implies standardization of controller training.

Standardization also applies to communications. English is the language of international air traffic control. Controllers everywhere are expected to adopt some similar procedures and to couch instructions in standard forms. There has to be international agreement on radar and its specifications, and on the allocation of radio frequencies for air traffic control purposes. Potentially ambiguous concepts have to be clarified, such as flight level which refers to a standard pressure setting and applies above 3000 feet, compared with altitude which refers to actual minimum pressure and applies below 3000 feet, and with height which refers to actual pressure at an airfield and applies during descent to it (Ratcliffe, 1991). Future international agreements must cover the automated transponding of air traffic control data between air and ground, and the integration of satellite-derived data into air traffic control.

Common forms of computer assistance are evolving for different air traffic control systems. The automated detection of an impending violation of separation standards between aircraft exemplifies general conflict detection principles. Aids to problem solving and decision making, such as automated proposals for conflict resolution, must be compatible with other forms of computer assistance and other equipment in the system, and therefore they tend to evolve towards standardized forms. Predictions about future traffic on the basis of flight plans, radar data or satellite-derived information also tend towards standardization.

Costs may force nations to buy off-the-shelf equipment and to adapt standardized items to meet local needs. As technological advances become available world-wide, international consensus is sought on their air traffic control applications. Many of the human factors problems associated with advances such as colour coding of electronic displays, direct voice input, touch-sensitive input devices, and computer assistance for cognitive functions, are general issues. When advances have been efficacious in one context they are likely to be so elsewhere, with any necessary modifications.

Some standardized human factors guidelines can be applied to air traffic control (Dul and Weerdmeester, 1993; Boff and Lincoln, 1988). Anthropometric recommendations for air traffic control consoles and seating, based on reach and viewing distances and comfort requirements, can be applied, modified according to the anthropometric data for the controllers in each nation (Pheasant, 1986). The installation of standardized equipment should encourage the extension of standardization to relevant physical aspects of air traffic control workspaces, such as the spectrum and intensity of the ambient lighting. Ergonomic evidence promotes standardization of the layout, positioning, spacing and functioning of input devices, of the portrayal, codings and legibility of the displayed information, and of the means to prevent glare and reflections. Other kinds of standardization are also applicable, such as those derived from medical and occupational health research on safe radiation emissions from displays.

An influence contrary to worldwide standardization is the increasing interest in cross-cultural differences (Kaplan, 1991). A few of these can readily be accommodated by applying existing anthropometric data, but for many there is insufficient evidence on what the best approach would be or on the degree of accommodation needed. For example, air traffic control workspace designs are imbued with Westernized notions about collective responsibility, team functioning, the delegation of authority, and leadership. Nations with other concepts of leadership, respect for authority and delegation of responsibility may prefer air traffic control systems that are more readily taught and understood in terms of their own traditions and legal systems.

The notion of standardization can have connotations of a different kind within a particular air traffic control workspace such as a centre or a tower if any practices which are non-standard elsewhere have become standard there, because they provide a safe and effective way to satisfy a local requirement. All controllers in the workspace adopt these practices and expect their colleagues to adopt them, whether they are officially sanctioned or not. Such local practices can become standardized as long as air traffic control remains largely manual, but their evolution and adoption become less feasible in more automated air traffic control environments where the mechanisms required to generate and implement them may no longer be present. If these practices have been genuinely beneficial, their removal by automation may impair rather than enhance efficiency, especially if the existence of such practices cannot be officially acknowledged.

2.5 Flight progress strips

A paper flight progress strip in its holder is the traditional tool for presenting in a standard format the information about each aircraft needed for air traffic control. The flight progress strips, one for each aircraft, are arranged in one or more columns on a flight strip board, in the most appropriate order for the tasks, such as by flight level or sequence of arrival. Their usage is illustrated in an air traffic control sector (Figure 1), in a terminal area control room (Figure 2), in an oceanic control room (Figure 3), and at an oceanic workstation (Figure 4).

Figure 1. A sector controller with horizontal radar display and paper flight progress strips

Figure 2.   Terminal area suite with paper strips

Figure 3.   An oceanic control suite

Figure 4.   A flight strip board for controlling oceanic air traffic

The flight progress strip was originally hand-written and updated manually, and in many places it still is. The controller, having talked to pilots by radio telephone and to colleagues by telephone or directly, formulates and checks a safe and efficient course for each aircraft that fits its known flight plan and objectives and meets other air traffic control requirements, and issues instructions to the pilot accordingly. The controller records instructions accepted by the pilot by writing them in a standardized format on the flight progress strip. A series of strips for a flight can be prepared in advance as soon as the flight plan details have been filed. Each strip is delivered to the appropriate controller a few minutes before the aircraft will establish contact, and before the controller assumes air traffic control responsibility for it. After the control responsibility for the aircraft has been handed over, the controller discards its strip, but it is kept as the official record of the control actions on the flight. In some cases, particularly in air traffic control towers, there may be only one flight progress strip for each aircraft, and the physical action of handing it on successively through several controllers denotes the transfer of control responsibility for the aircraft.

The preparation of paper flight progress strips in advance is quite easy to automate, combining the automatic printing of strips with computer storage of the information on them. Various attempts have been made to supersede paper strips with electronically generated strips which can be manipulated and amended in ways which are perceptually, cognitively and functionally equivalent to those that have served so well with paper strips (Vortac *et al.*, 1993). To capture their full functionality has proved to be a major challenge.

The paper flight progress strip exemplifies some of the attributes of air traffic control itself. It is incomprehensible to the layman for it contains no obvious information about what it is, what it is for, or how it is used. Initial impressions of it can be deceptively simple. It does not look complex, but seems to contain quite a small amount of information, all of which could be quantified and presented electronically. In fact, the paper flight progress strip has evolved into a complex and subtle tool (Hughes *et al.*, 1993a). While its most apparent features have obvious electronic equivalents, its full range of functions is far more elusive, complex and difficult to capture electronically than was initially realized (Hopkin, 1991c).

The full meaning and functionality of strips are learned from training and experience and are not self-evident. Paper flight progress strips are immediately comprehensible to a controller and contain all the information needed to control air traffic, given appropriate communications. They are still the main tool to control air traffic beyond radar coverage, though the radar display is now the main source of air traffic control information when it is available.

2.6 Radar

Of all the technological advances introduced into air traffic control since its inception, the most significant has been radar. A radar display provides a plan view of air traffic. Initially, primary radar displays depicted the current position of each aircraft as a blip, and a new blip appeared every few seconds as the radar head rotated. A long-persistence phosphor could show the recent track of each aircraft as a series of progressively fading blips on the radar display. The original primary radar displays contained no identity information, which was added as a "shrimp boat" for each aircraft, a small transparent plastic plaque on which its identity was written. The visual association between the blip and the identity of the aircraft was maintained by repositioning its identity plaque by hand beside its current blip on the display, which therefore had to be horizontal.

When secondary surveillance radar replaced primary radar, the displays portrayed synthesized information processed for human use. The position, heading and relative speed of each aircraft could be depicted, visual clutter and permanent echoes could be removed, adequate contrast ratios between dynamic and background information could be provided, appropriate visual information coding could be applied, and aircraft identity could be shown, usually by its callsign. Secondary radar displays with identity information made the plaques redundant, and radar displays no longer had to be horizontal. During an intermediate phase, some vertical displays could be pivoted to the horizontal if plaques had to be re-employed because of the equipment failures, but vertical or nearly vertical radar displays are becoming the norm in current systems and will be the norm in future systems.

Secondary radar displays can include further categories of relevant information, such as danger areas and areas with flying restrictions, geographical features such as coastlines, air routes and the positions of navigation beacons, and display scale and distances in the form of range rings. Slowly changing information, such as regions of severe weather or precipitation, can also be shown. Information based on computations can be introduced, such as whether an aircraft is straying from its planned route.

The convention evolved to depict the location of each aircraft by a symbol, and attach a transparent label of alphanumeric information that moved with it. On the label were the aircraft's identity, and often its height and whether it was climbing or descending. Larger labels could contain further information, but led to clutter and increased the probability of label overlap and consequent unreadability, which had to be resolved, manually or automatically, by reducing or restructuring the label content or by moving or suppressing one or more labels. A further development updated the information on the progress of a flight automatically; for example, the effects of a delay early in a flight on all its subsequent stages could be computed and displayed. In the future, satellite-derived data will provide plan views of air traffic beyond radar coverage, and could upgrade some plan views on radar. Air traffic control is expected to include plan views of the traffic for the foreseeable future.

A tabular information display of flight progress strips, whether electronic or manual, is difficult to integrate cognitively with a plan view of the air traffic. The consequent problems of cross-referral between radar and strip information are aggravated if traffic is heavy: there are more data to search through whenever there is less time to spare for searching. Windows of tabular information within the radar display do not wholly resolve this problem. A continuing challenge is the integration of these different kinds of information into a single practical format.

2.7 Air traffic control and the pilot

Spoken messages between controller and pilot were the main original means of communication between air and ground, and the basis of agreed air traffic control instructions and actions. Much of the information about aircraft needed for air traffic control purposes will in future be transponded automatically from air to ground. The pilot and controller will become aware of this transponded information only if it is displayed to them. The associated reduction in speech may save workload but may impair understanding or memory. Other pilots and controllers will glean less incidental information about the air traffic situation when there are fewer spoken messages to overhear. A recent development is the provision in some cockpits of a display of nearby air traffic as a short-term collision avoidance aid. The unaided pilot cannot reliably detect another converging aircraft in good visibility, and certainly not in darkness, poor visibility or cloud.

The concentration of human factors research and development resources on aircraft cockpits rather than air traffic control has had several consequences. Many problems have been recognized and solved in relation to cockpits first. Other subsequent applications, such as air traffic control, have followed this pioneering work, particularly in terms of methods and measures rather than findings, but it has sometimes been too tempting to extrapolate from cockpits to air traffic control without sufficient verification. Relationships between human factors work on cockpits and on air traffic control have been uncertain, to the extent that it has not always been conceded that the latter belongs in human factors texts on aviation. The texts edited by Wiener and Nagel (1988) and by Jensen (1989) included air traffic control chapters, though the fact that the former did so was cause for remark, but some recent texts about human factors applied to pilots scarcely refer to air traffic control at all (Edwards, 1990; Hawkins, 1993). There are specialized texts for pilots about air traffic control (Illman, 1993).

Human factors studies of cockpits or of air traffic control systems have relied extensively on simulation, which can be a costly research tool in terms of resources and funding. It is a major commitment to simulate either a cockpit or an air traffic control system, without trying to simulate both together and the interactions between them. Most studies of cockpits or of air traffic control have each included only those limited aspects of the other which seemed essential to obtain valid findings. Developments such as data link demand more human factors consideration of communications between air and ground, which must be fully compatible with the equipment and procedures in cockpits and air traffic control systems and must foster safety and efficiency.

The transmission and understanding of information have been studied quite extensively in research on communications in aviation. Topics have included errors, information presentation, tasks, language and vocabulary using auditory or visual data, and appropriate levels of detail. Speech between air and ground fulfils many functions. The judgements and assessments that pilots and controllers make about each other, concerning their professionalism, ability, confidence and apparent familiarity with tasks and messages, are based largely on the content, pace, phraseology, consistency, standardization, courtesy, and felicity of expression of the spoken messages between them. Rightly or wrongly, pilots make judgements about the competence and reliability of the air traffic control service they are receiving, and request clarification, confirmation or supporting evidence accordingly. Similarly, controllers make judgements about individual pilots based on what each says and does, and they may check more frequently that their instructions are being obeyed or require more transition states to be reported if they believe that a pilot is inexperienced or unfamiliar with local procedures. The validity of such judgements is sometimes unproven, but speech between pilot and controller conveys much more than the quantitative content of the spoken messages. If speech is curtailed by data links, so is the basis for such judgements.

The pilot is legally responsible for the safety of the aircraft and its passengers. The controller is legally responsible for the safety of the air traffic control instructions. No matter how the boundary between these responsibilities is defined, ambiguities can arise. When both the pilot and the controller are implementing air traffic control instructions that are presented on screens in the cockpit and the air traffic control workspace, but have been derived from software in the air or on the ground, the issues of legal responsibility can become quite complex. The ultimate reason for the retention of humans in aircraft cockpits and in air traffic control systems may be their legal responsibilities rather than considerations of air traffic control, of human factors, of technology, or of aviation. As long as controllers and pilots have these responsibilities, they must have the wherewithal to exercise them.

Air traffic control provides a service to pilots. In some regions of airspace, that service is mandatory and pilots cannot legally fly there without air traffic control agreement. Air traffic control has a commensurate obligation to provide an efficient service in an efficient system. These are not synonymous. An outmoded and inefficient system staffed with a highly dedicated and professional workforce can sometimes achieve an acceptable air traffic control service. An air traffic control system can be safe and highly efficient as a system, but its users will not think so if the service that it provides is not the service that they need.

3

The human as a system component

3.1 Limitations of the concept

A system in which numerous human operators collaborate with machines to fulfil the designated objectives of the system as a whole was originally called a man–machine system (Parsons, 1972), a concept now deemed sexist. This text generally employs the term "human–machine system", although it lacks universal acceptance. All the humans within the same system do not normally have identical tasks, jobs, equipment or functions, but they may share the same professional knowledge and qualifications and sometimes the same training. Nowadays, the machines almost invariably include computer hardware and software.

Studies of human–machine systems can be of real systems, of simulated or prototype systems in real time or fast time, of planned systems, or of hypothetical systems. They may aim to assess feasibility, efficiency, safety, practicality, reliability or capacity. They can be exercises in modelling, quality control, operational research, risk analysis, optimization, quantification, validation, or cost-effectiveness. They may examine procedures, innovations, changes, errors, or failures. An air traffic control system is one example of a human–machine system.

The description of an entire functioning system must employ language, notations and constructs that can be applied to all its components, to the interactions between them, and to the system structure and integrality (Wise *et al.*, 1993). Exclusively human concepts cannot suffice, no matter how anthropomorphic the descriptions of machines become. Engineering concepts, mathematical concepts and computer concepts have all been enlisted to describe human–machine systems, and have seemed feasible in so far as they can accommodate times, events, probabilities, associations, dependencies, and error and failure rates. Occasionally, they may apparently describe simple human functions within the system adequately for limited purposes, as when the human performing a manual tracking task resembles a servo-mechanism, but it can be very difficult to describe the products of human cognitive functions satisfactorily using such concepts.

The employment of machine language and concepts to describe the human as a system component has seemed a matter of expediency rather than choice. If concepts applicable to both human and machine are essential and human concepts are ruled out, machine concepts must be tried. They can be productive within their limitations, and it is not contended that no machine concepts can ever describe the human as a system component adequately for any purpose, but they cannot deal satisfactorily with human functions and attributes with no apparent machine equivalent, and they lack universal validity. Awareness of their limitations may prevent their misapplication. Most air traffic control questions are posed in system terms and the answers are expected to refer to system and subsystem functioning, although many questions about controller functioning cannot be answered by system measures alone.

One distinction can be fundamental in human terms but not in system terms: does the human drive the machine in that humans initiate most actions and the machine functions as an adjunct of human activities, or does the machine drive the human whose primary role is to respond to system-initiated events? The progressive introduction of computer assistance can tilt this balance of initiative in favour of the machine and against the human. An approach couched exclusively in machine language may conceal this tendency or underplay its significance. The replacement of a human function by a machine function can have profound human consequences, yet seem a trivial change when expressed in system concepts. Computer assistance for tasks can obscure the delimitation of human functions so much that they can no longer be measured separately from machine functions. Descriptions of the human in system terms render the boundaries between human and machine functions less discernible.

In principle, the system approach offers a means to integrate separate techniques and methods of study through common concepts and language. For example, activity analyses, real-time simulation, fast-time simulation, and modelling can be linked within a system framework. System studies can demonstrate feasibility, and can represent in system terms the benefits that accrue from such changes as reduced delays, increased capacity, fewer communications, and enhanced safety, but they may fail to quantify the improvements that are attainable in practice rather than in principle unless all the main consequences of human intervention can be expressed in system terms. System measures yield evidence about system performance, and individual measures about individual performance. The relationship between these two kinds of evidence has been of air traffic control interest for a long time (Buckley *et al.*, 1969).

It might be claimed, though usually it is not, that the human controller is being treated as a system component whenever the physical environment is adjusted to human needs, such as acceptable temperatures or noise levels. Parallels might also be drawn between human and machine as system components by considering activities applied to the humans within the system that could correspond to the system activities of routine testing, maintenance, and replacement schedules for components.

3.2 The allocation of functions

The process of allocating functions to human or machine seems attractive and logical. Fitts' (1951a) original allocation criteria assigned functions to human or to machine depending on which performed them better. Only when this simple process led to practical difficulties were some of its underlying assumptions and implications recognized and questioned (Jordan, 1968). Among the assumptions were that every function can be fulfilled somehow either by human or machine, and that relative judgements should remain the basis of allocation even of those functions which both human and machine could fulfil very well or very poorly. Among the implications were that advances in technology would progressively require the assignment of more functions to machines, that humans must do whatever machines cannot do, and that the only functions considered for allocation are those which a machine could conceivably perform for otherwise the question of allocation does not arise. This allocation principle seemed to exclude some options altogether, as it implied that human and machine should not closely share a function, alternate in performing it, or duplicate it by performing it independently in parallel.

During the evolution of the designs and specifications for an air traffic control system, the necessary functions and the means to fulfil them are developed progressively, and expressed in terms of tasks, instructions, human–machine interface designs, and communications facilities. Ideally, all human functions should be well within human capabilities after training, and meet all the conditions for effective human performance, related for example to equipment, knowledge, procedures and timescales. Human functions have to be treated collectively, not singly, so that they are mutually compatible and can be amalgamated into jobs. The matching of the results of task analysis with human capabilities should lead to human functions which are feasible and efficient.

So much for theory and principles: real life is different. The actual allocation process is driven mainly by capacity demands, technology and costs. If the benefits of an innovation cannot be stated in such terms it is unlikely to be introduced, no matter what human factors advantages it may confer, but if its benefits can be so stated it will be considered seriously, and may be adopted regardless of any consequent reallocation of functions between human and machine. For example, decisions on whether future air traffic control systems will include satellite-derived information, or data links, or computer assistance for prediction and problem solving precede consideration of the allocation of functions and therefore cannot be treated as primary tools to achieve an optimum allocation of functions. Rather, functions have to be allocated within the constraints of such decisions. This is a statement of fact and not really intended as a criticism, though it could be so construed. The practical point is that the allocation of functions to human or machine is largely technology driven and seldom the human factors process that it should be.

For technical and legal reasons, a function commonly allocated to humans is to deal with non-standard situations and emergencies. Whether the human has the means to fulfil such a role depends on the system design. Detailed knowledge of the emergency and of its preceding circumstances is needed immediately to resolve it quickly and safely. This seems to imply that the human must know the current state of the system at all times, which in turn implies monitoring. Yet machines are usually better than humans at monitoring. In current and proposed systems, the functions most suitable for the human have not necessarily been allocated to the human, and only some of the functions most suited to machines have been allocated to machines.

As technology advances, it continually extends the range of functions that could be allocated to human or machine. At first, many functions associated with the gathering, compilation, storage and presentation of data were human functions, but they are now generally machine functions at major air traffic control facilities. Functions often discussed at the present time in relation to human and machine roles include problem solving, decision making and prediction. As machines evolve, their roles in such functions will increase (Falzon, 1990). Beginning to appear are machine functions requiring flexibility, innovation, adaptability or intelligence, hitherto the exclusive prerogative of humans but now becoming machine attributes also. A future challenge will be the harmonious reconciliation of human and machine when both can be flexible, innovative, adaptable and intelligent (Hancock and Chignell, 1989).

A discernible trend is to reduce human participation in the tactical interventionist control of specific aircraft and to cast the controller in the roles of monitor, supervisor, manager of resources and strategic planner of traffic flows rather than of single flights. The criteria for allocating functions to human or machine are becoming blurred. As the objective of maintaining the controller's picture and understanding of the traffic at current levels of detail becomes more difficult to attain when there is more traffic but less direct intervention, the purposes of the objective have to be reappraised. As further forms of computer assistance enhance traffic-handling capacity, manual reversion in the event of system failure and return to the more automated state after manual reversion become less feasible and ultimately impossible. Long-term extrapolation from present trends leads to highly automated systems, but these raise issues not only of acceptability and technical feasibility but of legal responsibility.

Although the advance of technology sometimes seems to have restricted the human–machine relationships within air traffic control systems, in fact it has greatly extended the range of relationships that are possible between human and machine. There is much talk of human-centred automation, which bears a multiplicity of interpretations but always implies retention of the human in a pivotal role. Feasible kinds of human–machine relationships within air traffic control now or in the foreseeable future are listed below in the approximate chronological order of their appearance:

1. the human adapts to the machine;
2. the human competes with the machine for functions;
3. the human is replaced by the machine where possible;
4. the human is complementary to the machine;
5. the human supports the failed machine;
6. the human is adapted to by the machine;
7. the human and machine have a hybrid relationship;
8. the human and machine have a symbiotic relationship;
9. the human duplicates the machine in parallel with it;
10. the human and machine are mutually adaptive;
11. the human is interchangeable with the machine;
12. the human and machine have fluid interrelationships;
13. the human and machine form a virtual air traffic world;
14. the machine overcomes any human incapacitation.

As new options are added, the original ones do not disappear but remain available, so that the range of possible human–machine relationships expands.

Some of the above relationships render the preallocation of functions to human or machine inappropriate or irrelevant. For example, one conceivable option is for the human to choose what to do and leave the machine to do the rest. The human could choose to be idle or very busy. Such enticing technological options introduce new human factors problems and complications. Nevertheless, the allocation of functions to human or machine, which always was a process honoured more in theory than in practice, may be overtaken by events that render function allocation unnecessary and undesirable. In future air traffic control systems, the human controller will be different as a system component from the controller in past or present systems.

3.3 *The human–machine interface*

The interface deals primarily with the interchange of information between human and machine, and links the human to the system and to other components. All information from the human to the machine and from the machine to the human must pass through the interface. Information is transmitted from machine to human primarily by visual displays, with some ancillary information channels such as auditory warnings and communications. Information is transmitted from human to machine through controls and input devices, with some possible auditory channels such as direct voice input. Information transmitted directly between machines within the system and not via the human–machine interface remains unknown to the human. Information transmitted directly between humans by speech or gesture and not via the human–machine interface remains unknown to the machines and to the system. Although the interface is the only means for transmitting information between human and machine within the system, not every kind of information can be transmitted through it. Its design is crucial for system functioning (Wallace and Anderson, 1993; Dubois and Gaussin, 1993).

An obvious fact, too easily overlooked, is that the only functions which the human can fulfil within the system are those for which provision has been made in the specification and design of the human–machine interface (Hopkin, 1989a). Human actions require appropriate means to implement them. Lacking such means, they cannot occur, however apposite or praiseworthy they may be. It is futile, for example, to rely on human innovation if the human–machine interface vetoes innovative actions as invalid or makes no provision for them to occur. Spoken instructions can often be innovative in ways that instructions which have to be keyed in cannot be.

Most of the main attributes of the interface design and specification that can influence the efficiency of the human as a system component are known. They include the provision of appropriate equipment for every task and require-ment, derived from task or job analysis or by other methods. The interface specification will determine the skills which humans can acquire from per-forming tasks, the pace at which tasks can be done, the consistency and reli-ability of performance, and the kinds of human error that are possible. Some aspects of the interface design are straightforward. Displays and input devices should be located and matched so that the associations between them are self-evident. Input devices should facilitate the quick, accurate and error-free performance of tasks in their correct sequence, and provide feedback and encourage learning. Information displays should employ codings, levels of detail, layouts and formats appropriate for each task. When the same informa-tion is used for several tasks it must be satisfactory for them all, though it may be optimum for none. Its portrayal should foster the acquisition of skill, take account of the time available for each task, and acknowledge relevant charac-teristics of the user population such as minimum eyesight standards. The inter-face should help the user by showing plainly the facilities and options available, by issuing reminders, by assisting data input sequences, by support-ing planning and scheduling tasks and by aiding control interventions. With a well-designed interface, the controller should always be able to ascertain whether required information is contained in the system, and to discover easily how to gain access to it.

Decisions on the methods of displaying information, the specifications of input devices, and the provision of communications are also decisions on the kinds of human error that are possible. The interface design must seek to prevent human errors and to detect any that cannot be prevented. This requirement is much simpler to state than to achieve, but it is vital to identify all the sources of human error that are predictable because they are inherent in the human–machine interface design, particularly if they relate to pro-cedures which also contain system-induced errors. All human–machine interfaces include some predictable sources of human error. The notion of the perfect interface where humans cannot err is a potentially dangerous myth, for all human activities have the propensity for error. Identifying the potential human errors in any given interface design is the first step towards recognizing any that do occur and mitigating their effects.

It is important to recognize all the consequences of any changes in the human–machine interface design. A common incidental consequence of introducing new forms of quantitative information is the removal of qualitative information. One example is that the processed secondary radar displays that replaced primary radar displays retained little or no information about signal-to-noise ratios, fading, clutter from permanent echoes, the quality of the sensed data, or equipment serviceability. Another example is that the controller's displays of digital data transponded automatically from an aircraft lack the qualitative information about the pilot's confidence and competence contained in the human speech which the transponded data have replaced. Qualitative as well as quantitative information must pass through the human–machine interface. If there is no provision for it, qualitative information within the system remains inaccessible to the human. The human cannot discover how trustworthy the portrayed information on displays is, and the machine accepts inputs by humans who may not appreciate their full significance. Human–machine dialogues through the interface must be efficient as dialogues. An effective interface design should always present all the available menu options unambiguously.

One issue is the extent to which the machine should be able to adapt to differences in human proficiency. Much of the functioning of the machine components of the system is pre-programmed, with fixed machine responses to each human action. The human acts under comparable constraints. This raises the question of which human actions should be prevented by the interface design. Many already are prevented indirectly, whenever the controller's preference is not among the available options on a menu or is ruled to be invalid by the system when the controller tries to implement it. This seems sensible until a non-standard situation or emergency requires a human action which is normally invalid or cannot be implemented. A contentious aspect of human–machine interface design is any provision for the human to override machine instructions or constraints in the interests of safety under exceptional circumstances, and the precise specification of all such circumstances. Ultimately this dilemma poses the stark choice of whether the machine or the human is in control and has the legal responsibility for actions taken and solutions adopted in emergencies. These kinds of issue must be addressed as part of the specification of the human–machine interface. If they are not, the human may devise ingenious and safe solutions to problems, only to be thwarted because they cannot be implemented.

The human–machine interface as an entity has reached its present state through evolutionary processes. Most plans envisage its continuing evolution to incorporate further technological advances and accommodate increased traffic demands. Changes in the specification and design of human–machine interfaces should bring major intended benefits that can be predicted, quantified and absorbed, but they can also bring unwelcome and unintended disadvantages that can be difficult to foresee. The benefits of any change have normally been examined far more thoroughly than the disadvantages. Any change or innovation can be expected to remove some familiar kinds of human error and to introduce some new ones. These latter must be acknowledged and accommodated by revision of the interface design.

In air traffic control, the human–machine interface is complex and is used for numerous tasks. Any change in it will bring benefits (often predicted) and disadvantages (often not predicted). For a change to bring unalloyed benefits for every task would be rare. Whether a change should be adopted depends on judicious weighing of the balance of its advantages and disadvantages, which requires them all to be identified beforehand. In this weighing, all the tasks that would be affected have to be considered, and all the human factors consequences of the change recognized. Even apparently simple aspects of the air traffic control human–machine interface such as paper flight progress strips are revealed as complex by proposals to change them.

The human–machine interface is open and observable in traditional air traffic control workspaces. An incidental consequence of many proposed changes would be to render it less observable and thereby reduce the opportunities for team collaboration. Tasks can be shared only to the extent that the human–machine interface design permits task sharing. Functions can be observed only to the extent that the human–machine interface design allows them to be observed. Monitoring a controller's actions through a separate terminal is not functionally equivalent to direct observation. Failure to recognize such consequences can lead to unpredicted and unwelcome effects of changes in human–machine interface designs.

3.4 Automation and computer assistance

These two concepts have sometimes been treated as interchangeable when applied to air traffic control, but a useful distinction can be drawn between them. Automation implies that a function will be fulfilled, and perhaps also initiated, by a machine. Computer assistance implies a human hub or focus for the function, with machine help in task performance. Although there are exceptions, automation applies most often to processes concerned with the gathering, handling, storage, compilation and retrieval of the abundant data within a modern air traffic control system, whereas computer assistance applies to the human tasks and roles that make use of such data.

There is a considerable literature on the human factors aspects of automation and computer assistance in air traffic control (e.g. Hopkin, 1989b, 1994a; Wise *et al.*, 1991, 1993). The actual and proposed usages of computers in air traffic control have many kinds of human factors implication (Parsons, 1985). Before there were computers in air traffic control, the main database was the controller's professional knowledge, experience and situational awareness. A computer has its own independent machine database, containing more extensive and more frequently updated data than the human can perceive, process, assimilate, understand or recall. Most current functions need both human and computer databases. Acknowledgement of this clarifies some basic requirements and potential problems.

Some essential principles have gradually emerged on the conditions for successful human relationships with any form of automation or computer assistance that is intended for human use or can influence human tasks. The human and computer databases must be compatible and matched to the extent of sharing some concepts and terminology and some commonality of structure. There must be means to transfer data and communicate between the human and computer databases. Each database must have some knowledge of the content of the other database and of what it does not contain. This knowledge should not be confined to quantitative information but include some qualitative information about accuracy, precision, consistency, reliability, frequency of updating, and perhaps confidence and trust.

Each database should also include information on how to access the other database, and on its objectives. Perhaps most crucially, any effective collaborative relationship between the human and the computer, whatever its detailed nature may be, must presume that significant changes in either database will entail corresponding changes in the other database, in order to rematch the databases. This implies that computer changes that affect its content, processing or accessing, will incur human changes in required knowledge, procedures, strategies or training. It also suggests that computers should not necessarily respond in the same way to every controller but adapt to each controller's strengths, weaknesses, knowledge and experience. Neither the human nor the computer is wholly passive as a system component, for each initiates some activities while they are working together. Any particular air traffic control job usually incorporates several of the possible relationships between the human and computer listed earlier in this chapter in connection with the allocation of functions, including competitive, collaborative, adaptive and complementary ones.

Automation and computer assistance influence almost every aspect of air traffic control. This is not a unique feature of air traffic control but is also true of aviation and numerous other applications. Many controllers are already familiar with computer assistance, and future controllers may be selected partly for their ability to work well in more automated systems. The main previous impact of automation on the human as a system component has been in the reduction of data to manageable forms, with codings, timing, formats and levels of detail that accord with human tasks. Future forms of computer assistance will support human roles that require higher mental functions, such as problem solving, decision making and prediction. Increasing collaboration between human and computer will entail more mutual adaptability between them. Various forms of automation and computer assistance will be extended to air traffic control functions other than those of the controller, for example concerned with maintenance, quality control, and system planning and design.

Different kinds of relationship between the human and machine components of future systems, at first mainly within training contexts but eventually in operational ones, may include applications of artificial intelligence, neural networks, expert systems, common consoles and prototyping (Fabry and Lupinetti, 1989). As the range of possible relationships between humans and computers expands and more diverse ways to match them evolve, human factors studies must provide the principles and guidelines to be applied throughout system evolution and development to optimize human–machine relationships.

3.5 Human reliability

Although for some purposes the human can usefully be considered as a system component in air traffic control, most of the reliability assessment techniques applied to other components have not been extended to humans because they are not valid for humans. In safety-critical systems such as air traffic control, components with very high reliability are sought. Several safeguards are introduced to achieve high reliability, including software proving techniques, the duplication of critical components, and backup and standby facilities in the event of system failures. Maintenance schedules may include the replacement of essential components before they fail, and quality assurance procedures and system integrity checks seek to ensure system reliability. If a component failure with potentially hazardous consequences occurs, investigatory procedures are instigated to ascertain its causes and to prevent any recurrence, and the resulting recommendations are normally implemented.

Techniques to assess the reliability of functioning systems as entities have lagged behind techniques applicable to their components. The most difficult reliability issues concern the relationships and interactions between system components rather than the components themselves. Constructs applicable to the whole system are lacking, especially if human reliability is included in any reliability assessment of the system. The complexity and integrality of many future air traffic control systems are beginning to impose more consideration of reliability during the system design and of the development of links between theory and case studies (Wilpert and Qvale, 1993).

Human unreliability, human error or human inadequacy of some kind consistently emerge as the primary causes of a substantial majority of the incidents in air traffic control systems that are serious enough to warrant some form of official enquiry. It might therefore be expected that a preponderance of the resources expended on the improvement of system reliability would address human functioning, as the main source of unreliability in the system as a whole. In reality, only a very small proportion of the work on reliability deals directly with human reliability. Often it is not even considered whether human reliability could be enhanced significantly by reassigning or redesigning tasks. The comparative neglect of the influences on human reliability, and the consequent ignorance about them, seem strange.

A few explanations can be offered. The pessimistic one is that the study of human reliability is just too difficult. Another is that the origins of human unreliability seem so diverse, complex, wayward and intractable that attempts to reduce human unreliability at source seem futile, and more oblique approaches are preferable. A further explanation is that many causes of poor reliability in humans originate in exclusively human attributes, and therefore remain unrecognized because only human attributes with machine equivalents are studied. A different explanation is that humans are so complex that the techniques for assessing and quantifying their reliability seem rudimentary compared with those for other system components, and the evidence about human reliability yielded by these techniques seems too speculative a basis for recommendations to improve it.

Common experience proves that humans can achieve very high reliability in some tasks but only low reliability in others. Experienced car drivers are highly reliable in turning the steering wheel of a forward-moving car clockwise to steer to the right and anticlockwise to steer to the left. The achievable human reliability in this task seems comparable to those for some software and hardware components (though not when reversing the car!). In contrast, those over-familiar with conference presentations despair about the poor human reliability in the task of arranging a few slides or viewgraphs in their correct order and orientation. There is now evidence about human reliability in performing many tasks, so that differences between tasks in the reliability of their performance are often predictable. Keyboard data entry tasks have probably been studied most, and error rates of between one error per thousand and one error per ten thousand keystrokes or better are attainable by experienced operators using ergonomically designed input devices suited to the tasks.

Though quite impressive in human terms, this is not an impressive error rate for a system component. For many human tasks within an air traffic control system the reliability levels specified for other system components will not be consistently attained by the human. Therefore the dual aims are always to do everything possible to improve human reliability, but to circumvent any adverse effects of poor human reliability on the system. The reliability of the air traffic control system as a whole must exceed the reliability of its human components. This implies that the system and task designs, the procedures, and the forms of feedback must all be utilized to raise human reliability and to compensate for its deficiencies. Various techniques are available to assess human reliability in complex systems (Kirwan, 1994).

Existing system designs have not normally been influenced much by the reliability of the human as a system component. If inappropriate system or task designs prevent reliable human performance, the human cannot be moulded into a reliable component. Either the relevant aspects of the system or tasks have to be redesigned, or the system must detect and compensate for human inadequacies and failures. This apparently theoretical process also has to be practical.

Many functions in air traffic control require sustained attention by the controller and continuous monitoring over long periods. Sustained attention without active participation is unattainable reliably, and human monitoring is unreliable in many circumstances. Some functions in current air traffic control systems seem vulnerable to human unreliability, not deliberately but because human reliability has not influenced their design. Often the main sources of human unreliability can be predicted, and designed out or minimized. For many tasks such as data entry through a keyboard, errors are not random but follow a pattern that is broadly predictable from the layout of the workspace and the input devices in relation to the tasks, instructions and priorities. System designs that can compensate for the predictable human errors associated with the chosen data input devices and display–control relationships will be more reliable. Which controller will fail in which way at what time under what circumstances cannot be predicted, but general types, rates and incidences of human error can be predicted quite well, and such predictions provide a tool to improve human reliability.

The reliability of some functions is difficult to establish. Although the literature on accidents and incidents has been categorized and can be examined as a whole for patterns and significant relationships, the most striking impression from the accounts of many incidents is often the absence rather than the presence of any patterns or common factors. While general indices of human reliability could be derived, the prospects for increasing low general reliability by addressing particular problems seem poor because no single factor or cluster of related factors can account for much of it. All causes of potential human unreliability are not necessarily adverse in their effects. Human flexibility, adaptability and capacity to innovate are rightly construed as advantages in many circumstances, but when expressed in terms of reliability they can constitute unpredictable system functioning that lowers general system reliability.

A neglected aspect of human reliability is the human capacity for self-monitoring. Typically the human makes errors that are noticed as or soon after they are made, and errors that remain unnoticed. The former can be corrected if there are means to do so, whereas the latter are not recognized as a source of unreliability unless their adverse consequences become apparent and can be associated with the original errors. If a machine makes the same kind of mistakes as its human operator, it is not usually possible to convince the operator that he or she, and not the machine, made the mistake. People seem able to discriminate well between their own errors and identical errors made by a machine, yet be fallible in detecting their own errors. This aspect of human reliability and self-monitoring warrants further study.

Knowledge about human reliability is not always expressed in terms of reliability, but concepts unrelated to reliability are ascribed to it. Reliability should not be equated with attention: a boring and undemanding task may be performed either reliably or unreliably while attention wanders, and an attention-demanding task may or may not be performed reliably. Over-learned skills may become automatized and require no attention, yet be highly reliable. Estimates of confidence are a poor guide to reliability, for confidence may be misplaced and changes which should induce changes in confidence often fail to do so (Sen and Boe, 1991). Beliefs that reliability has been enhanced or degraded may be wrong, and unsupported by any evidence. The reliability of the human as a system component can be influenced by such diverse human characteristics as a dislike of idleness, inability to recognize incipient task overloading, and a propensity to confound the recency with the importance of information. Human motivation can influence reliability, but not necessarily in rational or predictable ways. The human may use disliked equipment unreliably to prove imagined deficiencies in it, even though it could be reliable if properly used. A human may achieve highly reliable performance with unreliable equipment if it poses a professional challenge, uses existing skills well, is cherished, or makes the human determined to master it. Although many specifically human attributes of the above kinds can and do influence human reliability, the extent, consistency and effectiveness of their influence and the conditions for it have not yet been fully established.

4

Human cognitive capabilities and limitations

4.1 Functions of psychological theories and constructs

Psychology is the scientific study of human behaviour and experience. Human capabilities and limitations are part of its subject-matter. Its theories and constructs should therefore be applicable to the human factors aspects of air traffic control. Although the human factors contributions to interdisciplinary work and discussions on air traffic control have to be couched in plain language, the human factors work itself, while eschewing jargon, employs the technical language of the discipline, characterized by its precisely defined concepts and their accepted technical meanings and connotations, and by implied theoretical frameworks and relationships. Much human factors terminology comes originally from psychology. The influence of psychology also pervades its methodology and many of its measurement tools and techniques.

Psychology and human factors deal with influences on human attributes and with influences that human attributes exert. Most topics included within psychology have potential human factors implications. Psychology can therefore contribute a substantive though incomplete list of issues to be considered in the application of human factors to air traffic control. Numerous theoretical and practical, laboratory and field-based measures developed in psychology usually provide the bulk of the evidence from human factors studies of air traffic control, although measures originating in other disciplines are needed too. These psychological measures that have evolved to be applied to human beings are associated with experimental designs and statistical techniques suited to the multiplicity of interacting influences that typically affect human thoughts and actions, and they thus seem apposite for the study of air traffic control.

General psychological theories and constructs fulfil further roles in relation to air traffic control. If they are applied to interpret or explain air traffic control findings, they can help to validate those findings by demonstrating their compliance with the theories (Bisseret, 1981; Narborough-Hall, 1987). When compatibility implies that the findings exemplify a more general psychological theory, that theory can be used to generate hypotheses on the circumstances under which the findings will remain applicable, and hence the theory may predict the generalizability of the findings. In such instances, theories are applied to evidence already obtained. A further role of theories is to identify combinations of variables and conditions that would be crucial tests of them, and hence to pinpoint the most productive kinds of additional evidence. This integration of theory and practice can be a cost-effective tool to define the boundaries within which human factors recommendations apply, by providing explanations of them and a rationale for the valid extrapolation of the findings beyond their original conditions.

The above roles are feasible wherever human factors findings about air traffic control are compatible with independently derived psychological theories. The air traffic control evidence and the theories are mutually strengthened. Conversely, incompatibility between air traffic control evidence and psychological theories may warrant reappraisal of both, for it could be that the theories are wrong or less general than they are claimed to be, or that the air traffic control evidence is invalid, misleading or incomplete. As a human endeavour, air traffic control must be explicable at some level in terms of general human capabilities, limitations and characteristics (Bisseret, 1971).

The proclivity to look at technological innovations and developments in software and hardware with the expectation that some applications of them to air traffic control can be found, has its human factors counterpart in a tendency to examine each new psychological theory, construct and tool for air traffic control applications with the expectation that it will be relevant and capable of yielding new insights. But even if new psychological developments have wide validity, which is rare, there is no particular reason why they should be able to provide insights into functions of controllers which were designed and have evolved without reference to them.

Many human factors concepts that originate from practical contexts such as air traffic control systems have no exact equivalent in psychological parlance or psychological theories. Some confusion and contention over the correct psychological interpretation of the concepts can result. Current examples of such human factors concepts include situational (or situation!) awareness (Endsley, 1994a; Smith and Hancock 1994; Hopkin, 1994b), readiness to perform (Gilliland and Schlegel, 1993), the controller's "picture" of the traffic (Mogford, 1994), and tacit understanding (Hopkin, 1994c), all of which await more refined definition and agreement as psychological constructs. The value of applying psychological theories to air traffic control may be to offer new concepts and approaches to old problems (Logie, 1993), to identify different limiting factors (for example, attentional rather than perceptual ones), to integrate factors previously treated separately (for example, through theories of motivation), or to distinguish and categorize different phenomena previously treated as equivalent (for example, different kinds of human error; Reason, 1990; Stager, 1991; Senders and Moray, 1991).

If the human is treated as a system component, obvious human factors issues concern the effects on system functioning of the interface between the human and the system through which human influences are exercised. These influences on the conduct of air traffic control depend on the information presentation, its cognitive processing, and the subsequent human actions that constitute task performance. Human cognitive attributes are important because they prescribe the capabilities and limitations of the human as a system component and the conditions for the successful matching of the human with other system components. Human cognition can be studied broadly (Solso, 1991), through individual differences (Carroll, J. B., 1993), through mental imagery (Logie and Denis, 1991), in industrial settings (Lord and Maher, 1989), and through human activity (Still and Costall, 1991).

This chapter is oriented towards the human as a system component to the extent that it focuses on human cognitive attributes and characteristics that are not grossly modifiable and to which the system must therefore adapt as a condition for successful matching. These processes are considered in the approximate sequential order of their normal occurrence, beginning with the cognitive processes applied to presented data and ending with the cognitive precursors of human control actions, although it is an artifice to separate these processes that constitute an integrated entity. Subsequent chapters consider some broader characteristics of the human as a system component, and examine more extrinsic and less cognitive human attributes associated with the interface design, the workspace or the physical environment.

4.2 Perception

Although many basic studies of perception which seek to elucidate human capabilities and limitations (Marr, 1982) may seem remote from the practical concerns of air traffic control, their findings must not be ignored for they define the human perceptual requirements that must be met in all air traffic control recommendations. While performing air traffic control tasks, the human controller synthesizes two broad kinds of information, namely, what is known beforehand and what is being presented. A primary perceptual issue concerns the feasible and desirable compatibility and integrality between the presented information and the already known information which is mainly in the form of professional knowledge.

The requirements for data presentation are also of two kinds. The first refers to the sensing of data, which in practice means that they must be visible or audible to the controller. The second refers to the perception of data, which means that the controller must interpret their structure, coherence, unity and meaning correctly. Successful information presentation entails correct sensing and correct perception. Data must be sensed and must make sense. Evidence of both kinds is adduced to confirm the successful presentation of air traffic control information. Failures of either kind mean that the data affected are not available to the controller, who therefore cannot use them.

The correct application of well-established psychophysical principles will ensure that data can be sensed. This does not imply that they will always be detected, only that they can always be detected. The psychophysical principles that are usually applied to guarantee detectability can also guarantee lack of detectability of unwanted sensory variations provided that the associated changes in presentation all remain below sensory thresholds. Perception depends on the magnitude of the minimum discriminable changes and rates of changes of stimuli. For most dimensions of psychophysical stimuli which can be differentiated perceptually, the minimum absolute magnitudes that can be sensed are known, and also the just noticeable difference thresholds which specify the minimum differences required between stimuli before they can be sensed as different. Normally, all presented information must be far from all these thresholds and minimum magnitudes, to prevent frequent and incurable human errors due to insufficient discriminability.

In practice, the perception of useful air traffic control information is confined to the two sense modalities of sight and hearing. Data from other senses—touch, taste, smell, heat, cold, pain or kinaesthesis—may occasionally obtrude, but they are more likely to be distracting than informative. Most air traffic control information is visual, a trend likely to be accentuated, at least until any more widespread adoption of direct voice input or automated speech synthesis. A busy controller may use both visual and auditory information continuously.

Some broad differences between sight and hearing as sense modalities can serve as criteria if the technology permits a choice between them in air traffic control. Auditory stimuli are essentially temporal in that their presentation occupies time. Visual stimuli are spatial and their presentation occupies space. Two or more visual stimuli can be presented sequentially or simultaneously, but auditory ones must usually be sequential. Visual stimuli offer good referability through continuous presentation but auditory stimuli are transient. There are fewer coding dimensions for auditory than for visual information. Most visual stimuli and some auditory ones have to be coded in advance, but auditory speech may offer flexibility and less pre-categorization. Visual stimuli may have to be sought, whereas auditory stimuli may have to be remembered. Higher rates of information transmission both to and from humans can usually be achieved with visual than with auditory data. Visual stimuli depend on direction of viewing and this can affect greatly their capability to command attention, whereas auditory stimuli tend to be more demanding of attention and are less dependent on their direction of origin. On the whole, hearing resists subjective fatigue better than vision. Reaction times to simple auditory stimuli can be slightly faster than those to simple visual ones. Noise can blot out auditory information, and glare, reflections or inappropriate ambient lighting can blot out visual information. Successful perception depends on the absolute intensity of the visual or auditory stimuli, and on their relative intensity in context, often expressed as a signal-to-noise ratio. To be perceived, visual data have to be of appropriate size, duration, intensity, contrast, shape and colour, and must not move too quickly. Different perceptual criteria apply when tasks require different discriminations, such as an absolute judgement that a particular symbol, specified by its shape, size, colour and intensity, is the one portrayed, or a relative judgement that two portrayed symbols are the same or different.

In perception, the whole is often not the sum of its parts. It is vital to be guided by this principle when choosing visual coding dimensions, symbols and icons. Perceptual data can be distorted or greatly changed by other perceptual data alongside them, near them or superimposed upon them. For example, if two colours have different meanings their combined meanings cannot be conveyed by superimposing them for the outcome will be a third colour and neither of the original colours may remain recognizable. The same applies to shapes. If two are superimposed, the combined shape may be perceived as a different shape unrelated to the other two. It is common to employ one visual coding dimension for one attribute and another dimension for another attribute but unwise to presume that the combined codings will be perceived as the combined attributes. This can place severe constraints on the design of icons which often must portray combinations of information categories. Superimposed sounds may not be heard as their constituent sounds either.

The psychological study of illusions can illustrate principles of perceptual structuring through exceptions where the principles are insufficient or inappropriate. The demonstration of an illusion reveals some discrepancy or anomaly which can amuse or bemuse, but if an illusion that is present is not perceived as such the human remains unaware of the resulting misperception. In applied contexts such as air traffic control, this could be dangerous, and it is vital to check that none of the information presented could induce gross misperceptions of this kind. Visual data which cannot be structured or which seem to have no meaning may not be perceived as information but treated as irrelevant or as clutter even if they are actually essential.

There have been many laboratory studies on the visual perception of forms. Some of their findings can provide guidelines on discriminability or legibility but not all laboratory findings hold true in real life. Psychophysical principles establish what cannot be discriminated, and such evidence is applicable to displays of air traffic control information, but these principles are less successful in providing practical recommendations on how information should be depicted. Some of the variables in laboratory studies are too crude: for example, many ergonomic recommendations on alphanumeric character designs lack the subtlety of traditional typeface font designs which consider criteria of legibility and aesthetics.

From psychophysical principles, checklists of factors relevant to information displays can be compiled (Hopkin, 1982a), and these factors should be included in their specification and in evaluations of them. Meeting such perceptual criteria is a necessary but not entirely sufficient requirement to optimize displays for human use. For example, displays should not flicker perceptibly and the appropriate psychophysical evidence can be applied to prevent perceived flicker, provided that it takes account of all other relevant factors such as eyesight standards, other luminous displays within the workspace, and the spectrum and intensity of the ambient illumination. However, this may not obviate problems originating from subliminal oscillating luminances which, being subliminal, are not among the perceptual factors studied. Luminous information, for example on a cathode ray tube, and non-luminous information, for example printing on paper, are not perceptually equivalent.

Perception is a function of the perceiver. Whether individuals can discriminate a symbol depends partly on their corrected or uncorrected visual acuity, on the retinal location of the symbol, on the state of adaption of the pupil, and on where the eye is focused. If the individual perceiver is tired or distracted or if there are gross differences in luminous flux as the perceiver scans the workspace, perception can be impaired.

Normal human perception is not experienced as an analytical process. It is immediate and continuous, and highly selective without seeming so. It incorporates structuring of the sensed data, and the addition of meanings and interpretations according to prior knowledge, experience, expectancies and needs. Awareness is of the products of perceptual processes and rarely of the processes themselves. If the products are flawed because faulty processes lead to misperceptions or failures of perception, the human will rarely become aware that this has occurred. The human can never monitor independently of perceptual processes but only through those processes, and therefore cannot recognize errors and omissions which are intrinsic to the perceptual processes themselves, for such recognition would require an independent mechanism.

If the same air traffic control information can be portrayed in alternative forms, such as alphanumeric, symbolic, graphical or pictorial, these should be appraised in perceptual terms. Technological innovations may permit less rigid divisions between sense modalities so that information that was originally auditory may be perceived visually, or visual information may be converted into sounds. Certain basic questions about presented data can be answered at least in part from evidence about psychophysical perceptual principles. Examples include:

1. What are the basic human sensory limitations?
2. What is the pace at which information can be perceived?
3. How can different codings affect this pace?
4. What kinds of human error do alternative forms of perceptual information induce?
5. What influence do alternative perceptual principles have on learning, memory and understanding?
6. How can perceptual principles help to integrate information?
7. How can they be used to depict qualitative values as distinct from quantities?
8. What are the respective merits and disadvantages in perceptual terms of alternative codings such as colour, shape, size or intensity?

The application of perceptual principles takes account of the following considerations:

1. relevant characteristics of the viewer;
2. the task requirements;
3. the level of detail of the information portrayed;
4. alternative methods of portrayal;
5. relationships among displays and between displays and input devices;
6. the interrelationships between information categories;
7. distinctions between what has changed and what has not;
8. the requirements for precision, clarity, contrast, accuracy and stability of information.

There is also the need to allow for other human characteristics. Unreliable information portrayed with crystal clarity can mislead if clarity of portrayal is interpreted to signify reliability. Qualitative as well as quantitative attributes of the portrayed information should be matched with task requirements. Perhaps information that is degraded should be portrayed so that it is perceived as degraded.

In air traffic control, the controller builds a mental picture of the traffic under direct control and other relevant traffic such as pending traffic. Generally, this picture is mainly perceptual, although it may also rely heavily on time information since time rather than space may be the better guide to the relative urgency of tasks. An outstanding issue is how far displays such as radar displays and tabular displays could or should be specified to match the controller's mental picture of the traffic so that the information portrayed does not have to be recoded before it can be integrated into the mental picture. The integration is primarily a cognitive function, but the success with which it can be achieved must depend in part, and perhaps mainly, on the perceptual attributes of the information. An associated issue is the extent to which controllers' mental images and models can be manipulated by differences in the portrayed information, in training, in procedures or in tasks. Individuals differ greatly in their innate mental imagery and in the kinds of mental picture that they are capable of forming. At one extreme some individuals may be able to generate and retain very vivid and detailed visual images, whereas others have stronger auditory than visual imagery. Technological advances that could make it feasible to portray information in alternative forms to match individual differences in imagery raise the issue of whether this is worth attempting.

4.3 Attention

Air traffic controllers, and others with jobs in large human–machine systems, are presented with a great deal of information, in auditory and visual forms. A common characteristic of the human as a component of such systems is that the limitations imposed by attention are more severe than perceptual limitations. Far more information can be presented than can be attended to. Attending occupies time, and switching attention occupies additional time.

The design processes of air traffic control systems should employ task analyses or equivalent techniques to describe the envisaged tasks of the controller and to adduce the information required for those tasks. Appropriate forms, contents, codings and levels of detail of the information can then evolve, and the information requirements of the complete range of tasks at each workspace can be specified. This process is driven primarily by tasks and requirements. Many tasks are performed at each workspace. Very few items of information are needed for every air traffic control task. When all the needs of many tasks have to be met, it is inevitable that far more information has to be provided or available than would be optimum for any single task. Almost every item of information is needed for more than one task, though its relative importance will be task dependent. The indispensable information for some tasks is an irrelevant distraction for others. Attention is disrupted by distractions.

The provision in usable form of all the information necessary to perform a task cannot guarantee that the task will be performed as intended, or performed at all. The controller has to perceive and understand the information in order to use it. Any surfeit of information entails some guidance of attention towards what is pertinent, so that time is not wasted attending to irrelevant information. The direction of attention requires some delicacy and finesse for it is counterproductive to treat attention like a bludgeon. Some of the most effective means to attract, direct or sustain attention cannot be adopted at all in air traffic control, or only very sparingly. Highly intense visual or auditory stimuli can always attract attention, at the unacceptable cost of interrupting or disrupting all current tasks. The most noticeable sounds or visual stimuli have to be reserved for the direst emergencies when the performance of all other tasks can be sacrificed. Since no items of perceived information have the same relative importance for every task, no routinely used item of auditory or visual information should become so dominant and attention-demanding as to obtrude whenever it is presented.

Codings that are different from those employed to code information for tasks may be needed to direct attention and curtail unproductive searching. The most appropriate ones generally involve movement or changes. Distinctive movement can be highly effective to attract attention but quickly becomes distracting and irritating if it lingers after attention has been attracted. Changes can be in size, shape, colour or intensity, can take the form of deletions or of additions such as asterisks and underlining, or can appear as oscillations or flashing. Commonality of coding can also direct attention effectively, as in coding two potentially conflicting aircraft in the same colour, size, shape, intensity or flashing, so that the controller perceives and attends to them as a pair.

Extensively debated theoretical and practical issues about human attention include the ways in which it is analogous to single or multiple information channels, its capabilities for serial and parallel information processing, and its channel capacities. With experience and familiarity, automaticity of human tasks can become widespread, whereby information is perceived, interpreted and acted upon with little or no attention to it. Many skilled functions of the air traffic controller possess this characteristic, and for some of them it is intrinsic to skill acquisition. This bears a complex relationship to situational awareness, a recently accepted and popular concept in air traffic control and other contexts. Intuitively, it seems inherently unsafe to perform tasks while remaining unaware of them even if they are performed well, but evidence to support this view is sparse, and commonsense may be a poor guide here. The important practical implication is that direct measures of the skilled performance of complex tasks must not be construed as implying attention to those tasks.

Attention switches from one topic to another. Subjectively, this switching sometimes seems controllable and sometimes involuntary. An approach to the study of attention and information processing that used to be in vogue employed a secondary task, to be done whenever attention could be spared from the main task. Time spent on the secondary task was equated with spare mental capacity. The approach fell out of favour for several reasons. It treated total mental capacity as a constant which it does not appear to be; it took insufficient account of the additional time and resources required to switch attention; it neglected the need to keep monitoring the primary task in order to know when to attend to it again; and it presumed that the primary task could be resumed as if its performance had not been interrupted.

The characteristics of human attention limit the amount of information that the controller can deal with. The main means to optimize attention in air traffic control are:

1. to code and present information in familiar forms so that the controller needs to attend to each item for as short a time as possible, can interpret it immediately, and does not have to recode any of it;

2. to train the controller so that all the presented information can be understood correctly;

3. to use appropriate codings to direct attention to the most pertinent items in the best sequence;

4. to present information at the optimum levels of detail for the tasks so that the controller does not have to seek more detailed information nor attend to irrelevant information only to discard it;

5. to eliminate the need to spend time searching for information, which employs attention unproductively;

6. to ensure adequate discriminability of all information so that attention to it is not prolonged by poor discrimination;

7. to use codings that make apparent and thus draw attention to significant associations between displayed items;

8. to minimize the distraction or disruption of attention from extraneous sources;

9. to select codings to attract attention to what is relevant, but select other codings to divert attention from what is irrelevant.

4.4 Learning

An air traffic control workspace is nearly meaningless to anyone with no knowledge of air traffic control. It contains very little information about its nature, objectives, functionality, tasks, or forms of feedback. There is no information at all on a paper flight progress strip about what it is or what it is for. Air traffic control therefore depends entirely on human learning. Even in the most sophisticated systems now being planned, with extensive automation and computer assistance, without the presence of the human air traffic controller there would be no air traffic control.

The human–machine interfaces in most current air traffic control systems are not designed primarily to support or extend the controller's learning but rather to enable the controller to apply what has already been learned. The functionality of future human–machine interfaces in air traffic control could be made more self-evident through labelling, changes in menu and dialogue design, the formulation and presentation of options and reminders, and the direction of attention. Technical developments that could support human learning in air traffic control better have revealed some ambivalences of policy about whether the functioning of human–machine interfaces should become more self-evident because they are also designed to function as teaching aids.

Just as the controller's tasks must not make excessive attentional demands, so they must not impose excessive learning demands either. Tasks that are so complex or cumbersome that only those who originally devised them can ever learn how to perform them properly are of no practical use in air traffic control. It is not sufficient for a task to be feasible; it must be teachable so that all or almost all current and future controllers will be able to learn to perform it consistently to an acceptable standard. This implies that the task must meet the following requirements:

1. it must make sense in air traffic control terms;
2. it must, after learning, be performed to the required standards, and must not take too long or impose undue demands on the controller's time and resources;
3. it must enable all who have to do the task to learn how to do it without protracted training or high failure rates during training, and without any modification of controller selection procedures to accommodate the task requirements;
4. it must be fully compatible with existing learning, knowledge, skills, procedures and experience;
5. it must provide sufficient knowledge of results for the performance of those learning it to improve with experience;

6. it should have definable required levels of learning and achievement, with objective measures to prove that each controller has attained those levels;
7. it should preferably incorporate a hierarchy of skills so that learning can progress beyond procedures, rules and knowledge, to improved situational awareness and understanding and to appropriate forms of automaticity.

Many learning methods can be applied in air traffic control. Training usually benefits from employing a combination of techniques. These include traditional classroom instruction, demonstrations, practice of the basic procedures and skills, simulation exercises of increasing complexity, worked examples and case studies, supervised training with the pace and content adjusted to each student's responses, and standardized tests and assessments as indices of progress. Much air traffic control involves teamwork, and controllers must learn during training how to be effective and collaborative team members. Stand-alone training packages may provide extra learning of tasks, skills or procedures for students who need more practice.

It normally pays to overlearn the most fundamental and important aspects of air traffic control, by continuing to practise them for rather longer than seems necessary. When learning employs examples that increase progressively in complexity, as it must often do, extra examples of the most difficult problems should be given, lest the student gains least experience of what is most difficult and most experience of what is least difficult, which is not a sound principle. Criteria with which to judge the efficacy of teaching methods and techniques should cover the long-term retention of learning; frequent relearning should not be needed.

Knowledge that is never needed is more likely to be lost. Skills that are never practised gradually atrophy. Procedures that are never followed may eventually be forgotten. Learning requires continuous maintenance in the form of frequent opportunities to apply it. Refresher training may be essential where this condition is impractical, as in rehearsing the correct responses to emergencies. If much of what must be learned is never used, the necessity to learn it at all should be questioned. It is dispiriting to possess extensive hard-won professional knowledge that seems irrelevant and wasted.

Air traffic control is learned initially at a specialist air traffic control or aviation school, college, academy or university. This learning is entrenched and enriched and becomes practical through on-the-job training, controlling real traffic under the close supervision of a qualified controller. Professional competence and prowess are confirmed before the newly trained controller is licensed to work with the normal level of supervision.

There is a distinction between what is taught and what is learned in air traffic control. Prior to on-the-job training, the rules, procedures, processes and knowledge required for air traffic control are formally taught so that the controller understands air traffic control and acquires the requisite skills. Much of what the controller learns on-the-job is never formally taught yet is of great importance. It includes the norms, standards and ethos of the air traffic control profession; acceptable attitudes to colleagues, pilots and management; the requirements for full acceptance as a team member; how to gain and keep the trust and respect of colleagues; what is tacitly presumed and what must be overtly expressed; and specific universally accepted local practices. This kind of learning is often a professional hallmark, familiar to every experienced controller.

4.5 Memory

Because few aspects of air traffic control are self-evident, each controller relies on memory to interpret what is displayed or available for display, and to control the air traffic. In a sense, the objective of all learning is to remember, as the influence of learning is at best indirect and at worst non-existent without recall of what has been learned. Most tests of learning take the form of tests of memory. There is therefore much concern in air traffic control with the causes of fallibility in human memory, with practical means to sustain and support it, with its guidance and direction, with circumstances when it seems prudent to provide automated reminders rather than rely on human memory, and with ways to influence and improve memory.

The requirements of memory and the influences on memory do not seem to have received much consideration in planning and design decisions about air traffic control systems, although many such decisions, such as the choice of visual or auditory codings, can affect memory greatly. This is not always recognized, and any cognitive considerations that influence such decisions usually relate to perception rather than memory. For example, when colour coding replaces monochrome on a radar display, the main emphasis is on its visual benefits, which are important. However, information tends to be structured and categorized by its most visually dominant codings, and entered into memory, stored in memory and retrieved from memory according to this structuring. Colour is a dominant coding, and its introduction can influence what is remembered, how it is remembered, and the attributed relationships between remembered items. In principle, codings that structure information appropriately for tasks and render it more memorable in ways that support task requirements could be a powerful tool to help the controller to recall the most apposite information in directly applicable forms that require no recoding. The extent of this potential benefit is not yet known because the differential effects of alternative codings on memory have not been quantified sufficiently to permit recommendations about it.

Some of the roles of memory in air traffic control do not fit the main theories of memory very well. Theories tend to emphasize timescales of a few seconds for short-term memory or relative permanence for long-term memory, or refer to active task performance for working memory (Baddeley, 1990; Logie, 1993; Stein and Garland, 1993a). The controller relies on a mental picture of the traffic that is based on a synthesized integration of radar, tabular and communicated information, interpreted according to professional knowledge and experience (Whitfield and Jackson, 1982; Rantanen, 1994). Although a simplified form of the picture can be built in a few seconds, as is routinely done at watch handover, building the complete picture requires more processing (Craik and Lockhart, 1972) and typically takes about fifteen to twenty minutes, by which time the controller knows the full history and intentions of all current and pending traffic and can plan accordingly.

Practical steps to compensate for the fallibility of human memory are taken routinely in air traffic control. One safeguard is a policy of widespread redundancy in procedures and communications. Another is to repeat, check and read back information rather than to rely on memory of it. A further safeguard is to update information regularly. The compulsory recording, by annotation or other means, of all actions incidentally reinforces memory. The need to reappraise the traffic whenever a decision is taken or new information is added also tends to entrench memory. Standard requirements for positive actions and acknowledgements support memory also. Openly accessible air traffic control workspaces and information increase the likelihood that a colleague will notice or remember anything that a controller has forgotten. As far as possible, controllers are not interrupted while working, lest they fail to remember to complete a task or recall incorrectly what they have done. A positive action signifying the completion of each task helps the controller to remember that it has been accomplished.

Various forms of computer assistance are intended to aid human memory. They depend on similar principles to those adopted by controllers in more manual systems when they offset a flight progress strip holder sideways as a reminder of an uncompleted action or a future check. Computer assistance can aid memory by signalling a need to confirm, initiate, contact or act, but the signals must be self-evident and familiar so that their significance is remembered correctly. If they remain unheeded, it must be possible to render them more obtrusive by increased intensity or other means. Codings intended initially to attract attention, such as of a pair of aircraft in potential conflict, can double as memory aids if they are in a suitable form to be retained until the conflict has been resolved. They denote the presence of a conflict rather than a solution of it.

Most relevant psychological theories employ the concept of memory, and related constructs such as recognition and recall. In air traffic control and aviation generally, the term "memory" is used to refer to applicable theories. This is unfortunate in some respects, because many practical problems concern forgetting rather than memory, and the theoretical relationship between memory and forgetting is confused and disputed. The purpose of most studies has been to aid memory. The nature and causes of forgetting have seldom been studied directly. Forgetting as a boon rather than a bane has scarcely been studied at all. Yet it is not always an advantage in air traffic control to be able to recall all the details of what happened previously, as this could encourage unwarranted presumptions that any intervening changes of circumstance are trivial and that previous solutions can be adopted again, whereas it might be better to work out fresh solutions without such remembered preconceptions. Some limited guidance on how to code air traffic control information to make it more memorable can be offered, but there is no comparable practical guidance on how to code air traffic control information so that it is easy and efficient to use while it is present but is readily forgotten after it has served its purpose and there is no benefit in remembering it. Given the perennial problem of too much information in air traffic control, recommendations on how to render the useless forgettable would have real practical value.

4.6 *Information processing*

As the forms of computer assistance in air traffic control have expanded beyond the gathering, storage and selective retrieval of data to support human cognitive functions, the controller as an information processor has been studied more closely. The design of the human–computer interface must facilitate not only the transmission of information from machine to human through displays and from human to machine through input devices but also the controller's processing of the information, since many of the products of that processing are reconverted via the human–machine interface into system functioning. The effective matching of machine capabilities and human information processing capabilities has become a major issue in air traffic control. The actual or impending advent of many feasible kinds of computer assistance, expert systems, and machines that can adapt and be flexible will demand the productive matching and reconciliation of human information processing characteristics with these advanced developments.

Many measures related to human information processing are feasible. The information presented to the human can be varied and controlled in specified ways. The resultant human actions can be recorded, together with the lapse of time between the information presentation and their occurrence. The effort involved in information processing can be gauged by psychological or biochemical indices. Any additional information obtained by the human can be noted. A technique such as eye-movement recording can indicate what information is looked at as distinct from presented, though it may be more effective for revealing what information is never looked at than for proving what is. Subjective evidence can be obtained about information processing, though it can be devalued by well-known deficiencies of human memory, or by difficulties in expressing in words the full subjective complexity of human information processing.

There is a mixture of data and resource limitations on air traffic control information processing (Norman and Bobrow, 1975). Information processing that precedes task performance by the controller involves the following:

1. the gathering of information;

2. the collation of information of different kinds or from different sources;
3. the categorization of information according to its meaning and implications;
4. the selection of the relevant information for each task and the discarding of the irrelevant;
5. the recognition of the interrelationships among information categories;
6. the definition and appraisal of options, choices and alternatives;
7. the condensation, summarizing and integration of information;
8. the identification and search for any essential information that is lacking;
9. the processes of reviewing the relevant information available.

The conversion of data into information involves the imposition of meaning. Information that remains meaningless or irrelevant will not be used, whether it should be or not. Information that is not initially meaningful may become meaningful during information processing, and may acquire additional meanings, for example if the controller detects in it a potential hazard or problem. Information processing does not necessarily lead to overt actions by the controller, who may conclude that an apparent hazard is not a real one and take no action or that actions should be postponed pending clarification by further information.

Human information processing is subject to several characteristic kinds of bias. The most recent events may exercise too preponderant an influence. Expectancies and assumptions may lead to interpretations of the perceived information that go beyond the factual evidence. A high level of confidence in human judgements may be impervious to strong evidence that it is misplaced. Human notions of probability, chance and randomness are not in full accord with the corresponding mathematical concepts. Humans may seek patterns, correlations and commonalities where there are none, and exaggerate the importance of any that are found. The imposition of preconceived classifications may lead to inappropriate categorizations of events. Humans have great difficulty in processing and weighing correctly information from several disparate sources, and any attempts to do so take a long time. Information in such forms, which are common in air traffic control, has to be processed by collation and integration to become usable. Given a dozen information sources, the human tends to rely almost exclusively on a favoured three or four; subjectively this may be described as the adoption of favoured and proven stratagems or tactics, but it is inefficient and biased in terms of information processing. Any essential information processing of this kind should be a prime candidate for computer assistance. The problems of weighting several information sources are compounded in air traffic control by disagreements about what the weightings should be. This can sometimes account for observed differences between controllers in their methods and tactics, even though the outcome is similar in terms of traffic flows and safety. Air traffic control has multiple criteria—the safe, orderly and expeditious flow of air traffic—but different controllers may weigh the requirements of orderliness and expedition differently whenever they conflict. The consequences for air traffic control of individual differences in cognitive style have been neglected (Streufert and Nogami, 1989).

Changes made for other reasons may influence human information processing more than they were intended to do. Information provided automatically is seldom processed as deeply by the human as information that has to be actively sought and obtained. The passive monitoring of changes made by the machine differs in information-processing terms from active intervention by the human to make the same changes. The practicality of the user-centred forms of computer assistance often advocated for air traffic control, and the quest for user friendliness so that computer assistance seems helpful rather than obstructive, both imply that computer assistance must support and be compatible with human information processing. Any products of that processing that cannot be implemented are defeated. To point out that the human does not know what the human does not know can seem trite and tautologous: what it means is that the human remains ignorant that pertinent information is available unless able to recognize its pertinence. Human information processing can be supported effectively by training and experience so that the controller is aware of all the information that should be processed, and by queries and reminders if relevant data seem to be ignored.

4.7 Understanding

The concept of understanding is not mentioned much in air traffic control objectives or evaluations, yet the desired culmination of all the foregoing processes of perception, attention, learning, memory and information processing is understanding. A common assumption is that the better the controller's understanding is, the better the air traffic control will be.

Although understanding is not emphasized, other concepts dealing with aspects of understanding are widely discussed and studied. Foremost among them is the notion of the controller's picture. This is a more dynamic construct than the conventional mental model, but the reported phenomenon of loss of picture is primarily a loss of understanding. The rebuilding of the lost picture and of understanding is not an instantaneous restoration of it as a whole but the painstaking recapitulation of the sequence of cognitive processes described above.

The controller's picture consists of all that is perceived and is meaningful, interpreted in the context of recalled events preceding the current situation, anticipated events predicted from the current situation, and professional knowledge and experience used to maintain control over the air traffic through sanctioned rules, practices, procedures and instructions. The picture is not static but evolves continuously with the traffic flow. Potential loss of the picture worries the controller because the accompanying loss of the meaning and understanding of what is perceived is usually accompanied by a strong subjective impression of temporary loss of control.

The circumstances which can lead to loss of picture provide clues about the conditions for maintaining the controller's understanding. Sometimes the controller is aware beforehand that the traffic picture is liable to be lost and tries hard to prevent it, but occasionally the loss of picture is sudden and without warning, though it is not random. Circumstances associated with it include a combination of high task demands and time pressures with heavy traffic and no prospect of respite, or premature relaxation following the successful resolution of a difficult air traffic scenario, or the discovery of unexpected additional information or of missing information. Following loss of picture, the controller has to rebuild the picture step by step, in the knowledge that, when this has been accomplished, similar circumstances to those which induced the original loss of picture will again prevail.

A more recent concept, originating in other aviation contexts but applicable to air traffic control, is situational awareness. This has affinities with understanding, although whether it may be measured in terms of performance or only subjectively has yet to be settled (Gilson *et al.*, 1994). Situational awareness includes knowledge of the present, past and pending situation, of system functioning, of human roles and tasks, and of roles, procedures and objectives. It may also include automaticity, the fulfilment of functions by the human without apparent conscious awareness, and in this respect it is broader than the notion of the controller's picture. It remains somewhat ill-defined (Sarter and Woods, 1991), but is no longer neglected (Garland *et al.*, 1993), and the effects on the controller's situational awareness of future automation (Garland and Hopkin, 1994), of technological advances (Hoshstrasser and Small, 1994) and of artificial intelligence and human-centred automation (Endsley, 1993) have all been considered. Full situational awareness is considered beneficial as an indication of better understanding, but whether it is possible to have too much situational awareness depends on how the concept is interpreted (Hopkin, 1994b). As a concept, situational awareness seems to attract more interest now than the older concept of imagery, though there is some evidence that the mental imagery of air traffic controllers may be more vivid than normal (Isaac and Marks, 1994).

Understanding can be impaired if active human involvement in tasks is replaced by passive human monitoring of a machine performing them. Whether proposed system changes will affect understanding can therefore be predicted to some extent by deducing their consequences for human involvement. The activities that are most crucial for understanding may remain indeterminate, but a link between level of involvement and understanding has been established (Narborough-Hall, 1987). Sources of errors in air traffic control associated with misunderstanding or inadequate understanding may be removed if they can be traced to deficiencies in perception, attention, learning, remembering or information processing.

Although some aspects of understanding affect performance, indices of understanding may relate only indirectly to performance measures. A controller may understand a form of computer assistance well enough to use it effectively according to its designer's intentions, but not well enough to recognize when it is malfunctioning. The meaning of evidence formulated by computer assistance may be understood correctly, but not its full implications. Improved understanding of computer assistance on the part of the controller may be inferred from ability to adapt to it, to use it innovatively or flexibly, or to act intelligently in conjunction with it, rather than from its application to the performance of standard tasks. A tidy and well-ordered workspace may be a sign of better understanding of the work and its requirements.

The importance of another kind of understanding in air traffic control, hitherto underestimated, is becoming more apparent because some forms of computer assistance affect it. Air traffic control has usually been a team activity conducted in an openly observable workspace. Each controller gains a tacit understanding of what immediate colleagues know, of what they can be relied upon to do, of what they prefer, and of how they will respond to the controller's own actions. Effective air traffic control teamwork may involve little overt communication between controllers but extensive tacit mutual understanding. Task analysis, modelling, simulation and evaluation may all neglect this tacit collaborative understanding, because there is nothing to measure. Computer assistance that fails to recognize it or to compensate for induced changes in it can undermine the mechanisms by which tacit understanding is acquired and curtail the opportunities of benefiting from it. The controller's understanding therefore encompasses the air traffic control itself, and also tacit understanding of how it is conducted.

4.8 Planning

In this section the traffic-planning activities by controllers, as distinct from the planning of air traffic control systems, are discussed. A complication is that jobs dealing with the planning of traffic flows are sometimes described as traffic management, which can be confusing when traffic flow managers who deal with air traffic also have managers with managerial responsibility for them. As air traffic control evolves from the tactical control of single aircraft towards the organization of traffic flows in advance, longer-term planning becomes a more important feature of the controller's work, and human cognitive capabilities and limitations in regard to planning functions assume greater significance. The emphasis swivels from solving problems to preventing them. Residual problems that still have to be solved tactically may become an index of planning efficacy if they can be construed as a consequence of planning failures, although some tactical manoeuvring will always be required to deal with the unexpected and unforeseeable. Widespread tactical manoeuvring can invalidate retrospectively plans based on the presumption that aircraft will follow their intended paths.

Some current air traffic control jobs such as that of the oceanic planner consist mainly of planning functions. They deal with actual or expected traffic and traffic flows rather than with aircraft singly, and either plan future capacities and routes or allocate aircraft to flows or slots. Their timescales are longer than those for tactical air traffic control, usually at least half an hour and often considerably more. Where they do deal with single aircraft, they may exercise some control concurrently over larger numbers of them than more tactical controllers deal with. Some planning functions envisage more negotiation with pilots than occurs in tactical air traffic control.

Several technological advances have extended the range of planning options. The most immediately applicable ones are prediction aids. These compute future system states from current states plus known intentions. They may present the predicted appearance of the traffic at some future time, or may indicate the implications of proposed alternative actions before the controller is committed to them. The indications may state possible problems or rule violations, without showing the predicted traffic situation in full detail.

The validity of predictions depends on the quality of the data available to compute them, in terms of precision, accuracy, reliability, consistency, and frequency of updating. In general, the more specific the predictions become and the further into the future they reach, the less valid they are likely to be. At some stage, perhaps beyond about 20 minutes ahead for the detection of conflicts in congested en route traffic, predictions lose their value because their benefits are outweighed by too many false alarms or, more seriously, because the data become inadequate to detect every potential conflict. Intervening factors include the closeness of the conformity between each planned flight and the actual flight, and the extent to which flights, both planned and actual, remain in a steady state without any changes in flight level, speed or heading. Such manoeuvres increase prediction errors through greater variability between aircraft and reduced data consistency. Technical aids can provide evidence about the validity of predictions and the confidence that should be placed in them.

Some forms of computer assistance for planning tasks are crucially dependent on the controller's correct understanding of their intended planning functions. Plans, whether based on the current traffic situation or on pending flights in the form of filed flight plans, require extrapolation, which in turn relies on professional skills and knowledge, on experience of how situations are likely to evolve, on possible circumstances that could invalidate the plans, and on judgement of appropriate solutions for predicted problems and of their optimum timing. Timing that is seriously awry can invalidate plans and require the formulation of alternatives.

The successful progress of flights in the future may depend far more on planning than it has in the past. Capacity limits on popular routes, or restricted access to airspace or to airports because of severe weather, may result in traffic demands that exceed handling capacity and lead to restrictions or diversions, or some form of rationing. This requires planning, whether it takes the form of traffic flow management which redistributes, smooths or curtails the total traffic within large regions, or the allocation of slot times which impose a maximum capacity on a particular route. Traffic flow management is generally applied before aircraft become airborne, and incurred delays are on the ground rather than in the air. Slot times are planned and allocated for every sector of a flight before aircraft departure. Much of the workload formerly associated with tactical adjustments to the progress of an airborne flight becomes incorporated in the preplanning of the flight within a traffic flow. The allocation of a slot time puts pressure on the pilot to meet it or incur a long delay, for any unoccupied slot wastes system capacity.

Crucial for the planning of traffic are appropriate forms of feedback to the controller on the outcome of plans. Although learning from experience is contingent upon feedback, planning tasks often lack adequate forms of feedback. It is more difficult to provide appropriate feedback for strategic air traffic control planning tasks where the consequences of decisions may not become apparent for several hours during which there have been many intervening tasks, than it is to provide feedback for tactical tasks where the controller implements decisions and becomes directly aware of their consequences very soon afterwards. It may be quite easy to ascertain whether plans were implemented as intended and to judge if they were safe and satisfactory, but not easy to establish whether the plans were near the attainable optimum. Plans that seem good may discourage efforts to make them better. Plans which are poor but the best available may induce fruitless efforts to improve them. Better tools and measures to assess air traffic control planning, and models or other indices to establish the theoretical or attainable optimum, are needed urgently to demonstrate where further improvements are possible, to define the nature of those improvements, to indicate where further improvements cannot be achieved with the same forms of feedback, and to devise alternative forms of feedback to reveal where further improvements could be made. Common claims concerning a gradual but significant improvement with experience in the performance of air traffic control planning tasks require confirmation and quantification, for if they are correct there are forms of feedback in planning tasks that have eluded the training for them, but if they are incorrect the common belief in the value of experience is unwarranted. The pending transition towards more planning functions in air traffic control may require new and better forms of feedback to allow achieved performance to evolve from the merely satisfactory towards a definable optimum.

4.9 Problem solving

Problem solving is an intrinsic feature of air traffic control. Its specific nature varies according to the traffic and the means to control it. Problem solving will remain a vital aspect of the controller's work but it should become less pervasive as the emphasis of the work changes from the solution of problems to their prevention. Greater reliance on predictive information and on planning tasks implies that more of the controller's time must be allocated to the prevention of incipient problems before they can develop into actual ones. To the extent that this is successful, less time should be needed for the safe resolution of problems that have actually arisen, for there should be far fewer of them.

Traditional problem solving has been associated with the following air traffic control functions:

1. air traffic demands that exceed the traffic-handling capability of the system;
2. the planning and continuing verification of safe separations between aircraft;
3. the detection and resolution of specific potential conflicts between aircraft;
4. the crossing or amalgamation of routes and traffic flows;
5. the resolution of the disparate air traffic control requirements of various air users;
6. the successful control of particularly difficult or complex patterns of air traffic;
7. acceding to users' requests;
8. the provision of an efficient air traffic control service with minimum delays;
9. the recognition and resolution of inadequacies or ambiguities in the information provided;
10. the maintenance of safety;
11. the optimum usage of the available traffic-handling capacity;
12. the scheduling of work activities in relation to the current and pending traffic;
13. the quick detection of any sudden and unpredicted aircraft manoeuvres or emergencies and the safe resolution of their consequences.

An unforeseeable set of circumstances may arise at any time and demand urgent action. It is prudent to plan all other activities so that their safety would not be jeopardized in such an eventuality. In practice this means that the timescale for problem solving must always be sufficient, and preferably allow a small surplus to permit some flexibility. Problem solving in air traffic control can never afford to become totally time constrained by pressures of demand, for there may be difficulties in contacting a pilot or controller through busy communication channels, in resolving ambiguities, in reaching agreement on the best solution, or in responding to changed circumstances. The timing of problem solving must be flexible enough to accommodate such difficulties; if it is not, the probability of errors or of non-optimum solutions will increase because of excessive time pressures whenever such difficulties actually arise.

The reliability of data decreases progressively as the timescales of predictions lengthen, until their misleading aspects outweigh their predictive benefits. How far ahead it is worthwhile to anticipate and resolve problems is therefore a contentious issue. A preferred solution implemented at once might not be preferred if it could be shown that it will incur future problems that are undetected and perhaps undetectable by the controller at the time of its implementation. A future conflict between two aircraft currently in different sectors under different controllers may be predicted to occur in a third sector. The consequent human factors problems are not only about the resolution of the conflict, but also about the allocation of responsibility for resolving the problem and implementing the solution.

Future problem solving must emphasize the planning of the maximum traffic-handling capacity of the system since this will become a more frequent requirement. This kind of problem solving requires a different understanding of the traffic since planning is applied to traffic which is currently the active control responsibility of others. The further ahead that plans are made to handle traffic already airborne, the less independent the plans can be of the intervening tactical problem-solving activities of the controllers currently handling it. The efficacy of future plans depends on the predictability of flights, which implies minimum tactical adjustments to flight plans. The effective coordination of problem solving at the strategic planning level and at the controller's tactical level presents a challenge, which includes the best ways to represent different kinds of problem (Woods, 1991).

Technological advances can assist many problem-solving functions. Their optimum utilization depends on the controller's understanding of how they function, and their capabilities and limitations. In principle, a computer can derive a future traffic scenario from the current one better than the human controller can, and therefore it may formulate for the controller solutions to problems that can be foreseen when this is done. The acceptability of the products of such problem-solving aids depends on how well their functioning is understood, how far they are trusted, whether the controller can interrogate them about the factors contributing to or excluded from the proposed solutions, and the legal status of the solutions offered. A particular issue is whether the controller retains the legal responsibility for the safety of the solutions.

Because most forms of computer assistance for air traffic control are intended to be used by individual controllers, team roles in problem solving tend to decline and new categories of problems often cluster around the interactions between each controller's own activities and those of others. A controller who cannot readily see or discover what colleagues are doing is poorly placed to judge the full effects of the solutions of problems on others or to assist others to solve their problems. The combination of planning and problem-solving activities extends to the allocation of work among controllers and the consequences for traffic handling if misallocation results in inequitable workloads. The optimum allocation of future work depends on gauging correctly how busy each controller is and will become. Computer assistance removes some of the existing means to base such judgements on direct observation, while offering new means based on prediction and problem solving.

4.10 Decision making

Most actions initiated by the controller represent the implementation of decisions that have been made, and should be explicable in terms of appropriate theories (Klein *et al.*, 1993). Decision-making ability has been sought in air traffic controllers for a long time, though satisfactory measures of it in the selection procedures for controllers have proved difficult to devise and validate. Controllers need to be able to choose and assimilate the relevant information quickly, weight it correctly, reach safe decisions, and act upon them confidently. Ideally it is always desirable to make the optimum decision, but it is not always possible to agree on the correct weighting of the various factors sufficiently to specify an optimum. In practice, each decision must be safe, effective and acceptable. Time to mull over alternatives is often limited, and dithering over decisions is obstructive. A requirement to write software to support the controller's decision-making functions can enforce the tighter specification of all the definable rules and influences contributing to a decision, but leave unresolved the respective merits of alternative air traffic control decision strategies because of a lack of objective criteria by which they could be compared with each other or with an independent optimum.

Although tactical decisions about the control of air traffic can appear to be unrelated, this is seldom the case except in very light traffic. The implementation of any tactical decision normally has significant consequences for other decisions, to the extent that the purpose of each decision is not merely to ensure the safety and expedition of the flight or flights to which it applies but also to contribute to the orderliness of the traffic flow and to the provision of an effective air traffic control service. The controller has to understand the extent to which each decision must conform fully with normal practices. Decisions that seem wayward or arbitrary, even if they are safe and efficient, are unacceptable in system terms if they constitute unpredictable behaviour in relation to other aspects of the functioning system. The limited scope for each controller to adopt a favoured controlling style and preferences, as defined by that controller's decisions, must be exercised within strict bounds of conformity, to avoid such unpredictability and remain compatible with others' decisions.

Controllers who work together as a team or who belong to the workforce at a particular place often evolve their own local practices which are defined most clearly by the decisions they take. Some local practices are officially sanctioned. Others may lack formal recognition but are tacitly understood and universally followed so that the decisions by all controllers are expected to conform fully with them. They can cover such practices as variations of normal procedures, short-cuts, expected omissions, the reservation of a flight level or portion of airspace for emergency use, unorthodox routeings to accommodate unusual but pre-defined traffic configurations, and the implementation of local policies on such issues as requests by the pilots of light aircraft to transit the control zone of an airport rather than fly round it.

Decision making in the course of the active control of air traffic is a continuing activity that utilizes the controller's comprehensive understanding of information at several levels of detail, together with the controller's knowledge of rules, procedures and instructions and their permissible flexibility. Tactical decisions about the air traffic are influenced by the controller's competence, experience and skill, by local customs and conditions, and by the planning of decision making. This planning embodies the individual controller's style, which influences not only the decisions themselves but also their sequence and their absolute and relative timing. Some of the criteria for assessing decision making go beyond the safety, efficiency, timeliness and appositeness of the decision itself and include its compatibility with other decisions, to the extent that the main benefits of a particular decision may be on the flow of traffic, the system capacity or the quality of the air traffic control service to the user rather than on the aircraft most directly affected by it, though no aircraft should be consistently and significantly penalized by a series of decisions that are for the benefit of all. Classifications of decisions that ignore such factors cannot explain or evaluate them fully and cannot be used exclusively to derive optimum criteria for air traffic control decision making. Likewise, satisfactory measures of decision making in air traffic control must not ignore such factors.

The most obvious cognitive aspects of air traffic control decision making are those that relate directly to the control of air traffic. Nevertheless, these tactical cognitive decisions are made within the constraints imposed by a hierarchy of previous managerial and organizational decisions, about the planned system capacity, its equipment and data sources, the staffing levels, the practices and working conditions, the training, and many other matters. Much of the literature on management decision making is applicable to air traffic control and its management. The history of air traffic control encourages retrospective criticism of managerial decision making in many nations, but this can be ill-informed and unjust and is almost invariably unproductive. What is important is to recognize that the quality of tactical air traffic control decision making depends not only on the controller and the traffic but on previous generations of managerial decision making.

Decision making may be impaired by fatigue, by unsuitable procedures, by inadequate data or equipment, by poor communications, or by inappropriate system designs. It may also be impaired by deficiencies in the individual controller, in management, or in conditions of employment. Some decision-making problems may become better recognized but aggravated, as tactical decisions yield to decisions about the deployment of resources, the planning of flows, stratagems to increase system capacity, and the furtherance of cost-effectiveness. If some of the decisions required of future controllers have a greater managerial and organizational component, this may raise their awareness of managerial decision making that affects their jobs and may encourage them to scrutinize such decision making more impartially. However, the associated need to select controllers with potential abilities for strategic decision making could result in a workforce with more insight into the decision-making processes of their own management and more drive to participate in management decisions.

4.11 Motivation

For most air traffic controllers, being a controller is important to them. They identify closely with their profession and with their place of work which they believe to be the best of its kind. Camaraderie develops among the controllers there. They believe that the excellent air traffic control service that they provide is primarily a product of their own high skills and professionalism. Any disparaging remarks by controllers about their colleagues at other facilities who do not quite match their competence are confined within the profession, and controllers unite to refute criticism of their profession from outside it. Further prevalent beliefs are that their skills are insufficiently rewarded, that their management fails to appreciate their achievements, and that they deserve more modern equipment.

A hallmark of a profession is that its norms and standards are often generated internally rather than imposed from without. Air traffic control conforms with this. Controllers are sensitive to peer judgements, for each controller strives to gain and keep the professional trust and respect of colleagues and is highly motivated to attain or exceed high professional standards at all times. Any lapses in performance reflect poorly on the individual responsible for them and also on the profession, and can cause anxiety whatever their circumstances and whether they are common knowledge or known only to the controller concerned. Controllers try hard to perform their tasks well. The demonstration of outstanding professional competence gains self-esteem and the esteem of colleagues in so far as it is directly observable by them or deducible from the handling of the air traffic. The combination of high motivation and team identification means that controllers will normally support colleagues in difficulty as well as they can, to sustain the high reputation of the air traffic control service at their place of work.

The motivation of controllers and its cognitive consequences have not been widely studied, although the need for appropriate research has long been acknowledged (Nealey *et al.*, 1975). Many factors are known to contribute to motivation (Kleinbeck *et al.*, 1990), though few of them have ever been validated for air traffic control. The reasons for the generally high level of motivation have not been established well enough to predict how proposed changes, such as new equipment, revised procedures, different responsibilities or computer assistance, might now affect it, though the topic has been studied and a methodology established (Crawley *et al.*, 1980). A main driving force lies in the high professional standards expected of each controller by colleagues, and ultimately imposed by them if necessary. It behoves those who advocate changes to ascertain their consequences for motivation beforehand, so that the conditions essential for high motivation are not undermined inadvertently, yet this is rarely done. Even those most in favour of confirming that motivation can be sustained balk at the work entailed and wonder if the available techniques would suffice.

It can be quite straightforward to gather evidence of high motivation among controllers, and it may seem obviously desirable to maintain it. However, the cognitive consequences of any significant loss of motivation among controllers, and the consequences for air traffic control in terms of performance and safety, are largely speculative. Commonsense suggests that air traffic control must benefit if controllers try harder to achieve high standards, but commonsense is not evidence and has been known to be wrong. Perhaps a requirement, albeit self-imposed, to sustain a continuous high level of effort is not the best guarantee of safety, or there may be no relationship between safety and effort as a result of motivation, or it may not be possible to perform tasks better by trying harder. The motivation to cope somehow, even when the traffic is imposing grossly excessive demands on the controller, may contribute towards job satisfaction but hide a problem that it would be safer to reveal. It is not axiomatic that safety must be enhanced by high motivation and impaired by low motivation. What is needed is evidence, whether supportive or contrary.

Many of the forms of computer assistance that effect the controller cognitively could also affect motivation. Although some of the cognitive consequences have been explored, the implications for motivation have not even been recognized. The successful resolution of a specific tactical air traffic control problem by active participative means which are fully observable by colleagues differs cognitively from the mainly passive monitoring and maintenance of a smooth flow of air traffic by means which are so unobservable that colleagues may not even know who is responsible. The difference is also large in terms of motivation and job satisfaction. Reduced team roles that change the mechanisms through which professional norms and standards are inculcated in newcomers to the profession can thereby change controllers' motivation, but the latter changes may remain unheeded until they have become so gross as to be irreversible. Some of the professional hallmarks of air traffic control cannot be perpetuated if the work consists mainly of unobservable manipulations of human–machine interfaces.

Although it is clear that many of the changes associated with computer assistance could affect motivation, the nature of the effects, and their consequences for performance and safety, are far less clear. Their effects on motivation are not a major feature of the future plans for air traffic control. The effects could therefore be particularly serious because they are unintended and unidentified, and the designs of future systems do not allow for them.

4.12 Human error

Human errors are very important in human factors. They are the subject of theorizing, of the development of taxonomies, and of attempts to predict, reduce and prevent them and to limit their consequences (Reason, 1990; Senders and Moray, 1991; Reason and Zapf, 1994). This general work has been adapted to air traffic control contexts (Stager, 1991; Reason, 1993). Also actual human errors in air traffic control have been studied, in relation to system operations (Danaher, 1980), to tracing errors (Empson, 1987), to cognitive failures (Langan-Fox and Empson, 1985; Empson, 1991), to factors related to errors (Rodgers, 1993), and to broader issues of risk management (Wilpert and Qvale, 1993).

In air traffic control contexts, most human errors are not random. The types of human error that occur are usually a function of the system design, the tasks, the equipment, the procedures or the traffic demands, in association with particular attributes of the individual controller making the error and with the conditions under which it was made. When decisions are made about the design and specification of the workspace and the human–machine interface, these are also decisions about the types of human error that will or will not be possible. An implication is that efforts to remove an identified source of human error through design changes will not merely remove the identified types of error if they succeed, but will also introduce some new types of human error that previously were absent. The human–machine interface that has no potential for human error is an unattainable ideal. Each particular interface design contains the seeds of its own human errors, on its surface in characteristic kinds of display misreadings and familiar failures in data input, and deep within it in flawed latent functioning that may remain dormant for a long time. Which seeds of error germinate depends on the tasks and training, but some always will. Because so many of the general types of human error can be identified from the designs, it is often possible to build in safeguards against them and to predict general circumstances that will change the probability of their occurrence, but it is rarely possible to predict the exact time and place and circumstances in which an actual error will occur.

One of the reasons for standard procedures is that they have evolved partly through efforts to prevent the repetition of errors that have occurred. Innovative short-cuts in procedures, adaptive stratagems that seem to fit peculiar circumstances, the abbreviation of messages, and the omission of time-consuming and apparently redundant acknowledgements when under time pressure can all seem sensible and safe until their unstandardized form leads to a new and unsuspected human error. Even redundancy of information or procedures is there for a purpose, and that purpose is often error prevention. The acquisition of good habits, the scrupulous adoption of standard procedures, and the adherence to the prescribed wording and not to some paraphrase or condensation of it, all contribute on balance to safety, but even these contributions to safety pose potential sources of error. Habits can become dangerous if a habitual procedure is interrupted and not resumed. Acknowledgements cannot prevent errors if they become so routine that they are not listened to. What was once safe can become a source of errors if changes in equipment, procedures or instructions invalidate established habits so that the controller has to remember not to do what was once familiar. As far as possible, system designs must preclude such inherently unsafe circumstances. Sooner or later a controller will lapse into a familiar habit, especially when under pressure and least able to spare the time to notice or correct an error, and it must not be dangerous when this happens.

Air traffic control is safety conscious, and must pay great heed to human error. There is much concern to ensure that the human and machine roles are mutually supportive in error prevention. The more successfully the kinds of error that a human could make can be identified in advance, the more feasible it becomes to devise forms of computer assistance to detect and prevent them. If computer assistance fails or cannot provide the normal support in unusual circumstances, it may nevertheless alert the controller and help to ensure that the human and machine together do not make errors. Already it is quite common practice in air traffic control for certain human actions to be interpreted as errors by a machine, which queries them, treats them as invalid actions or does not implement them. Unfortunately in emergencies normally invalid actions may occasionally provide the only viable solution, which implies some provision for the human to override the machine's veto and for human–machine dialogues to establish the causes of errors and to circumvent their consequences. In this context, the most pressing need may be for the human to be able to convey human intentionality to the machine in order to enlist machine support in achieving human objectives.

Often the controller lacks information about the occurrence of any machine error or the extent of its ramifications. After all, the correction of most machine errors and failures is a maintenance matter and not a controller function, and therefore the evidence provided about the failure is often directed towards the maintenance tasks. A difficult aspect of human–computer relationships is the provision of sufficient information to the controller about any machine malfunction or failure for the controller to know what remains unaffected and can still be used normally. This is what the controller really needs to know. Evidence about the human errors in other large human–machine systems can provide pointers and forewarnings for air traffic control, and a broad knowledge of parallel developments in other contexts should therefore be maintained.

5

Matching human and system

5.1 Principles of successful matching

In air traffic control, computers are used as tools at work, but also as things to learn from and as media to interact with other people (Carroll, J. M., 1993). Knowledge about the principles, procedures and objectives of air traffic control, about basic characteristics of the human as a system component, and about human cognitive capabilities and limitations can be combined to educe principles for the successful matching of the human and the system (Garland, 1991; Woods and Sarter, 1993). These principles suggest the kinds of relationship that are possible, permit predictions of the conditions for their success or failure, explain failures in terms of mismatching, and identify human influences that are relevant to matching. If these influences concern human attributes with no machine equivalent, they may be excluded from comparisons between human and machine since they can seem irrelevant to them.

Two of the principles that can be employed to achieve matching involve adaptation of the system to the human or adaptation of the human to the system. A further option, involving mutual adaptation of the human and system to each other, is becoming more common because of the rapid expansion of technical means to exercise it. Recognition of the principles for matching human and system should precede both the detailed definition of the human factors contributions during air traffic control system evolution and the detailed planning of air traffic control jobs and tasks, because the contributions of human factors include successful matching and the designs of air traffic control jobs and tasks must take account of the requirements for matching.

Possible kinds of relationship between the human and the machine were listed in Section 3.2. In principle, successful matching can be achieved through any one or any combination of these relationships, but it has to be planned. To plan effective matching, those who design and procure future systems need to know and be able to apply the extensive available human factors data, or need to work closely with those who possess this knowledge and can apply it. The nature and timing of the main human factors contributions during air traffic control system evolution are described in Chapter 6, but the most vital condition for successful matching is to introduce all the human factors contributions early enough during the system procurement process. If this is done, it should always result in matches that are adequate, and often in matches that are near optimum. At the very least, potential mismatches, and the constraints responsible for them, will be identified in time to prevent them. Practical examples of the removal of constraints might include the avoidance of poorly performed human functions such as continuous passive monitoring, or the inclusion of reasons and constructive guidance whenever a machine rules a human action as inadmissible.

Various guidelines for successful matching can be adduced. It is necessary to define at the outset the kind of matching that is being sought, and the criteria and measures that will be applied to judge its efficacy. There are two main reasons for separating measures of what is proposed or has been achieved from independently derived criteria of what is attainable, namely, to establish realistic expectations so that resources are not wasted on trying to improve the unimprovable, and to ensure that matching is not accepted merely because it is adequate when further substantial improvements in it are possible. Theories, models, fast-time and real-time simulations, and practical experience may all provide independent evidence about realistic matching objectives. Matching implies some adaptability, whether of human or machine. Hybrid systems (Hancock, 1993a), which show promise for integrating machine functions with each other and with their human users (Smith and Wilson, 1993), nevertheless pose problems of task allocation (Hancock, 1993b). The favoured human-centred approach (Norman and Draper, 1986) applied to system design (Rouse, 1991) has been advocated for aircraft (Billings, 1991) and has wide applicability (Preece *et al.*, 1994). A crucial human factors issue is whether the envisaged outcome of matching is a single solution applicable to everyone or multiple matchings that can accommodate some individual differences in behaviour, task performance, skill, experience, knowledge, procedures or instructions (Coeterier, 1971).

A further condition for successful matching is that the decisions that affect matching are all reached and acted upon at the appropriate stage during system evolution, and are not postponed until some of the best means to achieve matching are no longer feasible. Successful matching implies the pre-identification and acceptability of the full range of consequences of the forms of matching proposed. The commonest cause of previous mismatchings has been the failure to recognize all their consequences beforehand. It is necessary to assess the importance of various consequences and to distinguish matching criteria that should be mandatory from those that are merely desirable. To match the human and system successfully, some form of veto may sometimes become essential wherever a proposal that is acceptable to one side will remain utterly unacceptable to the other, no matter how it is modified or adjusted. On the human side, it may be impossible to select and train controllers to perform an envisaged task reliably to the required standards. On the system side, it may be impossible to provide the kinds of data that humans would need for satisfactory task performance. If either the human or the machine has been assigned functions which are fundamentally unsuitable, successful matching becomes impossible. The penalties can be high (Perrow, 1984; Wise and Debons, 1987; Broadbent *et al.*, 1990).

Matching is not only about the compatibility of human and system capabilities, but covers well-being and health. The outcome of mismatching could be a loss of efficiency which must be avoided or a hazard to safety which must be prevented, but it could also be high attrition rates or occupational health problems in the human, or increased unserviceability or misuse of equipment in the system. Neither the human nor the system can be required to achieve what there are no means to achieve. Flexibility, if properly planned, can be an effective tool for successful matching, but flexibility, and particularly human flexibility, does not always denote good matching for it can also be a substitute for adequate planning and foresight. In any case, true flexibility is unachievable unless the matching process has included means to be flexible.

Matching of human and machine should be viewed in the broader context of air traffic control objectives, resources and management. It occurs within an organizational and management structure which has to provide sufficient latitude for effective matching processes to be formulated, tested and implemented. Over-narrow policy guidelines may inadvertently curtail successful matching. A further criterion of successful matching is its acceptability to those concerned indirectly with it as well as to those most directly affected by it. For example, a match between human and system may initially seem satisfactory but actually be far from satisfactory if the consequent procedures are unteachable or the software requirements are unattainable. It is prudent to confirm that any ostensibly successful matching does not lead to difficulties elsewhere.

5.2 Common mismatches

The range of potential and actual mismatches between the human and some aspect of the air traffic control system is broader than is commonly supposed (Hopkin, 1989a). The most obvious and easiest to resolve mismatches are physical: examples include temperatures, humidities and airflows that are inappropriate for humans, background noise that masks speech, furniture that leads to postural problems or has poor accessibility, seating that lacks adequate adjustment or support, input devices that are too near the user or too close together, or are too far apart or out of reach, glare, reflections or gross variations in luminous flux within the visual environment, and displayed information which has to be peered at or is at different viewing distances. A combination of knowledge of the tasks, task conditions and the characteristics of controllers with the application of ergonomic standards and handbook recommendations during system design and procurement can prevent such mismatches, and cure them if they arise. Specially commissioned research is seldom necessary, as the application of existing evidence, with occasional verification, should suffice.

Another common mismatch that is quite easy to redress occurs in relating the duration of work to human capabilities. This applies to short-term durations where the available evidence about work–rest cycles, rostering, hours of continuous duty and overtime should be used, and to long-term durations where the applicable recommendations cover retirement age and changes of duties because of ageing or burnout. Though the human factors evidence about recommended work durations is less strong and consistent than the corresponding evidence about physical factors, nevertheless its application should prevent gross mismatches.

A further mismatch concerns human jobs and task demands. Its commonest practical form is a mismatch between task demands and human capabilities and needs. Such mismatches attract most attention if the task demands have become excessive enough to lead controllers to complain about too much workload and perhaps about stress, but they can also be of the opposite kind when an air traffic control service must be provided for little or no traffic, resulting in boredom and in shifts that seem never-ending. In principle the initial remedy is the same for all gross mismatches between human capabilities and task demands for it relies on changes in staffing levels. Very large variations in the amount of air traffic to be controlled are often endemic in air traffic control, and usually follow predictable daily or other cycles. The extent to which mismatches can be corrected by staffing adjustments depends on how far the system specifications allow jobs to be split or amalgamated.

Within these gross adjustments in staffing, more sensitive matching of the human and system may be achieved through forms of computer assistance that provide some flexibility in functioning or some control over task demands. Flexibility can result from computer assistance that can take over some functions when the controller is busy, or from the reversion of some automated functions to the human who has too little work. Matching of this sort entails the planning of systems to permit such flexibility, and the provision of means to prevent potential mismatches associated with task sharing, with training limitations, or with vagueness in the timing of interventions.

At a different level are mismatches that originate in task designs and concern particular tasks. Mismatches of this kind are very common, especially when the nature, level of detail, or format of the information provided is inappropriate for the tasks or for their timescales. Many of those identified by Bainbridge (1987) are familiar in air traffic control. Searching and monitoring are generally more appropriate as computer functions, but are often assigned to humans who perform them slowly and inefficiently. Human actions may be overridden or prevented by the system without adequate explanation. Desirable human actions may not occur because the human does not know that they are possible, or cannot discover how to perform them. A policy of manual reversion in the event of system failure may lack adequate consideration of the associated mismatches, both in the reversion itself and in the subsequent return from manual to automated mode. There may be no means to convey human intentionality to the system in order to enlist system support for human objectives. There may be inadequate means to initiate and conduct constructive dialogues either when the human does not understand what the system is proposing, or when the system cannot accommodate what the human wants to do. Functions that seem simple may require complex communications or cumbersome procedures. Major cognitive or social consequences of apparently minor changes associated with computer assistance may remain unrecognized. Information may be in an inappropriate form for human use or altered by the transmission processes.

When the manual form of a task or function is replaced in whole or in part by a more automated form which in system terms is functionally equivalent and similar to the manual form but in human factors terms may be very different from it, mismatches can arise where there were none before. Other mismatches derive from facilities or procedures that seem incompatible with responsibilities: a controller with control responsibility for an aircraft may be expected to accept the computed solution of a problem concerning it, yet retains personal responsibility if the computer recommendation is unsafe. A contrary kind of mismatch occurs if computer assistance can conceal human inadequacy. Latent mismatches are introduced whenever changes intended to fulfil defined objectives bring other changes that were not predicted: for example, computer assistance for a single controller changes the feasible roles of supervision and assistance. Section 5.3 contains further examples of this. Whenever system changes modify human roles significantly, mismatches result unless the revised human roles are rematched with the system. The expanded human roles in resource management resulting from the evolution from tactical to strategic air traffic control were recognized only belatedly.

Some potential for mismatching is associated with any system changes that require human adaptation. Current incipient mismatches usually refer to computer assistance but common future mismatches will refer to artificial intelligence, expert systems, neural networks, virtual reality, and adaptive interfaces, which must be reconciled with human needs for status, self-esteem, job satisfaction, professionalism and motivation. Demographic trends point to smaller workforces in the future. When jobs become more interchangeable throughout all complex human–machine systems, those who can provide satisfying jobs that still require and use human knowledge and skill may have most success in recruiting and retaining qualified workforces with low attrition rates and retraining costs. Future mismatches may result from ignoring such requirements. The ultimate penalty of severe mismatching is that able people leave and cannot be replaced. A contribution towards matching would be refinement of the selection and training procedures for controllers, based on improved understanding of the factors that fit individuals for working in large human–machine systems and motivate them to do so.

5.3 Human characteristics with no machine equivalent

Some technological advances can produce system components that seem to
possess human-like attributes. They may aid traditional human functions such
as problem solving, decision making and prediction, or even fulfil these functions without human participation. They may be adaptable, flexible, intelligent
or innovative in ways that apparently resemble humans. Such developments
pose new problems of how to match human and system when attributes are
shared, but at least they ensure that the relevant issues are addressed in relation to air traffic control. No comparable assurance can be given about human
characteristics with no machine equivalent (Hopkin, 1982a). As they are
human and not machine characteristics, they cannot be expressed in systems
language and a systems-centred approach cannot be applied to them directly.
They cannot influence the allocation of functions to human or machine if the
allocation criteria are confined to factors applicable to both, yet many human
attributes without machine equivalents are affected greatly by the allocation of
roles, tasks and functions.

The gulf between these human characteristics and any machine equivalents
can be illustrated by posing some human questions about machine components. For example, do electronic components dislike protracted periods of
idleness and inactivity, and fidget or devise activities to fill the void? Does an
electronic component wonder if it fulfils a function as efficiently as other
similar components elsewhere in the system, and does it initiate actions to try
and find out? Does an electronic component obtain satisfaction and motivation by applying learned knowledge and skills to solve a difficult problem
successfully? Posed in this form, such questions seem and are nonsense.
Therein lies the problem. Some need for matching can readily be acknowledged if the human and system are collaborating to solve the same problem.
Even that does not guarantee effective matching, for apparently commonsense
prescriptions, such as assigning to humans the functions which humans
perform well and assigning to machines the functions that machines perform
well, are often not followed. Nevertheless, it is possible to formulate a few
principles for matching even if it seems impractical to implement them.

The problems are different when a match is sought between human attributes with no machine equivalent and the system. There is a long history in
human factors of ascribing specifically human influences to other causes
(Gillespie, 1993). Only three kinds of change are available to satisfy human
requirements with no machine equivalent. Human roles must be modified; or
systems must be modified for reasons which may seem irrelevant to their functioning; or the ways in which the human and system interact must be modified, again for apparently irrelevant reasons. Those who design or procure
systems need persuasive evidence to be willing to change the system or its
interfaces with controllers in order to satisfy requirements that do not seem
directly pertinent to system functioning. Without such evidence, satisfactory
matching to accommodate human characteristics with no machine equivalent

cannot occur. The list of human characteristics to be accommodated in this way is a long one. The problem has originated partly in the tacit precedence accorded to systems thinking. The flexibility and adaptability of many technological advances enhance the opportunities to adapt the system to match these human characteristics without incurring penalties elsewhere.

A central issue concerns measurement. Normally questions are posed and answers are expected in system concepts rather than human ones. The variables that are controlled to answer such questions and the measures that are employed to quantify the effects of those variables are naturally those that are obviously related to system issues. Measures of human attributes that are apparently unrelated to human functionality as a system component seem at best to relate indirectly to system issues, but mostly they simply appear irrelevant. Although some appropriate measures, for example of factors affecting or affected by attitude formation, have been developed in other contexts, they are not routinely applied in air traffic control. Some possible measures are qualitative rather than quantitative and still in the process of development, and need verification for air traffic control applications. None of them may seem particularly constructive to those charged with the development or operation of air traffic control systems, who need practical human factors recommendations that they can act upon and not tentative suggestions so hedged about with caveats as to be of little practical value, which reinforce suspicions that human factors recommendations are vague or equivocal.

Some of the ways in which the human is not a system component concern human needs and aspirations at work. Humans develop attitudes towards their jobs. For many controllers, air traffic control is a very satisfying occupation. They like the work, and are conscious of their professional knowledge on which air traffic control depends. The opportunities to develop skills are potentially satisfying, and the opportunities to exercise those skills provide actual satisfaction. Controllers work in a team environment and become conscious of circumstances when their skills will be noticed. Proof of skill is a condition for professional acceptance by peers, and for maintaining self-esteem. These skills are also a basis for the development of professional norms and standards and ethos.

To gain full acceptance in a team, the incoming controller must be able to gauge what is expected. Knowing what to do is not enough. Knowledge of the prevailing standards has to be gleaned, mainly by tacit means. This issue has recently risen in relation to on-the-job training, which, through practical experience, teaches the controller how air traffic control is actually conducted, how real as distinct from simulated problems are solved, how far incipient problems are anticipated and prevented, and what kinds of error cannot be tolerated. During on-the-job training, the controller also learns a great deal that is never formally taught, much of which relates to human characteristics with no machine equivalent and to ways in which a human is not a system component. The controller learns what the beliefs, opinions and attitudes of a controller are expected to be towards management, conditions of employment and equipment. The controller learns what is important to the professional controller and what is not, and whether various prerogatives lie with legislative or organizational authorities, management, a trade union or professional guild, the local facility, the watch, the supervisor, or the individual controller. Controllers learn how supervisors and peers judge individual acceptability and competence. They learn which issues are sensitive and require solidarity, and why they are important.

In common with many other jobs, air traffic control can be rewarding as an occupation, but this can also have a negative aspect. Mismatches can induce strain in individuals, if they seem unable to achieve the required levels of performance and safety consistently in the available time, whatever the reasons may be. Satisfying jobs require effort and present challenges. Ideally the controller should not be driven entirely by the system but have some autonomy over task demands and their peaks and troughs. Neither the work itself nor the work conditions or environments should ever harm those who do the work. The very conditions which in moderation are sources of challenge, interest and opportunities to demonstrate skills can, if they become excessive, impose unreasonable burdens on controllers, induce occupational health problems or generate anxiety.

Decisions about the human as a system component impinge directly on the ways in which the human is not a system component. To recruit, retain and get the best from an enthusiastic, competent, healthy professional controller workforce, it is necessary to acknowledge the ways in which those workforce members are not system components and to meet the associated human needs. This is a vital aspect of matching the human and the system. Many of the main guidelines can be formulated: jobs must not harm those who do them; job and task designs should grant each individual some autonomy over the broad level and the fine detail of work, should provide opportunities to develop and use skills, and should foster the development of professional norms and standards; the work and working conditions should engender favourable attitudes towards all aspects of air traffic control; and positive planning should promote high morale and provide opportunities to gain esteem including self-esteem. The above guidelines refer to specifically human attributes which should be encouraged if only because failure to do so could degrade the performance of the human as a system component. However, some human attributes are not worth preserving and the objective is to remove their influence. Examples include human inability or unwillingness to accept human deficiencies, a propensity towards wilful obstructiveness as a response to innovations, the use of computer assistance as an excuse for human inadequacies, and a reluctance to concede that, as air traffic control evolves, controllers must evolve with it.

Full matching of the human and the system implies recognition of the whole array of human attributes with no system equivalents, and of their current contributions. Informed judgements, based on sound evidence that has been specially gathered if necessary, have to be made about which attributes should continue to be influential, which may be discarded without incurring problems of matching, and which should be retained but in another form. Any major system changes provide opportunities to act on these judgements. Catering for these human needs must not necessarily always have the highest priority; nor should the system be adjusted to accommodate every human requirement, however slight. But it is in the interests of everyone that all these human needs are recognized in advance, long before the detailed design, planning and procurement processes for an air traffic control system take place, so that crucial decisions about the final form of the system are influenced by the need to allow for these specifically human attributes.

Matching is always a two-way process: sometimes the human must change, sometimes some aspect of the system, and often both. What must not continue is the past practice of failing to recognize the importance of these uniquely human attributes until it is too late in the system evolution to introduce modifications to compensate for the deficiencies they cause. In a system where human and machine are matched, the human is never fighting the system, is not at cross-purposes with it, is not being excessively driven by it, is not resentful of it, and does not feel helpless. The human is not being harmed by the system but is an advocate of it, an enthusiast for it, keen to use it to its maximum potential, and identifies with it closely and positively. Such are the goals of matching.

5.4 System adaptability

The notion of system adaptability is commonplace in closed loop systems, but recent or pending forms of adaptability in the more open loop air traffic control systems can have new human factors implications although they are in accord with the traditional ergonomic approach (Grandjean, 1988). A main issue is whether the human user of the system knows that the system is adapting to the user. In an adaptive air traffic control system, the system response depends on what the human does. The system may adapt through changes in the structure of dialogues, in the level of detail at which dialogues are conducted, or in the range or sequence of presented options. The human user is seldom aware that this is happening, and it is usually assumed or contended that the human does not need to know about it. In human factors terms, the human is driven by the machine without being aware of it, and the associated human factors problems differ from those that occur when the human is aware of being driven.

System adaptability can be employed to prevent, guide or query designated classes of human action. If a controller, through miskeying, allocates an aircraft to a level at which it cannot fly, the system can be programmed to treat the miskeying as an invalid action that cannot be implemented. Successful human–machine matching requires the adaptive system to notify the controller of the reasons why the action is invalid. Various helpful forms of computer assistance for the controller, including the automation of routine functions and the replacement of tactical problem solving by strategic planning, can be quite rigid in system terms and do not necessarily incorporate much system adaptability. More adaptability is entailed when these or other forms of computer assistance are optional, and when the controller can introduce alternatives that require an adaptable system to implement them. System resilience, the capability to accommodate change without catastrophic failure, implies adaptability (Foster, 1993).

System adaptability is an important means to fulfil chosen policies about matching the human and machine. A policy requirement that entails extensive system adaptability is the provision of forms of computer assistance that are most useful when most needed, so that most of the work can be done by the machine when traffic demands are heavy and most by the human when traffic is light. The controller then acquires substantial control over workload, but this is dependent on flexibility in the allocation of functions to human or machine, and on system adaptability in the exercise of functions. Sometimes there is a tacit policy disagreement over whether the system should control human workload by assigning tasks to the human only if the human will be able to cope with them according to predetermined machine criteria, or the human should ultimately control the work of the system by having the final say over which tasks should be allocated to the human and which to the system. In effect, the human factors issues become more complex because both the human and system components of such flexible systems have to be capable of some adaptation to each other. This evolves towards the problem of specifying the optimum match between human and system when both are adaptable. Although the principles applicable to this kind of matching are not yet clear, some of the potential problems are, such as instability if each is attempting to adapt to the other at the same time. Comparable problems arise when system adaptability takes the form of intelligent system behaviour in order to fulfil such traditional human functions as planning, problem solving, decision making, prediction or scheduling, because criteria to reconcile human and machine intelligence optimally when they are employed together are still being devised.

An obvious practical application of system adaptability is to the training of air traffic controllers. Basically, the student's progress during training is then assessed by the system, which adjusts the amount of training on each topic, the sequence and complexity of training examples, the degree of similarity and repetitiveness in the training content, and the scheduling and timing of training and assessments, in accordance with the needs and progress of each student as denoted by the student's responses. The training presented by the system is continuously adapted to the needs of the student as perceived by the system. When training is by stand-alone self-administering packages containing examples of a single air traffic control topic or type of problem, system adaptability becomes the primary means of adjusting the training content to student needs. In such applications, the role of system adaptability becomes similar to that of the human instructor whom the training packages supplement or replace. The system adaptability offered can vary greatly in its elaborateness and sophistication. There is some current uncertainty as to whether the effort required to provide such sophistication is always repaid by commensurate improvements in training outcomes or in training times. However, all stand-alone training devices that allow the student to learn from experience must incorporate some system adaptability principles as a condition for selective learning of that kind.

Any form of system adaptability in an operational air traffic control system poses a practical dilemma that has to be addressed, concerning the circumstances under which the human controller can override the adapting system. This should be a rare event. System adaptability is pointless unless it is reliable, safe, efficient, beneficial and cost-effective in normal circumstances. However, in emergencies or unusual traffic configurations some of the characteristics of system adaptability that are usually advantageous can become disadvantageous constraints if the controller ever has to adopt unorthodox procedures. Matching human and system when the human must override an adapting system is a difficult but inescapable problem for which acceptable solutions must be devised. Policies on whether and when the human may override the system, and on the associated legal implications, have to be formulated in advance, for they can be implemented only if suitable provision for them has been included in the design of the human–machine interface.

For the controller, the main issues concerning system adaptability are:

1. Does the controller need to know whenever the system is adapting?
2. Does the controller need to be able to override system adaptability?
3. Does the controller need forms of system adaptability that differ from those that would be implemented by the system if there were no controller intervention?
4. Does the controller need means to interrogate the system about the available range of system adaptability?
5. Is it possible to predetermine the limits of system adaptability acceptable to the controller?
6. What does the controller need to know in advance about the consequences of proposed forms of system adaptability in order to decide whether to accept or reject them, and how far ahead would this knowledge be required?
7. Should system adaptability be viewed primarily as a means of encouraging individual differences between controllers or as a means of preventing them, since it has the potential for either?

5.5 Human adaptability

The matching of human and system has traditionally relied on human adaptability. Fitts (1951a) treated adaptability as a human function in an era when it was not a common system characteristic. Even in quite highly automated air traffic control systems, human adaptability will remain the main basis for continued human involvement. The history of accidents and incidents teaches that our current technology and knowledge are insufficient to foresee and allow for everything adverse that could happen. Human adaptability is the final resource when all else fails.

As systems evolve and become more automated, system-induced constraints on human adaptability tend to accumulate, either by design or default. In particular, the feasible extent of human adaptability becomes largely predetermined by the human–machine interface whenever it is the main means to communicate with the system. Human adaptability is not usually a first priority in interface design. It gives more emphasis to the safe, efficient and prompt performance of standard human tasks and functions for which human adaptability may be a liability rather than an asset. For example, if the system formulates a series of proposed solutions to an air traffic control problem and presents them in a recommended order of preference for the controller to make the final choice, human adaptability, in the form of rejection by the controller of all the options offered in favour of the controller's own preference, may produce a solution that seems inferior in system terms, that may be treated by the machine as invalid, or that may trigger a protracted dialogue through the human–machine interface. Human innovations can be particularly unproductive if longer-term system planning processes, proposed or already implemented, presume that one of the system solutions that has been checked automatically to ensure its full compliance with longer-term plans will be adopted, and not a human solution that has never been checked in this way. In normal circumstances, human adaptability is not always a boon. It is not helpful when procedures have been standardized deliberately to ensure thoroughness as in the application of checklists (Degani and Wiener, 1993), and broader prescriptions for the application of human factors to systems do not generally encourage much human adaptability in doing so (Booher, 1990).

The means to introduce various forms of human adaptability may be very limited or cumbersome if the need for them has not been anticipated. The scope for human adaptability is far less in automated than in manual air traffic control systems. In the latter, where the controller evolves solutions to problems and implements them by instructing pilots, there is much more scope for human adaptability, but adaptability should not necessarily be equated with efficiency. The increased predictability of traffic handling often associated with reduced human adaptability can in many respects be advantageous, and may be a precondition for the successful strategic planning of traffic flows.

Controllers may resort to innovative behaviour for their own purposes, particularly if they cannot convey those purposes to the machine. The ways in which future systems will handle human intentionality seem linked to the role of beneficial human adaptability in them. Otherwise many attempts at adaptability will be thwarted by interface limitations or by rulings that actions are inadmissible. Adaptability in human terms can constitute unpredictability in system terms, without the intervening rationale of intentionality. Successful human adaptability requires considerable stability in those aspects of the system to which the human is adapting.

In the past, there has always been some human adaptability in air traffic control by individual controllers, and sometimes controllers have adapted to each other or teams of controllers have adapted to the system. When human interchanges have to take place via the system, this curtails mutual human adaptability, partly by reducing each controller's direct knowledge of colleagues that is applied in successful adaptation, and partly by reducing the functions that can be performed collaboratively by team members and hence can benefit from mutual adaptation. An important form of human adaptability in previous air traffic control systems has been assistance to colleagues which has flourished because it has been possible to recognize their need for assistance and because the system and interface have been flexible enough to enable assistance to be given. Neither of these conditions will apply as often in the future, and therefore mutual adaptability between controllers will become less significant unless a policy to retain opportunities for such adaptation is coupled with positive and practical planning to achieve it. It seems prudent to ascertain what the current benefits of human adaptability are before curtailing it too much.

The common presumption that technological innovations curtail opportunities for human adaptability within systems is not necessarily correct, but results either from a policy to discourage it or from the lack of any relevant policy so that it occurs by default. An alternative technically feasible option is to identify the needs for human adaptability within systems, to define human roles in those terms, to provide facilities at the human–machine interface for those roles to be exercised, and hence to retain in the system human functions that require human adaptability. What is needed is a thorough appraisal of the respective advantages and disadvantages of human adaptability which can sometimes be a strength and sometimes a weakness. An informed and coherent policy on the role of human adaptability in future air traffic control systems can then emerge and be applied with full knowledge of what it is intended to achieve.

6

Human factors contributions during air traffic control system evolution

6.1 The basis of contributions

The first condition for successful human factors contributions to air traffic control is acknowledgement that human factors as a discipline has a role in air traffic control. This is a necessary but not a sufficient condition, for it has to be accompanied by a willingness to act on human factors recommendations. Attitudes towards human factors are generally more supportive now than they formerly were, and scepticism and indifference are more common than outright hostility. The burden of proof has now switched from gaining recognition for human factors to delivering what it has promised. As a discipline, it has been lax in educating others about its roles, recommendations and achievements. Many who would confine its contributions to the later stages of system evolution have never been informed about its contributions at earlier stages. Contributions draw on basic human factors data and procedures (Wickens, 1992; Sanders and McCormick, 1993).

The sections of this chapter form a hierarchy of human factors contributions, approximately in their preferred sequence of occurrence. The earlier in the hierarchy the contributions are made, the more effective they can be. Unless the right basis for human factors contributions is present, many of the actual contributions cannot be sufficient or optimum. If a human factors issue should have been addressed during system planning or design but is not recognized until system procurement or evaluation, many of the best options for resolving it may no longer pertain. The most significant human factors contributions are always made early in system evolution, and can be fully effective only when made then. The sections of this chapter, being in the approximate chronological order of their application, are therefore also in order of their human factors significance. Decisions at earlier stages curtail later options. For fundamental human factors flaws identified too late in system evolution, no satisfactory remedy may remain. Human factors as a discipline is not a panacea. Being able to define a human factors problem and to prescribe means to tackle it cannot guarantee an acceptable answer to it. Sometimes no satisfactory matching of human and machine remains possible without gross recasting of human or machine functions or of both.

If human factors problems can be identified early, it is far easier to treat them as broad problems which affect the whole system because of their interactions than as specific problems of displays or input devices for example. Although human factors problems are usually phrased as specific problems initially, this is almost always an oversimplification. The full range of human factors implications seems relevant to broad system problems and can be acknowledged, whereas specific problems, for example of display contents or keyboard layouts, are confined within the terms in which they are posed, and it is too late to raise such broader questions as whether pictorial information might have been better than alphanumeric or whether a keyboard is the most appropriate input device. With specific problems, the human factors contributions can be reduced to detailed tinkering to improve the usability of something which can no longer be changed much but is fundamentally flawed beyond recovery in human factors terms.

Perhaps the most important practical human factors contribution during air traffic control system evolution is to educate those who conceive and plan air traffic control systems sufficiently to ensure that professional human factors contributions are made. Ideally these contributions should also be welcomed, but they can stand or fall on their merits, and are usually deemed to be beneficial and effective if there is an opportunity to make them. Human factors expertise is not synonymous with practical experience of air traffic control; both make essential contributions to an evolving air traffic control system, and it can be particularly effective if some of their work is done jointly through the close collaboration of a controller and a human factors specialist, each bringing different knowledge and experience to bear. Human factors generally supports the involvement of controllers as users in the design and development of future air traffic control systems (Levesley, 1991; Simolunas and Bashinski, 1991; Dujardin, 1993).

Those who advocate the introduction of a technological innovation into air traffic control may be reluctant to concede that it may have no significant human factors benefits, although it may bring other kinds of benefit. Some human–machine interfaces exemplify this. The first forms of computer assistance were treated as acceptable if they could be made to work, even though controllers had to learn to think like software writers or computer scientists in order to fathom how to use them. These were genuine technological advances, but they often raised major human factors problems. Controllers had to use in their own specialization tools developed for them by other specialists who did not know much about how the controllers would use them or what they would use them for. If the tools had been optimized at all, it was in terms of the other specialty, and they had to be adapted to assist and be understood by the controller. The very concept of user-friendliness testifies to the inappropriateness for human use of many early technological advances. Unfortunately, these early mismatches also induced much user scepticism about promised benefits.

To get the best from human factors contributions, they have to be planned and organized, with resources and funding for appropriate contributions at each stage. There is nothing specific to human factors about this. If the management, planning, hardware or software are inadequate at any stage during system evolution, their inadequacy also will affect all subsequent stages.

Human factors as a discipline encounters a further set of different problems when it is applied to air traffic control because it is the discipline that is most likely to reveal for the first time that fundamental human limitations will prevent the efforts of others from being fully successful. As a discipline it may seem to generate more than its share of findings that are politically unwelcome, administratively awkward, or financially prohibitive. It adopts a balanced approach to achieve the basic objective of the most efficient, safe and effective air traffic control system that can be devised within the criteria set, and one which fosters the well-being of those employed within it. However, the attitudes of human factors specialists in making a fair and balanced appraisal of respective benefits and problems can be construed as obstructiveness or intransigence by those who are promoting strongly a technological innovation on the basis of known and probably undisputed technical merits, without any comparable consideration of what might be wrong with it in human factors terms. It is almost unknown for its disadvantages to be explored as thoroughly as its advantages, but they must be to arrive at a human factors balance. As a consequence, human factors is sometimes accused of looking for difficulties and flaws when it is actually making a balanced and impartial appraisal.

6.2 Problem definition

The definition of problems is an anticipatory exercise. Its processes and their timing are crucially important in applying human factors to air traffic control. The approaches to problem definition may be either general or system specific, and actual problem definition is usually a mixture of both. Some human factors issues arise in evolving from system requirements to specifications (Wrightson, 1993) and in the exchange of information between the human and the system (Clare, 1993).

General human factors problems can be identified either because they commonly arise in complex human–machine systems, or because they are associated with other disciplines which interface with air traffic control, or because they can be inferred from basic human attributes. They all lead to the same question about what their significance is for air traffic control, whether they originate in developments in systems, in hardware or software, in technology or computer science, in input devices or displays, in psychological theories, in cognitive or social psychology, or in the relationships between psychology and other disciplines. They may also originate from revised national or international rules or regulations, or from management policies that affect organizational structures and socio-technical features of systems. They can arise from developments in sensing such as satellite technology, or in navigation such as area navigation, or in other contexts such as aircraft cockpits. In this approach, the human factors consequences of developments elsewhere are deduced in general terms. Being intrinsic to these developments, they are expected to apply wherever the developments are introduced, including air traffic control. This general approach combines deduction of the human factors implications of the developments with examination of the human factors consequences of previous applications of the developments elsewhere. From the methods, findings and experience of others, hypotheses about probable human factors implications are generated, to be verified for air traffic control.

The other more system-specific approach to human factors problem definition is quite different. It starts from experience with current air traffic control systems and from the initial proposals to build a new air traffic control system or recast an existing one, and attempts to identify deductively the human factors implications of the specific proposals. The human factors contributions must therefore be in phase with the contributions from all the other disciplines involved in the progressive evolution of the air traffic control system. The human factors implications have to be deduced iteratively at each successive stage of system evolution as plans become progressively more detailed and crystallized, the purpose being to influence all decisions about how the system should evolve so that identified human factors requirements are met. A role confined to the definition of problems is not constructive enough in a collaborative interdisciplinary work environment, and an essential further human factors activity is to propose solutions to the identified problems, based on all relevant and valid human factors evidence. A clear statement of each problem is required, including its significance, the range of its possible effects and the conditions under which it will arise. The human factors specialist should advise on whether there is likely to be a solution to each problem. If there is, the advice should cover the range and practicality of possible solutions, incidental benefits and penalties associated with each, and the familiarity of the problem. It should also cover the adequacy of existing evidence for providing a solution, any additional evidence needed, and the preferred solution according to human factors criteria. Further advice could be on how to verify the applicability to air traffic control of recommendations derived from other contexts, or how to expand evidence to add credence to recommendations. Advice should include the best means to fill identified gaps in knowledge in time to influence the requisite decisions during system evolution.

Human factors as a discipline always seems to have more work to do than time, resources, funding and staff to do it. Therefore, maximum use must be made of existing evidence wherever it is valid. Sources of evidence can include findings from other large human–machine systems, the consequences of the application of theories, constructs, technologies, or research programmes in other systems, other current or pending human factors studies in air traffic control, and current thinking about methods, measures and the interpretation of findings. Of particular importance is the range of the search for relevant evidence. It must be wide enough to differentiate between human factors problems endemic in all large human–machine systems, problems that characterize a subgroup of systems such as aviation systems, problems that are associated with a particular technology and recur wherever it is used, problems that occur in all air traffic control but not elsewhere, and problems that are specific to a single air traffic control system. The conclusions on this issue will determine where to seek relevant evidence, whether commissioned research should be theoretical, practical, general or specific, and if studies should seek specific findings valid for only one system or more general findings applicable to other systems.

The product of the problem definition phase should be a listing of human factors problems, indicating whether they are general or specific, new or old, and also describing how they are relevant to air traffic control, what is already known about them, what gaps in knowledge must be filled, and what the most effective, economical and timely means are to gather the required evidence.

6.3 Initial concepts

Human factors reasons feature prominently in the updating of current air traffic control systems and in the justification of new ones. Perhaps the commonest reason is that projections from present to expected future traffic reveal that controllers using existing equipment and practices could not handle the envisaged traffic. Most of the measurable evidence to support this conclusion deals with the behaviour, performance and achievements of controllers. Human factors evidence of other kinds may suggest that the well-being of controllers could be affected adversely by excessive task demands. An initial human factors role is to contribute professional judgements about what controllers can achieve now and in the future, what changes are entailed in evolving from the present to that future state, and what the human factors implications of those changes will be.

An initial step is to apply the products of the problem identification exercise to formulate operational concepts and to compile outline human factors requirements (Phillips and Tisher, 1985). In-house human factors staff, who understand the true needs because they have been associated closely with the compilation of requirements, can often do this most effectively. Alternatively, external human factors specialists may be employed, though this is likely to be less cost-effective at this stage. The human factors requirements must indicate how compliance with them by a contractor will be tested. Testing for compliance may rely on checklists, human factors standards and guidelines, or human factors handbooks. While it is possible to go into more detail at this stage, such general guidelines may suffice as the main purpose is to include the human factors requirements with all the other kinds of requirement. Similarly, if specifications or broad requirements are developed by competitive tendering, human factors guidelines may be given either in some detail or by citing a main source such as a human factors standard against which compliance will be tested. While the evidence in handbooks, standards and guidelines is often of a general nature and not specifically tailored to air traffic control requirements, nevertheless it is usually sensible to put the burden of proof squarely on those who prefer to ignore that evidence, so that they must justify their non-compliance with it. Sometimes there are satisfactory justifications, but proposed non-compliance more often offers a way of avoiding difficulties which must not be circumvented at this stage, for they will only re-emerge later in the system evolution when it has become much more difficult to resolve the problems that they pose. The main applicable human factors databases and suggested criteria for applying them should be identified during the development of outline system requirements.

A human factors role is to assess the feasibility of initial concepts in human factors terms. Many kinds of study can contribute to this: for example, anomalies in a proposed division of responsibilities may be demonstrated, with suggestions on how to resolve them. Another role is to evolve or recommend criteria: for example, if several data input devices seem appropriate for entering information and instructions into the database, one practical criterion for differentiating between them could be whether an experienced controller can learn to use them fast enough to keep pace with the spoken messages to which they refer, or is forced to jot down their gist and enter them into the system afterwards because the data input task takes longer than reaching agreement does. A further human factors contribution at the initial concepts level is to compile a complete list of functions that controllers fulfil so that none is inadvertently lost through ignorance of its existence. An example concerns paper flight progress strips (Hopkin, 1991c). Earlier attempts to evolve electronic surrogates for paper strips concentrated on attributes related to their visual appearance and manipulation, and failed to capture their full functionality. This must be defined in human factors terms as a basis for interdisciplinary judgements about which functions could be discarded, which must be retained on electronic flight strips, and which must also be retained but could take a different form.

Human factors advice at the initial concept stage of a new system may often draw attention to categories of human factors implications not otherwise considered because the proposed changes or innovations are being made for other reasons. At the heart of the human factors role is the interacting nature of human factors, so that most proposed changes have human factors consequences far beyond those that are immediately apparent. A few examples can illustrate this point. The procedures that controllers must follow in order to use a new device as it is designed to be used must not only be safe and efficient but sufficiently user-friendly to be teachable to controllers, although sometimes they are not. Proposed computer assistance must not impose a search task because this would introduce a wrong human factors principle since the busier the controller becomes, the less the time that can be spared for searching but the longer the time occupied by searching. If the successful use of an innovation depends on information not readily accessible to the controller, the way in which the controller uses the innovation has to be changed or the information must be provided in a more accessible form. Sometimes the consequences of planned new procedures for team roles and supervision have not been thought through, and a human factors role may be to question how performance can be supervised and whether under any circumstances others are allowed to intervene.

Another human factors role is to relate technical innovations and advances to air traffic control requirements and controller capabilities, and to make recommendations accordingly. Such advances may seem to have sufficient promise to develop and evaluate them for air traffic control needs, or they may have no evident air traffic control application despite being technically exciting. The recommendations rely on human factors expertise in matching the human and the system, taking account of what the system can potentially offer and relating it to innate or acquirable human capabilities and to avoidable or insuperable human limitations. A further human factors contribution at the initial concepts stage is to comment on any discrepancies between stated air traffic control policies and their translation into practice. This never courts popularity. Examples of such policies are that a future system is intended to remain controller centred, or that future air traffic control must still be sufficiently observable to permit effective supervision.

6.4 System planning

Human factors principles can and should be applied to the planning of air traffic control systems as well as to the products of that planning. Human factors contributions at the planning stage can be of two kinds. One represents further development of the previous problem definition and initial concepts stages whereby general human factors evidence is applied to the specific system being planned. The other involves examination of the plans as they are formulated, to deduce both the human factors problems that are specific to them and the particular forms that general human factors problems will take if the plans are pursued.

Human factors contributions to planning sometimes involve compromises, but they must always be informed compromises. All human factors recommendations depend ultimately on evidence. Even the poorest evidence should be superior to guesswork and to opinions with no factual support. The nature of the evidence determines the form of each recommendation, the confidence in it, its probable validity, and the extent to which it may be compromised to reconcile human factors with other requirements. The strength and quality of the supporting evidence for human factors recommendations are not usually apparent from handbooks, standards, and other sources of human factors data, so that the expertise of the human factors specialist must encompass the recommendations themselves and the strength of the supporting evidence.

Consider as a simple example a display viewing distance recommended in a handbook. Its interpretation depends on the number and incremental steps of the display distances for which data were obtained, the range of viewing distances covered, the quantity of data gathered, visual characteristics of the viewing population, the ambient lighting, the experimental material and the tasks, the procedures and the instructions, the measurements, and the magnitude of the difference between data for the optimum distance and for the second-best distance. If there were few measures, big increments and small measured differences in performance, the recommended viewing distance could probably be compromised considerably without serious penalty, but if there were many measures, small increments and large measured differences in performance between the recommended optimum and the next best alternative the prospects for compromise without resultant penalties in performance are poor. There are no general rules on this. Human factors data vary grossly in the strength of their supporting evidence. Some human factors dimensions contain step functions so that compromises without significant loss of performance can be made up to a point, beyond which sudden major decrements occur.

One useful role for the human factors specialist is to compile a list of human factors recommendations for the particular system being planned. These can be bald statements, or can specify a maximum, a minimum or a permissible range. The latter can be a useful indication for others of the potential for compromise. If each discipline participating in planning submits proposals and conditions in this form, it becomes much easier to see which planning proposals satisfy all requirements or which requirements are not immediately satisfied. Some of the latter will offer scope for reconciliation but others that seem to lead to incompatibility between human factors and other requirements should prompt some reappraisal of the initial concepts stage. The human factors specialist may be the first to detect incompatibilities because a necessary aspect of human factors is to examine others' proposals in relation to human factors proposals in ways in which specialists in other disciplines seldom have to do. The human factors specialist can then become the focal point or catalyst for resolving incompatibilities in so far as they affect the controller, even though such a role may have no formal status. The resolution of incompatibilities is not a human factors perquisite, however, but concerns the whole planning team.

At the problem definition and initial concepts stages, human factors contributions are often the main means to identify future research needs and to gather evidence related to those needs before it has to be applied. Nevertheless, if human factors problems have not been detected or the resources for research are inadequate, they may be recognized for the first time at the planning stage when existing evidence will be inadequate to resolve some of them. The planning stage normally offers the last opportunity to commission research on identified problems. Any that are not identified until the design stage or later must usually be resolved in a shorter timescale than properly conducted human factors research requires. Relevant evidence then has to be gathered by means other than research. This points to another human factors role during the planning stage, which is to clarify what the opportunities will be to introduce relevant human factors data at subsequent stages of the system procurement cycle. Methods for gathering such data include laboratory studies, simulations at various levels of sophistication, extrapolation from real-life environments, applications of theories and models, and the canvassing of professional opinion among those with relevant expertise.

Human factors contributions throughout the planning stage must be practical. Practicality depends on knowing which decisions become final at what stages of system evolution, and ensuring that relevant human factors evidence is provided in advance of those decisions. This is the overriding criterion for all practical recommendations. The most brilliant human factors recommendation is of no use if it comes too late in system evolution to be acted upon. This guiding principle must dictate the nature of any additional human factors evidence gathered to fulfil a requirement identified at the planning stage of system evolution. This is particularly applicable to any research, which must produce findings on time despite the fact that research programmes rarely run smoothly.

6.5 System design

At the design stage, system plans reach fruition and are converted into specifications and descriptions of facilities, functions and procedures. Originally the main envisaged role for human factors was as an aid to system designers. It should have a major role in this, but its efficacy depends on the identification and resolution of human factors problems at each previous stage during system evolution. The successful application of human factors recommendations and guidelines throughout the design process requires the application of existing valid solutions to any remaining human factors problems or the prior commissioning of research to obtain the requisite human factors evidence. Communications difficulties between system designers and human factors specialists have not really been satisfactorily resolved. Each generally has insufficient understanding of what the other can offer and does not interpret the other's specialist knowledge as a significant contribution to their own requirements.

The design process which in principle seems logical and progressive through increasing levels of detail often has several iterative stages as initial proposals are recast and refined to effect improvements and overcome obstacles. The human factors principles and checklists that may serve the early stages of design well may not be specific enough in its later stages. While they may still be helpful in indicating difficulties, they may fail to provide adequate solutions. Evidence from previous systems can be valuable, including negative evidence about what did not succeed and should not be tried again. It may also be appropriate to reconsider any previously discarded design options that have become technically feasible.

The full range of human factors contributions is relevant to system design. Some should be repeated throughout the design process whenever a significant design change is made or proposed. The following categories of human factors information should be treated in this way.

1. Anthropometric evidence relates the range of the body dimensions of future system users to the specifications of the furniture and workspaces. It covers reach and viewing distances, console profiles and seating, display–control relationships, consequences for posture and for head and eye movements of the tasks as designed, and workspace accessibility for all purposes including maintenance, supervision, and on-the-job training. There should be checks that the workspaces can accommodate the entire envisaged range of staffing levels, including those at handover.

2. The display designs relate the full range of human task and information requirements to the presentation of all the requisite information in usable form, coded optimally at the right level of detail and matched with the physical environment. The displays are also treated collectively in the sense that the ambient lighting suits them all and there are no confusions.

3. Controls and input devices provide the means by which the controller is able to influence the system. From the task requirements, the full range of human functions can be defined, and provision must be made to perform them. Contributions include checking that all the input devices collectively can supply the complete range of intended human functionality, and optimizing the displays and the communications facilities.

4. The communications requirements, especially those between humans and machines and between humans directly, are deduced from the task requirements and the task designs, so that judgements can be made about how well all the communications requirements can be fulfilled.

5. At the design stage, crucial decisions are taken on the feasibility and attainable proficiency of human tasks. Design decisions on displays, input devices, communications, workspaces, and the physical environment predetermine the ways in which the human can fail and the types of error that it is possible for humans to make while operating the system.

6. An aspect of the system design concerns the physical environment. It is important that this meets human requirements. Main physical factors, in addition to visual ones, concern heating, cooling, ventilation, air flow, noise levels, acoustics, decor, and room layout.

7. Although many of the conditions of employment are not yet directly relevant at the system design stage, some of them are. Obvious examples include such influences on fatigue as the normal period of continuous working without a break, the work–rest cycles, and arrangements for the handover of control responsibility. Provision must be made at the system design stage to identify any viable options precluded by system design decisions taken for other reasons.

An important function of the human factors specialist during system design is therefore to ensure that designers understand the relevance of human factors contributions enough to allow the human factors recommendations to be integrated into the design process. It is not the function of the human factors specialist to design the system, but to facilitate the design process so that it does not lead to subsequent human factors problems that could have been prevented (Day, 1991). The purpose is also to supply and apply human factors evidence to optimize the design for human use (HMSO, 1989). It is important to identify all the human factors consequences of design decisions as the decisions are taken, and not to wait until they have been overtaken by other events. It is also important to provide whatever evidence is available to resolve identified problems, and, if the evidence is insufficient, to give the best advice according to existing evidence while acquiring further evidence. The human factors specialist must possess or know where to obtain sufficiently broad and deep professional knowledge to be able to apply it judiciously in order to reach practical compromises that do not sacrifice fundamental human factors principles. The basic objectives are to ensure that the system as designed meets all the human factors requirements that can be deduced, and that the human factors implications of design decisions are recognized from the outset and not left to emerge by default much later in the procurement cycle when it is too late to deal with them adequately.

6.6 System procurement

System procurement follows the system specification and design stages. If the specification is very detailed, most of the human factors contributions should have been made in the form of requirements included in the specification. If the specification is broad and many design options remain flexible so that those who tender for system contracts can propose how the requirements should be met, the human factors contributions become rather different. An essential one is the examination of tenders for system procurement to verify that they meet the human factors requirements in the specification and also to draw attention to any apparently unrecognized human factors implications of the tender that could result in failure to meet the full system requirements.

Most human factors contributions during system procurement are not research and should not be treated as if they were. Sometimes human factors specialists tend to be research minded, and reluctant to concede that the opportunity to commission useful supporting research in time for its findings to be applied to the system development has generally gone before the procurement stage. The essential human factors role during system procurement is to apply existing evidence and recommendations so that human factors problems are not built into the system but prevented because the design and procurement processes together have taken sufficient account of them.

A vital human factors role is to get the balance right. Successful system procurement in air traffic control is a collaborative interdisciplinary effort. Human factors has no monopoly of knowledge, requirements or wisdom. Normally the conversion of the system design into system procurement reveals some human factors anomalies that must be resolved. If the human factors contributions at earlier stages have been adequate, system procurement should confirm that the design is practical and broadly correct, and the further human factors contributions should make minor improvements that aim to optimize rather than recast the system. A further human factors role during system procurement is to keep checking that all the controller's tasks will be within human capabilities and can be taught and that all the essential knowledge for tasks will be readily accessible in usable form. Any fundamental human factors flaws remaining at the system procurement phase must be remedied at once lest they delay the entire procurement and evolution process. At this stage, it may be necessary to settle for practical rather than optimum remedies.

It is important to know why systems have been procured as they are. Sometimes their proposed or actual form depends on design decisions which are correct for reasons other than human factors ones but are not optimum in human factors terms. Human factors specialists should question such decisions only if they would have serious and unavoidable human factors penalties. While examining human tasks and roles during the procurement process the human factors specialist must remain realistic and be willing to settle for what is achievable and acceptable rather than optimum, although the latter is always the goal if it can be defined and attained.

Another human factors contribution during system procurement is to state the human factors procedures necessary to verify that the procured system can meet its operational objectives, and provide guidelines for subsequent testing and evaluation procedures. These are also influenced by current operational systems and by the views of future system users (Dujardin, 1993). However, the system objectives, including its human factors objectives, are inherent in the procurement processes. A selective approach is needed during the procurement process to recommend which objectives require most rigorous verification and validation by subsequent system testing or evaluation procedures (Leroux, 1993; Stubler *et al.*, 1993). Descriptions of the future system become the basis for procuring it and for the specification of adequate replications or prototypes of it for testing and evaluation to proceed.

A human factors role during procurement is to relate the system to the required training methods, contents and procedures, with particular emphasis on any requisite human knowledge for which the training apparently makes no provision. The teachability of the envisaged tasks and procedures should have been confirmed in principle at earlier stages, but the system procurement stage must cover not only the detailed specification of the equipment but comparably detailed descriptions of how it should be used and how the tasks will be taught. Knowledge of why the system takes the form that it does and of how it is intended to be used has to be converted into appropriate equipment, procedures and instructions for which suitable training has to be devised, before the next phase of prototyping, simulation, testing and evaluation can begin.

The processes of system procurement often have to strike compromises, sometimes for powerful and inflexible reasons. It may become obvious that there is not enough money or not enough time to procure the system in the optimum form to meet the identified requirements. The resultant compromises may include less flexible or sophisticated specifications, lengthened procurement cycles, the abandonment of technically advanced aids, and the omission of features deemed to be desirable but not essential. Alternatively, the introduction of the system may become more evolutionary and less revolutionary, with a planned succession of changes whenever funding for each can be obtained and a series of interim system states cobbled together to keep it functional while it is evolving. Other compromises have to be reached when technological advances are not ready in time, encounter unforeseen developmental difficulties, or fail to meet their specifications.

If the system is procured through competitive tendering, the human factors specialist should provide relevant guidance whenever competitive tenders propose alternative ways to meet the specifications. These alternatives should be reviewed by the specialist in terms of their balance of human factors advantages and limitations, with a comprehensive human factors audit and a recommendation on the tender preferred on human factors grounds. Ideally those in other disciplines are following parallel procedures, and it will be necessary to hold meetings to resolve discrepancies and to weigh divergent recommendations since it is unlikely that the same tender will be preferred according to all criteria. However, agreement must be reached, and the objective must be to meet all the essential human factors requirements and most of the desirable though not mandatory ones. Knowledge of the mandatory, desirable or abandoned human factors requirements must be kept and applied to all subsequent stages following procurement, particularly testing and evaluation, since this knowledge points towards the human factors requirements most in need of verification.

Some uncertainty may persist regarding technically innovative equipment about which there is little human factors evidence. The human factors specialist must discover what evidence there is, and advise on its applicability and validity. A few brief experiments may confirm its suitability for the intended tasks, suggest its most practical forms, indicate what the user will need to know about it, and provide guidance on instructions and training. From an understanding of the technology and of the tasks, it should be possible to deduce the main kinds of human error and to devise appropriate measures to evaluate it for the envisaged air traffic control tasks.

During system procurement the human factors specialist usually has to rely on the best available information since there is no time or opportunity to obtain more. It should often be adequate, but if the specialist becomes conscious that the supporting evidence for the recommendations is less strong than it should be it can be opportune to identify during system procurement any topics on which research ought to be commissioned, not for the system being procured but for its first major updating or for the next generation of systems. The identification of further research needs is a normal spin-off during system procurement.

6.7 System testing

In one kind of system testing, a prototype or first production example based on the system design and procurement specification is evaluated. In another kind of system testing, preliminary evaluations of proposed innovations are conducted to establish if they seem promising enough to treat their potential suitability for air traffic control seriously, by seeing how they fare in a simulated air traffic control environment. Occasionally, testing may be done in a real operational environment if definitive data can be gathered there without disrupting operational tasks or if real-time simulation cannot yield valid data because some aspects of actual operations that cannot be simulated are crucial for validity.

One of the commonest applications of human factors to air traffic control in the past has been in system testing and evaluation, where human factors has several roles. The first is to identify human factors issues on which reliable and valid data can be gathered. Another is to recommend whether such data are worth gathering; if they are, further recommendations cover their scale and appropriate methods and measures, but if they are not, the reasons, which usually refer to insufficient validity, have to be explained. For example, the circumstances and restricted duration of evaluations usually preclude the emergence of typical short-cuts in procedures. In the past, the human factors contributions at the system testing phase have sometimes seemed large because insufficient contributions at prior stages have postponed the identification of major human factors issues until then.

Human factors knowledge about human attributes that are relevant to system testing includes measurements of their effects on the achievement of air traffic control objectives and on the system performance as mediated through the human–machine interface, and also measurements of effects in the opposite direction, namely, effects of the system on the human, in terms of such factors as fatigue, trust, stress, occupational health, and the effort imposed by task demands. During system testing, the functioning of the system may be gauged in the following ways, among others:

1. Is it possible for controllers to function as teams, and has such functioning been included in the system design and objectives?

2. Are the envisaged work–rest cycles and rostering arrangements likely to prove satisfactory in relation to the task demands and responsibilities?

3. Is it possible for a supervisor to exercise the kinds of supervision envisaged, and to make sound judgements about staffing levels and the allocation of responsibilities on the basis of the available evidence?

4. Are the arrangements practical for flexibility in staffing and for the associated splitting and amalgamation of jobs, and does the equipment facilitate this flexibility?

5. Are the methods chosen to teach controllers how to use the system satisfactory, and how could training be improved?

6. Is the total package of equipment, facilities, tasks, procedures, instructions and work conditions acceptable to controllers, and will it satisfy human needs at work?

Much preparatory work should have prevented serious human factors problems of these kinds from arising during the testing and evaluation phase, but it may provide the first opportunity to verify the adequacy of the earlier work or to detect any unsuspected remaining problems.

During the system testing phase, it is important to gather and examine carefully any evidence about the ways in which the human can err. It is not sufficient to count and classify errors or to express them as proportions or percentages, although this should be done. Single errors should also be studied in detail in terms of inadequate understanding that could be corrected by changes in training, inadequate equipment that may require modification of the human–machine interface, excessive workload that may entail revision of the task demands on each controller, or inadequate diagnosis of error on the part of the controller, colleagues, or the system. If an error could have serious operational consequences, means to prevent or circumvent it must be devised. These need not be complex, but could include simple mechanical or electronic interlocks. The point is that the initial evidence about the occurrence of a particular type of human error may consist of only a single example of it during system testing or evaluation, and its significance must not be lost by consigning it to a broad preclassification of human error that does not recognize its unique importance as evidence.

It is futile to expect to prevent every human error, but it is vital to remove every identifiable source of human error. An objective of system testing is to discover if the system as planned would fail to prevent any kinds of error that could be dangerous, and to devise and prove checks that such kinds of error have been prevented. For less serious errors that can be tolerated, less drastic system modifications may be justifiable but should nevertheless be made. Having suggested measures of the human that are appropriate for system testing, the human factors specialist must recommend how to treat the resultant data. Specialist knowledge is required for the interpretation of findings, for example on their generalizability and whether they could be artefacts of the simulation or testing or will recur in real life.

Guidelines have been adduced on the role of real-time simulation testing (Hopkin, 1978), and lists compiled of human factors issues which can validly be tackled using system testing and evaluation methods and of those which cannot (Hopkin, 1990). An objective of testing is to confirm that the design and procurement processes have resulted in a feasible system. Testing may point towards further possible improvements, through modifications to system functioning or changes in training, in procedures, or in other aspects of the system. Some minor design details may be left to the testing phase, particularly if they are achieved by simple software changes or by the substitution of modular or interfaced components.

A common objective of system testing is to prove that the system can actually handle its planned maximum traffic, but this is difficult to ascertain by testing which is fundamentally more fitted for revealing why a system will not achieve all its objectives than for confirming positively that it will. For example, testing can show a major imbalance between the task demands on different team members, or a failure of team members to function as a team, or excessive communication, coordination and liaison demands, although the participants are seldom fully familiar with all the procedures at the time of system testing. The later stages of testing and evaluation must normally employ experienced controllers in current practice, who should also be familiar with the type of air traffic control and the region of air traffic control being tested.

System testing covers the briefing, the familiarization and demonstration processes, the methods of training, and the acceptibility of the system and its components to the user. This last point applies especially to innovations, whether of equipment or procedures. Sometimes simple feasibility assessments are made by controllers visiting the test and evaluation facility, who are given considerable freedom in their approach, but it is usually necessary to gather more formal data, particularly on issues which are intended to be resolved by comparative data on the practical options. Prototyping is a technique for gathering the comments and impressions of future users and others regarding the capabilities and limitations of a simulated air traffic control workstation or workspace while it can still be changed. It is not a substitute for testing, and its functions are more analogous to planning since it encourages the formulation of alternatives but cannot usually yield quantitative evidence about capacities or characteristic human errors and is weak in dealing with interactions. Prototyping can be a useful technique to discard fundamentally flawed options quickly, to identify crucial combinations of circumstances that require testing, and to discover main topics of agreement and of disagreement among controllers.

A further human factors input concerns the level of realism or fidelity required during system testing in order to extrapolate the findings validly to the operational system. The intention is to avoid any conclusions that are false because they are artefacts of the level of realism during testing. Knowledge of human capabilities and limitations, and especially of principles of learning and the transfer of training, gives an insight into the aspects of a system which are crucial for the human tasks and must be simulated faithfully, and into other apparently more obvious aspects of the system with little effect on task performance, for which much less fidelity may suffice. Sometimes the stance of human factors in debates about the required level of system fidelity for purposes of system testing and evaluation is criticized because it appears to contradict commonsense and the expectations of other disciplines. Given a fair hearing, the human factors specialist can usually demonstrate why the recommended level of fidelity is appropriate. Levels of fidelity that are very simple may be too far removed from reality, but extreme fidelity may also be invalid if the resultant complexity obscures the true explanations and if interpretations become biased by trivia and extraneous factors. If excessive complexity is ever equated with fidelity it may imply, wrongly, that every human factors issue can be tackled by simulation methods if only they are complex enough.

6.8 Operational systems

There will always be a need for some human factors contributions to operational systems because even the best planning cannot foresee every problem, but these contributions should be quite rare if competent human factors specialists have been on hand throughout the system evolution. When a specialist is asked to look at problems that appear to have arisen in operational systems, a common initial conclusion is that the problem as posed is not the real one but a symptom of it. Complaints about equipment may be due to excessive task loading and not to deficiencies in the equipment itself, which is performing to specification. Existing knowledge is the main tool for dealing with problems that have arisen in operational use. The experienced human factors specialist accumulates and learns to apply a mental checklist of likely causes of trouble.

Common physical causes of problems include the following. Postural problems may relate to poor seating that is not well matched with the console profile or with the tasks. Visual problems may relate to eye movements imposed by the tasks and the positioning of displays and controls, confounded with ambient lighting problems such as glare and reflections. The physical environment may be inadequate, with deficiencies in heating, ventilation or air flow, excessive noise, or mismatches between the displays, the spectrum and intensity of the ambient lighting, and the decor. Scanning may impose changes in pupil size because of gross changes in luminous flux within the visual environment. Although advice on operational problems must depend on known evidence, these operational problems can suggest the sufficiency of existing evidence and requirements for future research.

Operational experience reveals the functionality and efficiency of equipment and users' attitudes to it, and can furnish direct evidence about possible human errors. Much of the practical evidence from operational experience can be applied to remedy deficiencies. For example, if a procedure is not used at all, or not used as and when intended, additional instruction, training or demonstration may be necessary. Anticipated levels of skill may not be achieved, and the reasons must be discovered. If some individuals cannot sustain acceptable performance, the reasons must be found, whether in inadequate selection, training, skill or understanding. Operational experience provides evidence, in ways which simulation never can, on whether the work is enjoyable and satisfying or tedious and boring, and on whether tasks make excessive demands on controllers. In extreme cases, this can result in stress-related illnesses, absenteeism, high staff turnover rates, characteristic occupational health hazards, or frequent complaints and low morale. On such human factors topics, it is difficult to gather definitive evidence except from operational conditions.

Sometimes human factors as a discipline functions as an interface between other disciplines. If equipment is not satisfactory, the human factors specialist may not merely report this but explain the reasons for it and be able to suggest changes that would remove the source of complaints. Traditionally, human factors is an independent discipline, not aligned with management, users or suppliers, and this perceived independence is essential for some human factors functions.

Controllers performing operational tasks may occasionally commit serious errors unnoticed by others, which are nevertheless worrying to the controllers because they realize that they could have been potentially dangerous; or they may be worried by the actions of a colleague; or they may suddenly realize that an action they were about to perform could have been potentially dangerous but the system would not have prevented it by treating it as invalid; or they may perform a task safely in a certain way but realize that they could have tried to perform it in other ways that would not have been safe. In such instances it is important for aviation safety that the controller can bring the potentially hazardous situation to the attention of the relevant authorities without incurring personal blame or retribution. This cannot always be reported through line management without being construed as careless or unprofessional, or without putting the controller's career prospects at risk. It could even be a feature of line management that is causing the anxieties about safety.

The Aviation Safety Reporting System conducted by NASA in the United States, the United Kingdom Confidential Human Incident Reporting Programme under the auspices of the Civil Aviation Authority, and comparable focal points in a few other countries, provide confidential and disidentified channels for reporting potential sources of error in the interests of aviation safety where appropriate follow-up action can be taken. This kind of reporting facility benefits safety by tapping evidence not otherwise available, and is therefore in addition to and not instead of other more traditional means to improve aviation safety. Almost all the reported incidents originate in operational circumstances, and a vital bridge between operational systems and human factors specialists is their role in confidential incident reporting (Baker, 1993).

Current air traffic control operational systems are also examined to assess whether any remaining human factors issues are specific to each system and must be solved as air traffic control problems, or are endemic in large human–machine systems and may have been solved elsewhere. The human factors contribution is then to recommend if solutions obtained in other contexts can be adopted for air traffic control. A final human factors contribution to operational systems is to study them at first hand from time to time, to suggest further human factors benefits and to identify further possible improvements even where everyone seems satisfied.

6.9 System evolution

Human factors as a discipline can fulfil two broad functions in system evolution. One is to assist the planned evolution of the system during its lifetime, and the other is to define the human factors ideal towards which it should be evolving. Both functions are appropriate depending on their timing and circumstances, and both should feature in the human factors programme.

Air traffic control must change with increasing demands. The human factors implications of some evolutionary changes planned during the system lifetime have to be identified as part of the original system procurement. For example, if it is intended at some stage to replace paper flight progress strips with electronic ones, this change must not be impractical because of constraints built into the initial system design. Satisfactory human factors solutions must be feasible within the technical and operational constraints, which can be formidable, so that the human factors role can be to find a practical rather than an optimum solution.

The other role of human factors in relation to system evolution is to try to specify what would be the optimum air traffic control system in human factors terms, if necessary by looking far beyond current technical constraints with intentions to harness technology optimally to meet human factors objectives, to cast each human in roles which are most suitable for humans, to obtain the maximum efficiency and safety from human jobs and to make the work healthy and interesting. Potential benefits include a satisfied workforce, low job attrition rates, pride in profession, high morale, job satisfaction, and harmonious relationships between management and controllers. This approach requires the definition of goals and of the means to evolve towards them, rather than evolving more passively from the present state by benefiting from technical advances. Current and future human factors problems in air traffic control that can be identified, such as being overburdened or bored, are considered in relation to possible solutions such as the provision of more control over workload for the individual controller. Stress and its causes need to be understood and alleviated if they become excessive, although the issue of whether there could be an optimum level of stress also needs to be addressed.

Human factors as a discipline should become more active in driving the evolution of future systems in directions which take more account of human factors. At their simplest these require the human–machine interface to be compatible with human capabilities and limitations and to meet human needs at work. At more complex levels, they concern aspects of management, acceptability, status, roles, expectations from work, and its social climate, and they require the optimum relationships between the controller and the computer systems to be formulated. Many developments in expert systems, in artificial intelligence, in intentionality and in memory aids, as well as some of the traditional ones to do with attention and skill, offer prospects for designing interesting and satisfying jobs which make the best use of human strengths. The lot of the operator, as well as the efficiency of the system, has long been within the province of human factors (Taylor and Garvey, 1959). Previous criteria for system evolution have emphasized safety, reliability and efficiency rather than human needs and aspirations. A policy will be needed on how important human needs are. Advancing technology provides the means to foster human well-being positively, but the achievement of such a goal should be based not simply on vague humanitarian principles, though there is a role for them, nor on well-meant opinion, but on impartial human factors specialist recommendations resting on sound and valid evidence.

7

Air traffic control jobs and tasks

7.1 *Job description and allocation*

In its most general form, an air traffic control job description can be a brief and simple statement. Advertisements inviting applications for selection as controllers employ job descriptions of this kind. Appropriate human factors advice on advertising can cover addressable markets, publications with relevant readerships, and the principles by which advertisements in various media are framed to gain the attention of suitable applicants. This is partly a matter of image building and of conveying the rewards of air traffic control as a profession, but any disparities between the public perception of air traffic control and its reality should be dispelled by the recruiting literature so that the expectations of candidates are realistic.

Beyond these simple job descriptions for recruitment are a multiplicity of more comprehensive and detailed job descriptions for other purposes. Air traffic control job descriptions become voluminous if they cover the functions, duties and responsibilities of the job, the nature of the work, the individual qualities required for it, and some mention of conditions of employment including its organizational and managerial aspects. Many purportive job descriptions are actually only partial job descriptions because their narrower or more specific objectives do not need a full job description to achieve them.

The main reasons for compiling an air traffic control job description include the following:

1. To provide a comprehensive description of what being an air traffic controller entails and includes.
2. To delineate the key aspects of air traffic control jobs that distinguish them as a group from other jobs.
3. To reveal the ways in which air traffic control jobs differ from each other and the ways in which they are all similar, as a rationale for categorizing air traffic control jobs.
4. To provide a basis for selecting air traffic controllers and for evaluating selection procedures in relation to job requirements.

5. To provide a framework and policy for the training of controllers.
6. To develop a tool for the quantitative evaluation of different air traffic control jobs, in terms of their pay, status, gradings or responsibilities.
7. To classify or categorize measures that can be applied to appraise the performance of individual controllers.
8. To provide means to indicate and quantify the main effects of proposed job changes resulting from computer assistance, new forms of information, increased traffic-handling capacity or a new air traffic control system.
9. To reveal desirable career development paths in terms of commonalities and differences between jobs.
10. To generate a comprehensive list of the human attributes relevant to a job, and to trace interactions between listed attributes.
11. To show which jobs are more compatible with each other, or would be most easily split or combined.
12. To provide a tool for manpower planning.
13. To show how the hierarchy of job characteristics and responsibilities relates to organizational and managerial structures.
14. To devise jobs to be studied by simulation, evaluation, research or modelling methods, and to revise job descriptions accordingly.
15. To define in advance the aspects of a job that are most crucial in determining the attitudes to it of those who perform it.
16. To show how a job enables those who do it to meet their legal obligations.

The level of detail of the job descriptions that suit the above purposes varies greatly. Occasionally a job description includes details of specific tasks, but usually tasks are identified only broadly, together with their relative importance and prevalence and some mention of the skills, abilities and knowledge required for them. Job descriptions indicate the main human responsibilities and initiatives, relationships between jobs, whether the job is an individual one or done as a member of a team, the kinds of supervision and assistance provided, and desirable working conditions. An incidental advantage of job descriptions, though not normally a primary purpose of them, is that they must be couched in a common and reasonably non-technical language in order to include every aspect of them.

Successful job descriptions should reveal why jobs have different names and the differences that are crucial for classifying them. This is not as straightforward as it sounds. In air traffic control, apparently different jobs in different locations can share the same job title, or nominally different jobs can be very similar. To be useful, job descriptions must be objective and impartial, and they should employ the same methods consistently to achieve these objectives. Jobs may be classified by the type of airspace—en route controller, terminal manoeuvring area controller, approach controller, ground movement controller; by the equipment in use—radar controller, procedural controller; by functions—controller, supervisor, instructor; by type of user—military controller, civil controller; by experience or seniority—senior controller, grade 1 controller; or by strategic/tactical or planning/current roles—support controller, executive controller, oceanic planner.

Job description methods depend on whether the job already exists. Categorizations of factors relevant to the job have to be devised and adhered to. Most commonly, some of the numerous job analysis methods and procedures reviewed by Spector *et al.* (1989) are used. Most job analyses attempt to identify all the factors that are relevant to a job and to deal with their relative significance by some form of rating, since all factors are never of equal importance. Even if all the practical job dimensions seem to have been identified, and often they have not, the ratings of them are often suspect for they lack a convincing rationale and can seem too intuitive or biased. The products of ratings may also be flawed because of large differences in the abilities or criteria of the raters. The grouping and clustering of jobs to derive taxonomies of them can also be of suspect validity.

Air traffic control policies on similarities and differences between jobs have to be promulgated, with or without the support of job analyses and job descriptions. Current issues are whether a single selection procedure is suitable for all jobs as controllers or the differences between jobs justify separate or modified selection procedures for them, and whether a common selection procedure should incorporate some pre-allocation of controllers to jobs or no allocation should precede the completion of any common training in air traffic control fundamentals. Whatever wrangles there may be on these points, the selection of controllers must proceed whether they are resolved or not. The definition of the nature and extent of the differences among air traffic control jobs would permit judgements about their practical differentiation using tests and measures that could be included in the selection procedures. Although the validity of tests to select controllers for different jobs or to allocate controllers to different jobs would have to be proved, they should be better than random processes if the job descriptions differ significantly. The relationship between job descriptions and job allocation seems likely to be a live issue for some time. However, to compile a taxonomy of air traffic control job descriptions entails the commital of extensive human factors resources, for it represents a lot of skilled work. Jobs in air traffic control may become more different from each other in the short term while computer assistance is being applied selectively in the busiest regions, and then more similar to each other in the longer term as the forms of air traffic control automation become more universal.

7.2 Task analysis and task construction

When the system objectives and jobs have been specified broadly, the tasks that future jobs imply can be deduced and the tasks comprising existing jobs described. The objectives of task analysis determine the level of detail that is appropriate. Task analysis can be applied either to work being done or to work to be done, but not using the same techniques. A task analysis of an existing job employs descriptions, classifications, observations and measurements of the steps to accomplish each task and of the procedures followed by those skilled and experienced in performing it. A task synthesis for a future system employs deductive methods based on the objectives and facilities, on the rules, procedures and instructions, and on the envisaged skills, experience and knowledge. The practicality and standards of achievement are known for existing tasks, but can only be hypothesized for future ones.

In air traffic control and elsewhere, job analysis and task analysis share the characteristics that the literature suggests they are standard and routine, whereas both are actually quite rare. They represent a major commitment, and are espoused more in theory than in practice, for there are in fact very few of them. Alternative approaches to task analysis, including those that employ computer-based methods, have been reviewed (Edmondson and Johnson, 1990). There is a bewildering array of task analysis techniques (Whitefield and Hill, 1994), and textbooks may provide practical guidance on the cost-effectiveness of alternative techniques for particular needs (Kirwan and Ainsworth, 1992). Tasks have been classified in relation to human performance (Fleishman and Quaintance, 1984), and it has been claimed that some kinds of task analysis can be applied early in system development (Diaper and Addison, 1992). An early study concentrated on skills rather than tasks (Whitfield and Stammers, 1978), but task analyses were compiled as a tool to predict selectively the effects of automation on controllers (Crawley *et al.*, 1980). Broader compilations of air traffic control operations concepts (Alexander *et al.*, 1988; Ammerman *et al.*, 1988) have been converted into taxonomies of air traffic control tasks (Rodgers and Drechsler, 1993). Endsley and Rodgers (1994) have employed task analysis to determine the controller's situation awareness requirements for en route air traffic control. The most detailed air traffic control task analysis is probably that by Cox (1994). There seems to be considerable confusion about the role of task analysis in relation to other techniques, its functions, its objectives, and its strengths and weaknesses as a method. Time analysis as an alternative is a time-consuming and skilled process, requiring the combined knowledge of human factors specialists and air traffic controllers, but it has been tried (Soede *et al.*, 1971).

Air traffic control can give an initial impression that quite a simple taxonomy of tasks and functions would be sufficient for a complete description of it. Only when the full functionality of the tasks and equipment are considered in detail does its complexity emerge, whereupon the provision of a comprehensive task analysis begins to seem a much more ambitious and daunting undertaking. Writers of air traffic control software have often had the same experience, for the same reasons. When air traffic control task analyses have occasionally been conducted, there has been a tendency to treat them as completed objectives rather than as applicable tools, and they have been used surprisingly little considering the amount of compilation work that they represent. Perhaps those who commission them do not know how to use them.

As a technique, task analysis is most easily applied to human activities that can be observed and preferably recorded and classified. It may be more difficult to include cognitive tasks except by implication, particularly cognitive tasks with no overt activity at all, such as a controller deciding after appraising a traffic scenario that the best course is to take no action. Even where there is activity, deductions and inferences about its cognitive antecedents may be wrong. Evidence intended to be pertinent to tasks may not be used, and even methods such as eye-movement recording may fail to reveal this. Task analysis is usually most successful with simple tasks. It must be a passive process. Even the presence of an observer may be enough to change what is observable in some circumstances, for example by precluding short-cuts in procedures that are known to be safe but lack formal or legal sanction. Task analysis may reveal what a controller did, but not why. The reasons suggested by other controllers may not be the true ones. Reasons imputed by the task analyst may not be the real ones either. Task analysis as a technique possesses both significant advantages and serious limitations.

The basic principles and methods of task analysis are applicable to air traffic control, including such facets as the analysis of skills. It requires a lot of work, and for most purposes must be supplemented by other techniques. The partitioning process that is entailed in converting the continuous flow of air traffic control activities into discrete tasks that are treated as separate entities underestimates the importance of task sequencing, task interactions, the overlapping of concurrent tasks, and the compatibility of tasks that share the same resources. Most task analysis focuses on the controller in that it is individual rather than team tasks that are analysed or synthesized, yet this approach underplays collaborative joint activities and the dependence of each controller on the actions of others, neither of which can be described wholly as individual actions. Descriptions of the functions and performance of air traffic control teams that could serve task analysis purposes scarcely exist. The multi-tasking that is so typical of air traffic control is conspicuously absent from most descriptions of what the controller actually does. Task analysis is unable to deal adequately with the extensive tacit understanding between controllers that influences many of their overt activities.

7.3 The sequencing of tasks

Descriptions of the tasks within a job or of the subtasks within a task usually carry implications of their sequencing or ordering even when these are not formally or explicitly stated. There are several obvious constraints on task sequencing. Some relate to dependent functions: a conflict has to be detected before it can be resolved, and a message sent before it can be acknowledged. Some relate to divisions of work or responsibility: actions by one controller may have to be completed before those by another can begin. Some relate to priorities, such as performing emergency actions before routine ones. Some relate to the methods of information presentation, such as placing the most recent flight progress strips at the top of the board. Some depend on the tasks already being done or the equipment already in use, such as the availability of a communications channel. Training influences task sequencing through preferred or mandatory procedures. Informal but universal procedures that are practised within a particular air traffic control environment include specific task sequences.

Some forms of computer assistance curtail the flexibility of task sequencing, and this can be interpreted by the controller as being driven by the machine. Common examples are requirements to deal with different categories of information in a particular order or to make the correct response to signal the completion of one task before the machine permits the next task to start. In system terms this can seem sensible; in human terms it seems to curtail freedom or options. An everyday example is the disruption of other tasks by an insistently ringing telephone; options include restoration of freedom manually by leaving the telephone off the hook, or automated assistance in the form of an answering machine. The task sequencing demands of the telephone are aggravated because it is poorly matched with human needs, by employing the same signal for messages of every degree of urgency, and by providing no means to reduce its distractability short of switching it off.

Work allocation implies careful task sequencing within and across jobs, and the planning and integration of tasks so that controllers can work in parallel and sufficiently independently not to be delayed by each other. It is also wasteful and frustrating if some controllers are idle while others are busy. The main constraints on task sequencing in unautomated systems concern aircraft entering or leaving the sector at times determined by the aircraft position and not by the controller, who otherwise has considerable discretion over the sequencing of control actions within the sector. Because the human can be treated in some respects as a parallel processor of information and in other respects as a serial information processor, the sequencing of tasks should be influenced by considering which tasks the human could do concurrently and which must be done successively. In effect, decisions about task sequencing are also decisions about task compatibility.

7.4 The compatibility of tasks

The range of factors deemed relevant to the assessment of task compatibility has expanded with the increased realization of the cognitive complexity of human information processing. The compatibility of the tasks within the job of a single controller is discussed here. Comparable issues of compatibility do not arise in other contexts where people have far fewer tasks or only a single one, but they always arise in air traffic control where every job includes many tasks. The main influences on task compatibility concern equipment, task demands or human attributes, which are considered in turn.

Controllers can usually perform many tasks with the same equipment, make many data entries with the same input devices, and obtain many kinds of information from the same displays. Therefore a sign of incompatibility among the tasks of a single controller is if all the input devices or all the information displays cannot be housed to meet the ergonomic requirements for reach and viewing distances within the workstation, for a criterion of compatibility is that there should not be so much disparity among the display requirements or among the data entry requirements of the tasks of the same controller. When tasks are incompatible, it is difficult to code the same information satisfactorily for all of them. Codings that suit some tasks hinder others. The more the tasks differ, the greater the prepondance of displayed information that is irrelevant for any given task. The information portrayal becomes sub-optimum for every tasks because the extensive information not required for each task is not merely redundant but constitutes clutter. The input devices cannot be optimized for all the tasks either; nor can the relationships between the displays and the input devices. If any equipment is shared with other controllers, tasks done concurrently by different controllers within the same suite must also be fully compatible with each other, as all the foregoing considerations apply to them but with less flexibility for temporary adjustments or suppressions since information suppressed as superfluous to one controller's tasks could be essential for the concurrent tasks of others.

Task incompatibility may also be traced to inappropriate task demands, especially to excessive demands because there is too much work or the procedures are too cumbersome and time-consuming. Some tasks that could be concurrent cognitively cannot be concurrent physically if they require the same equipment or impose incompatible physical actions. Another form of task incompatibility occurs when a controller must await signalled events but cannot perform other tasks in the meantime because they would block the signal or because they share equipment which cannot be utilized for another task during a lull in an incompleted one. This source of potential incompatibility becomes particularly serious when it applies to a shared communications channel during high system loading because the performance of all the tasks utilizing that channel is likely to suffer.

Human attributes are a fecund source of incompatibilities between tasks. Some requirements almost seem intended to be incompatible, so contrary are the human attributes needed. The attributes for continuously successful passive non-interventionist monitoring tasks and for very occasional highly interventionist problem solving and decision making seem so nearly opposite as to be irreconcilable, yet both are demanded of air traffic controllers. A further potential incompatibility is between the numerous air traffic control tasks which require high levels of skill but occur relatively rarely, and the few routine air traffic control tasks which require low levels of skill but occur very frequently. Unfortunately, skills atrophy if not exercised, and knowledge that is never used may be forgotten. Tasks that collectively are tedious, boring and undemanding may be as incompatible with each other as tasks that collectively impose excessive workloads and stress. Compatible tasks collectively strike the right balance: some tasks are challenging and interesting; some tasks may be routine; some tasks require the maintenance and exercise of skills; some tasks provide opportunities for esteem and pride and satisfaction. Some task flexibility is often essential to respond to differences in demand, but an associated incompatibility can be a fudging of the issue of who is responsible for task performance. A skills analysis may be helpful, by indicating tasks which share sufficient skills to be considered for inclusion in the same job.

In the past there has been some confusion over task compatibility and workload. When human information processing capacity was treated as a finite constant or maximum, attempts were made to measure how much of that capacity was not being used, either by filling the unused capacity with secondary tasks or by reallocating any unused capacity to other designated tasks. Although there were signs that all tasks were not equivalent, there was a failure to recognize either the nature and magnitude of the differences between them or that the compatibility of tasks is crucially dependent on whether they draw on different mental resources or compete for the same resources. Yet this concept is central for explaining incompatibilities between tasks and for recommending tasks that could be added without employing resources that are already fully utilized. This approach to task compatibility through the compatibility of mental resources offers considerable promise. As a tool, it could check that tasks are compatible, could specify forms of computer assistance that would conserve fully used resources, could describe air traffic control jobs in terms of the mental resources they employ, and could result in the controller fulfilling more functions more efficiently with no corresponding penalties. An example may clarify these points.

Suppose a highly experienced skilled controller to be very busy handling an intricate traffic scenario with many aircraft imposing high task demands that present difficult problems requiring complex decisions. Workload would be rated as very high, and the controller might judge it impossible to do more. In one sense this is correct, and in another sense it is not. Probably the controller could not handle a substantial increase in traffic safely or resolve more problems of greater complexity as safely at a faster rate, for this would require more of the same resources that are already being stretched to the limit, but the controller could probably learn to do the same work as well eventually using a different language. There would be so much to learn that it would probably take years, but it could be done without any loss in ability as a controller because the mental resources in learning a language are not the same as those for being an air traffic controller, and the applicable forms of automaticity are different. So much is language taken for granted that measures of it are usually not even included among measures of workload, yet very large resources are required to learn and use it. The point is that tasks that do not require the same resources need not be incompatible, but tasks that do require the same resources have the potential to be incompatible for that very reason.

7.5 The grouping of tasks into jobs

How tasks should be aggregated into jobs and where the divisions of responsibility between jobs should be, are practical issues that have to be settled, but there are two complicating factors in air traffic control. One is that some job differences arise from differences in the air traffic control service offered, but the grouping of tasks into jobs must still achieve some consistency of responsibilities across jobs. The other is that in some air traffic control environments the differences in task demands are so large that they have to be accommodated by splitting and amalgamating jobs rather than tasks, a theme discussed in the next section.

An approach based on task analysis would advocate that single jobs should consist of tasks which share common skills and knowledge, can be performed sequentially rather than concurrently, and do not seem to require potentially incompatible human attributes. Various task analysis methods can group tasks according to these criteria, but some constraints intervene. Collectively the equipment for all the grouped tasks must be housed in a single workstation, and there must not be too many displays or input devices or communications channels. The grouping of tasks into jobs requires compromises. An approach based exclusively on task analysis is too narrow.

Jobs are not simply groups of tasks but require demarcations of responsibilities. It must always be crystal clear who has the control responsibility for the traffic. Each job must incorporate those tasks that permit the controller to be responsible for the traffic within a block of airspace and to plan the flows of traffic transitting that airspace. The grouping of tasks into jobs therefore must relate closely to the divisions of airspace. Responsibility is exercised through many tasks, including the following: monitoring, searching, planning, problem solving, decision making, predicting, communicating, discussing, coordinating, liaising, instructing, verifying, understanding, remembering, handling and structuring information, scheduling work and managing resources. These functions are fulfilled within constraints imposed by system and interface designs, international and national rules and regulations, professional norms and standards, and accepted practices.

Jobs must be compatible with each other. The handover of responsibility between jobs must be smooth and effective. One controller must not be thwarted by another controller doing another job. Many jobs are done by teams, and the grouping of tasks into jobs must take full account of team activities and their timing and coordination. Individual controllers must not disrupt team tasks, and teams must not disrupt individual tasks. The grouping and sequencing and flexibility of the tasks of each controller must therefore allow for the tasks of other individual controllers in the team, for tasks done by the team, and for the needs of pilots. A grouping of tasks that is acceptable to all controllers cannot be implemented if it would regularly disrupt pilots' tasks at inopportune times.

If different tasks requiring the same facilities must be done at the same time they must be within different jobs with duplicated facilities. Task groupings should reflect traditional responsibilities whenever possible. The groupings reflect the extent to which each controller can act autonomously, and must reflect policies and legal obligations on such matters. When tasks are grouped into jobs, the demands of each job should approximately be equated so that some controllers are not grossly overburdened while others are idle. Tasks should not be grouped so that two controllers could be attempting mutually incompatible tasks at the same time, such as contacting the same pilot even though they have different purposes. Task grouping should take account of the system as it currently is but also as it is planned to evolve, so that gross task regroupings do not result from the automation of certain functions or the introduction of new forms of computer assistance. Further requirements can be facilitated by suitable task groupings, which must be compatible with on-the-job training needs, with all available forms of assistance, with the planned forms of supervision, and with any planned splitting or amalgamation of jobs in response to gross changes in traffic demands.

7.6 The splitting and amalgamation of jobs

Air traffic control deals with a gross range in demand. At small airfields with only a few aircraft a day, all air traffic control duties may be fulfilled by a single controller who may also have managerial and administrative functions. At major control towers and centres which never close, there can be gross variations in task demands, often because of restrictions on flying at night when staffing levels can be much reduced. However, this pattern is not universal: for example, the eastbound transatlantic traffic peaks during the night, and oceanic air traffic control in the United Kingdom can be handling its peak traffic around dawn. The demands on military air traffic control can also vary grossly but for different reasons, such as an air defence exercise. Military air traffic control can impose much greater workload per aircraft on the controller if control responsibility is for a particular aircraft wherever it goes rather than for the aircraft within a designated region of airspace. General aviation may impose great air traffic control demands temporarily, associated for example with international or national events that attract many general aviation pilots. A minimally equipped and little-used airfield may then become extremely busy.

In principle, there are two ways in which gross changes in traffic demands, whether they conform to a pattern or not, can be accommodated by splitting and amalgamating jobs. One way is to combine jobs so that, for example, a controller might combine the duties of a planning and executive controller. The other way leaves the duties unchanged but changes the region of airspace within which they are exercised, so that, for example, an en route controller might be responsible for one sector in heavy traffic and for several sectors in light traffic. Air traffic control jobs and tasks have to be planned to facilitate the chosen form of splitting and amalgamation with no loss of safety or efficiency, and the human–machine interface, all the communications and all the functions have to be capable of appropriate expansion and contraction. Gross changes in demand are generally predictable because they are associated with particular events or weather states, or with familiar patterns related to the time of day or time of year.

The extent to which jobs can be split or amalgamated is set at the planning stage by the flexibility built into the workstations within the air traffic control operational environment. The total number of workstations largely determines the maximum traffic-handling capacity. The extent of possible amalgamation sets the minimum possible staffing. The methods of amalgamation determine which responsibilities can be combined. The more extensive the envisaged splitting and amalgamation, the more difficult it becomes to cater for all of them effectively within each human–machine interface. Communications facilities may become complex when traffic loading is light if a large number of communications channels are then needed. It may be acceptable for the communications to be a little more unwieldy under such circumstances since the system will not become overloaded. But the extent to which the splitting and amalgamation of jobs is required must be a matter of policy. To implement the policy, appropriate communications facilities and workspace designs must be provided, since they set the practical limits on what can be achieved.

7.7 *Staffing flexibility in response to changes in demand*

Different kinds of changes in demand require different kinds of staffing flexibility. A military air defence exercise may require new control positions to be staffed, rather than a splitting or amalgamation of existing ones, and so may the provision of an air traffic control service at small airports used primarily for leisure flying. This kind of change in demand entails not only the staffing of extra positions but more liaison, coordination and communications with normally staffed positions in other parts of the air traffic control system. In extreme cases, major demands for temporary additional air traffic control services can become difficult to reconcile with air traffic control as a full-time and not a part-time profession, since the provision of extra services in any part of the system is liable to result in the whole system becoming busier. More of those controllers whose work is normally spread to provide a continuous service may have to be on duty together to cope with transient peak demands, or controllers may have to be seconded to provide a service for a particular event.

The different daily, weekly or yearly patterns of demand which characterize much of air traffic control impose requirements on staffing flexibility. Means to meet these requirements include: sufficient operational workspace for the maximum requirement, a capability to split or amalgamate many existing workspaces including their communications and other facilities; flexible rostering and work–rest cycles to maintain the match between resources and demands; and staffing levels that can adjust to foreseeable peaks and troughs in demand. These considerations apply not only to controllers but to the supervisors, assistants and others who work with them, and sometimes also to those less directly affected such as maintenance staffs, although wherever possible equipment is not withdrawn for maintenance during periods of high demand.

Predictable changes in demand have further effects. If any new items of equipment or procedures are being tried or there are extra constraints on the system for any reason, these should not coincide with periods of peak loading. Research requirements can lead to particular difficulties because any innovation which is the subject of research must normally be checked under conditions of peak loading and yet these are the very circumstances when an operational system can least tolerate any potential disruption to meet research needs. In practice, the controller with the responsibility for the air traffic must have the option of dispensing with any innovation that seems to have any adverse consequences. Some of the most informative data on new devices may not concern how well they performed but the circumstances, such as peak traffic loading, when the controllers were willing to use them. Such difficulties may force the employment of simulation, or of other methods with less apparent validity than real-life data.

Staffing flexibility is not merely an adjustment of numbers. If controllers are accustomed to working in closely integrated teams, many benefits of increased staffing may be fully realized only if the additional controllers are fully familiar with the ways of the particular team. The problem is not intolerance of individual idiosyncrasies or preferred methods, though these must not become extreme, but rather of ensuring others' prior familiarity with them so that they do not constitute a distraction. Circumstances when new staff are introduced to meet a significant increase in task demands are not ideal for an incoming controller to have to learn a team's idiosyncratic ways for the first time. Such differences between teams can in some circumstances become extensive. They can cover the accepted pace and order in which tasks should be done, the particular formats employed for communications, the occasions when colleagues expect to be consulted, the expected level of mutual monitoring by colleagues, the accepted divisions of responsibility and sharing of facilities, policies on the air traffic control service offered when it is optional, circumstances when a supervisor expects to be informed or consulted, criteria for repeating or verifying messages or instructions, and so on. Staffing flexibility entails a high level of agreement on such matters among the staff who share the work.

Staffing flexibility also has implications for training. The extent to which training has resulted in uniformity of performance will affect the efficiency of teams of different composition. Training must also be employed to produce the smooth handovers of responsibilities, splitting of responsibilities and amalgamation of responsibilities associated with flexibility in staffing. Not only must the workplaces permit effective flexibility, but the training must equip controllers to make the necessary changes efficiently and safely.

7.8 *Team functions*

Most air traffic control jobs and tasks are shared among team members. In all but the smallest air traffic control workspaces, there is far too much work for only one controller. The division of duties and responsibilities is a main determinant of the constitution of each job and of the tasks within it. The allocation of functions between and within teams, the ways in which functions can be split or amalgamated in relation to team composition, and the communications within and between teams are all strongly influenced by team composition, by the planned interactions among team members, and by the intended interdependence or autonomy of various team functions. Therefore decisions about teams have major human factors consequences.

The members of a team who often work together become familiar with each other's work. They have realistic expectations about the reliability and competence of each team member, which is probably a bonus, but may tend to take the performance of colleagues too much for granted. Products of close teamwork can include the tacit rather than active acknowledgement of actions, the omission of routine confirmatory procedures, the development of nonstandard procedural short-cuts that all team members adopt, detailed knowledge by peers of each controller's strengths and weaknesses, the associated recognition of circumstances when a particular controller may need help, and the emergence of accepted norms and standards of performance applicable within the team. Most of these trends can aid efficiency and safety. If the team members get on well together and develop mutual respect and liking, the work environment can be happy and congenial.

In such circumstances, every controller identifies primarily with the team, believes in its superiority over other teams, and defends its procedures, practices and idiosyncrasies against any outside challenges. Solidarity becomes the watchword. The members of a team of high repute will do their utmost to maintain its high standards and reputation. When someone leaves such a team, it can take some time for an incoming replacement controller to earn full acceptance because the newcomer has to learn the team's preferred methods and be comfortable in adopting them before the full acceptance, respect and trust of colleagues can be gained.

In an alternative approach, such close teams are never formed because individual controllers are much more interchangeable. Everyone follows standard procedures closely and an incoming controller has little new to learn in order to conform fully with everyone else. There is much less opportunity in such circumstances for team idiosyncrasies or short-cuts to develop, but also much less general awareness of the particular strengths and weaknesses of individual controllers. Acceptance and trust are built by being predictable and being perceived by colleagues as able to cope. In this approach to teams, individual controllers tend to identify with their profession or with their work location such as an air traffic control centre, rather than with a smaller team or watch. From the point of view of the pilot, the air traffic control service depends very little on which controllers happen to be on duty. Divergent policies and practices between watches do not evolve. Beneficial but initially unorthodox innovations do not evolve either. More transient groupings and teams are less sensitive in recognizing and dealing with a controller experiencing serious difficulties or problems, and less able to provide constructive assistance quickly without fuss.

The mechanisms by which professional norms and standards develop depend on the functions of teams, on their organization and supervision, and on the degree of autonomy afforded to each team. Since system policies, plans and designs can provide some control over these variables, they could in principle influence the information and evolution of norms and standards among controllers, but norms and standards have in the past been generated mainly within the air traffic control profession and owe little to external influences. The mechanisms through which norms and standards are generated can be weakened inadvertently whenever the forms of computer assistance intended to aid the individual controller also curtail team roles and render the activities of a controller less visible to others. The threatened consequent weakening or removal of norms and standards enforces consideration of their roles and desirability.

If there can be no self-generated professional norms and standards, what could replace them? Could they be instilled during training if they cannot be learned on the job? Would professionalism perpetuate the highest individual standards among controllers or would the mechanisms for generating that professionalism also atrophy? Perhaps professionalism and norms and standards incidentally ensure greater uniformity of performance among controllers than there would otherwise be, but many forms of computer assistance could also reduce the incidence and consequences of variability in human performance. The very forms of computer assistance that reduce team functions and lessen the roles of professionalism and of norms and standards may also compensate for or disguise any resulting weaknesses or inconsistencies in human performance, so that some loss of norms and standards and professionalism occurs, but the consequences of the loss are also minimized. If this were so, the diminution of such team functions might not have many serious consequences for performance and safety, and the significance of its consequences would then have to be appraised in other terms.

Occasions for gaining the esteem of colleagues, for self-esteem, for camaraderie, for job interest and challenge, and for motivation and job satisfaction all derive from functions that teams fulfil. The importance that should be accorded to some of these factors is difficult to gauge now but will become even more difficult to gauge in the changed social climate of the future, when jobs in which an individual rather than a team interacts with a large complex system through a human–machine interface will be far more common, will have acquired their own set of skills, and will probably be much more interchangeable.

Misunderstandings about functions and about communications and divisions of responsibility that have been responsible for human errors and incidents in aircraft cockpits have led to extensive research (Foushee and Helmreich, 1988), and to the widespread adoption of team training in cockpit resource management in the furtherance of air safety (Wiener *et al.*, 1993). Corresponding failures of team functions in air traffic control may also in some circumstances be inherently unsafe, but his has not yet become so directly apparent. Even if team functions in air traffic control break down under very high loading, this has not necessarily resulted in any reallocation of functions within the team, provided that the safety and efficiency of the air traffic control service have not been impaired. What happens is that each controller is too busy to act as a backup for another or to take on some of the workload of another, but the diminution in team activities and in each controller's knowledge of what colleagues are doing may be an acceptable cost if the alternative is reduced traffic-handling capability, delays to traffic, and further incidental penalties such as increased usage of fuel. However, the roles of team functioning in air traffic control are now treated more seriously. Key issues include the extent to which functions for teams should be retained in more automated systems, and better identification of all the current roles of teams in air traffic control so that their functions are not discovered to be indispensable only after the means to fulfil them have been automated out of existence. Research on computer-supported cooperative work (CSCW) can provide an alternative framework to cockpit resource management for air traffic control team studies (Hughes *et al.*, 1993a, b; Diaper and Sanger, 1993).

7.9 Supervision

In most current air traffic control workspaces, a supervisor is in immediate charge of those who are on duty at any given time. The supervisor is concerned with the detailed planning of the work, including staffing, rest breaks, the allocation of duties, and the amalgamation or splitting of air traffic control jobs in response to traffic demands. The supervisor commonly has some say in which forms of optional assistance will be used and under what circumstances. To the supervisor devolves the responsibility for dealing with visitors, whatever the pretext for their presence.

The degree of participation by the supervisor in the actual control of air traffic varies greatly. It varies between nations, between military and civil control, between type of air traffic control, and between planning and executive jobs. It varies according to management practices and policies, and according to the norms and standards evolved by the controllers or by their representatives. Sometimes it can vary most of all according to the wishes of the individual supervisor. At one extreme, the supervisor's desk may not be near any of the controllers' positions and the supervisor may remain there and rarely initiate any dealings with controllers directly but expect them to draw their supervisor's attention to any matter which requires it; such a supervisor can be quite a remote figure. At the other extreme, the supervisor may seldom occupy the separate desk but mainly stand or sit behind or alongside the controllers at their suite, listen to them through a headset, and participate actively in the control team. Most commonly the supervisor avoids both these extremes, but is occasionally peripatetic within the control room, maintaining up-to-date knowledge about the traffic and its handling, noting broad problems, and attending sometimes to specific problems. A supervisor may annotate a flight progress strip or otherwise draw a controller's attention to a particular matter. The supervisor may either wish to confirm that the controller appreciates the full significance of some information on a display, or may wish to pass to the controller additional information that has been routed via the supervisor's desk.

Although these practices can be flexible and diverse, they all signify that some form of supervision is present. The supervisor has some appreciation of the air traffic and of how it is being handled, and judges progress and the adequacy of staffing. The supervisor forms opinions about the professionalism, tidiness and competence of each controller, and will probably have a duty to assess the controllers formally in terms of their competence, fitness for promotion, apparent weaknesses, retraining needs, and continuing ability to cope. The supervisor guards against any actions that could be illicit, that could not be defended, that seem potentially hazardous, or that seem unpredictable to others. The supervisor should be on the alert for any arguments or dissensions within the workspace or between controllers and pilots. Part of the supervisor's job is to foster good working relationships between team members, which includes following agreed practices, clear divisions of responsibility, mutual assistance where appropriate, and the satisfactory melding of functions fulfilled by different members of the team. The supervisor may also be best placed to detect if the demands of the job seem to be excessive or potentially harmful either for an individual controller or for a team of controllers. Controllers with difficulties expect sympathy, understanding and help from their supervisor, whether the difficulties originate within the work environment or outside it.

As air traffic control jobs become more automated, some traditional supervisory roles change and may ultimately vanish because they are incompatible with the forms of computer assistance provided. An itinerant supervisor can plug in a headset at a controller's position and listen to the interchanges, can look at the radar and the annotated flight progress strips, can understand the traffic situation at that work position, can judge how well it was being handled and assess current or pending problems, and can intervene directly if necessary. Much of the information that is most pertinent for such supervision is no longer accessible in this way to anyone other than the controller directly concerned, because it must be accessed through the dedicated human–computer interface that is designed for only a single controller to use. It becomes more difficult for the supervisor to observe what each controller is doing and hence to ascertain whether a controller's actions are correct, timely and in the right order. It is much more difficult to distinguish between what the controller is doing and what the machine is doing. The result of their combined endeavours may be visible but not their respective contributions to it. The supervisor has fewer means to assess the strengths and weaknesses of individual controllers in absolute or relative terms. The concept of the good controller becomes elusive because the evidence for such judgements has been undermined.

A problem is that supervision could become largely nominal. If someone with the title of supervisor is still present in the air traffic control environment, this can imply that effective supervision remains possible and continues to occur, whether it does nor not. There should not be anyone with the title of supervisor within the workspace if effective supervision is in fact impossible, for this could be dangerously misleading. It can fudge the issue of where the responsibility lies. It can suggest some form of validation of the controller's work when none exists. It can imply that intervention in tasks and functions is feasible when it is not. If intervention is possible only through the human–machine interface there may be no record, as there is now on paper flight progress strips, that intervention has occurred, and no proof of responsibility for actions.

Currently there has to be some basis for the supervisor's extra responsibilities. Most commonly it is longer experience, and presumed greater knowledge, skill or ability. Supervisors have often been judged by their peers to be very able as controllers. This raises the issue of whether the attributes of the good controller and the good supervisor are sufficiently similar for this to be a valid criterion for choosing supervisors. Normally the supervisor has access to additional facilities and communication channels, and participates in decision making at a higher level within the air traffic control organization. Most of the supervisor's additional authority relates to controllers rather than to their equipment. The less the direct knowledge that the supervisor can have of the information that the controller is using, the less the justification for direct intervention by the supervisor can be. Intervention can be justified easily if the supervisor's knowledge is different from the controller's, but not if it is likely to be less than the controller's. It was noted more than ten years ago

(Hopkin, 1982a) that traditional forms of supervision can become impossible in more automated systems, and that if reallocations of functions from humans to machines also recast the remaining team functions this can terminate the traditional supervisor's role. The supervisor's duties are then to manage the workspace and resources rather than to supervise controllers.

From the outset, the role of the supervisor must be specified as clearly as the roles of each controller within the team. If the supervisor is an integral team member, the supervisor's role must be included in the team structure. The supervisor's equipment and facilities are specific to the supervisory functions, and the supervisor can do only what they allow. If a planned role of the supervisor is to intervene actively within the suite to assist a temporarily over-burdened controller, then the supervisor must have the means to recognize a need for help and to respond to a request for help. The supervisor must be able to participate in activities within the suite without disrupting the work in progress and without sowing confusion over who is responsible or how tasks are being shared. When team members are very busy and need help, the efficacy of any assistance depends crucially on its smooth integration with existing activities. It must not result in the duplication or omission of functions, and must reduce rather than add to the burden of the overloaded controller. Within the next few years, the role of the supervisor is likely to change more than the role of the controller in many air traffic control contexts.

7.10 *Assistance*

Currently air traffic control teams often include air traffic control assistants. Their actual functions vary but share certain attributes. Assistants require significantly lower levels of knowledge and skill than controllers. Their tasks are less complex, carry fewer responsibilities, and often involve more routine and repetitive work applied to a smaller range of tasks and functions. An assistant may undertake some routine liaison, prepare and deliver flight progress strips, and perform some delegated functions such as setting up under the direction of the ground movement controller the lighting signals that route aircraft along taxiways within an airport.

Many of the traditional functions of air traffic control assistants are among those that are most amenable to automation. For example, paper flight progress strips can be printed by machines rather than hand-written, delivered automatically rather than manually, separated from their holders and stored mechanically after use, and ultimately replaced entirely by electronically generated and presented strips. Related ancillary functions employing a simple data input device may be incorporated into the controller's tasks. It may be little more work, or even less work, for controllers to perform such functions themselves rather than tell others what to do and check that they do it. The trends are for many assistants' roles to diminish and for the remainder to change.

The reasons are similar to those that apply to the supervisor. As more automated aids are provided, the controller's tasks become more self-contained and more directly linked with the machine so that assistance is mainly from machines rather than humans. It becomes more difficult for anyone, whether supervisor or assistant, to participate actively in the controller's tasks, to assist them directly, or to intervene without disrupting them. In manual systems, it is quite easy for an experienced supervisor or assistant to judge by observing the controller when the controller may be interrupted. It is much more difficult to choose opportune moments for intervention when the controller's activities consist mainly of interactions with the machine through the human–machine interface.

Assistants and supervisors share the further attributes that their functions must be designed in conjunction with the functions of all the team members, and clearly differentiated from and reconciled with those of the controllers. It must not be presumed that effective assistance is possible merely because someone called an assistant is present within the air traffic control workspace. The assistant must have a clearly defined job which actually is helpful; otherwise the assistance is nominal only. Although the assistant's job should be designed to be subsidiary to and at the behest of the controller in accordance with their respective legal responsibilities, an effective assistant must have some autonomy, as assistance which requires continuous and detailed scrutiny and cross-checking by the controller cannot really be construed as effective assistance.

As automation progresses and the remaining roles for the assistants become more constrained, some of the human factors problems encountered elsewhere in air traffic control spread to them. In many countries a few of the best assistants have traditionally earned the chance to apply for retraining as controllers, and this has been a source of recruits with realistic expectations. As jobs become more constricted, so does the evidence for defining who the best assistants are. Opportunities for assistants to become controllers may therefore diminish, and as the basis for selecting prospective controllers from assistants becomes more speculative a larger proportion of those selected may fail to qualify.

If the job of the assistant becomes more marginalized, assistants may not identify so readily with air traffic control as their profession and with its prevailing ethos and norms. They may see themselves as peripheral to it rather than part of it. This could be counterproductive. It is important that the remaining jobs for assistants in air traffic control are perceived to be essential by the assistants themselves and by all others, and that these jobs retain clearly defined responsibilities and skills.

A major reappraisal of the future roles for assistants is overdue. It must make sense in principle for as much work as possible to be done by those who require fewer skills and shorter training, provided that efficiency and safety are ensured. Assistants' jobs must be real jobs and not a mere bundle of ancillary functions that have proved difficult to automate but seem too routine for controllers. This is not the way to employ automation. There may be considerable scope for assistants rather than controllers to perform many future air traffic control functions. This does not seem to have been seriously addressed. As long as air traffic control combines a preponderance of routine activities with occasional demands for high level skills and abilities, there is a case for separating the routine from the higher level tasks. Disparate attributes and abilities are required for the optimum performance of such different functions, which implies that their job descriptions and their selection procedures should also differ. In air traffic control, the progressive whittling away of many of the traditional roles and responsibilities of assistants has been less a deliberate policy than an incidental consequence of decisions taken for other reasons, particularly related to the provision of forms of computer assistance that allow more traffic to be handled safely and efficiently. Nevertheless, technological options are becoming so flexible that such changes in the roles of assistants need no longer be an inevitable concomitant of technology.

At some stage it may be worthwhile to test different assumptions. An alternative premise is that there will be a similar proportion of jobs for assistant air traffic controllers in the future as in the past. The jobs, tasks and functions in more automated systems that could be identified and grouped to form the core of satisfactory roles for assistants could then be considered. If skills as complex as those possessed by controllers are likely to become scarce in the future, it seems sensible to formulate alternative and more attainable balances among jobs and roles within teams, which could alleviate future recruitment problems by continuing to provide satisfactory careers and prospects for more air traffic control assistants.

8

The selection of controllers

8.1 Recruitment

All successful procedures for the selection and training of air traffic controllers depend ultimately on the efficacy of recruitment. Effective selection is impossible without suitable applicants. There has not been much research on why people apply to become air traffic controllers. In many countries such research has seemed unnecessary because a plentiful supply of applicants has emphasized the practical problem of devising effective and fair sifting procedures to reduce the initial flood of applicants to administratively manageable numbers without discarding the best candidates in the process. Sifting should streamline the initial phases of selection, and avoid raising unrealistic expectations by calling candidates for interview whose prospects for acceptance seem poor. For successful initial sifting, the application forms must yield sufficient valid information.

In some countries there have been recent recruitment campaigns for air traffic controllers, usually with one of three objectives: a need to increase the supply of applicants, a requirement to fill specific job vacancies, or a policy to target under-represented groups such as women or certain minorities. The population of air traffic control applicants normally diverges demographically from the corresponding national population. Sometimes the reason concerns national policies based on human factors evidence, such as a maximum recruitment age, for example. Sometimes air traffic control requirements introduce biases because of a confounding of variables: insistence on normal colour vision is bound to favour women rather than men, for example. But often the demographic differences originate in the applicant population. Any national policy to redress a perceived imbalance therefore has to be tackled primarily at the recruitment stage. For example, the main reason why there are relatively few women controllers in most Western nations is that few women apply to become controllers, and any change must start with recruitment drives to encourage more women to apply. Air traffic control recruitment procedures, whether in the public or private domain, must actually be impartial and be demonstrably fair if challenged. Considerations of aviation safety have to remain paramount. This is why discrimination on the grounds of age is commonly permitted, but discrimination by gender is not.

One cause for concern is the effect on recruitment of ill-informed criticism or public denigration of air traffic control and controllers, especially by public figures or media representatives. Safety is not news, but lapses in safety are. The rarity of serious lapses in air traffic control safety ensures that the details of any lapse will be emblazoned in headlines in a crusade for safety, unless the lapse has to compete for media attention with the titillating peccadillos of public figures. The thorough and impartial investigation of the full circumstances of any incident follows formalized procedures rigorously, and those directly concerned must do nothing that could be construed as an attempt to prejudice or influence the investigations; in particular, they must refrain from public comment and must not reply to criticisms, however unjust they may be. Some air traffic controllers fret about the unfairness of uninformed and even malevolent criticisms of their profession, which they are forbidden to refute directly. Ironically, any human factors specialist originally trained as a psychologist can empathize with the controller, for any professional psychologist becomes accustomed to ill-informed gibes about psychology. Whereas the psychologist expects disparagement, the air traffic controller does not, and can be dismayed by it.

It is uncertain how far prospective applicants for air traffic control jobs are deterred by its image as a profession that seems to attract blame rather then praise and seems slow to defend itself. Air traffic control can also be a butt, the ever available excuse to passengers for delays, but with no opportunity to provide its version of events whether air traffic control has influenced them or not. It seems unlikely that the image of air traffic control among the general public has no influence on recruitment at all, particularly since many applicants have very limited understanding beforehand of the nature of air traffic control jobs.

If any potential recruits are discouraged by their perception of the public image of air traffic control, this is particularly important because the image is false. It surprises even well-informed people how much the typical air traffic controller enjoys controlling air traffic. Air traffic control is a very satisfying job for most of those who do it (Wise *et al.*, 1991). Many enjoy it so much they wish for no other job. There is therefore a gulf between the public image of air traffic control as a profession and the controllers' own perceptions of it, their pride in it, and their identification with it. The effects of this gulf on recruitment merit urgent consideration.

One application of human factors to the recruitment of air traffic controllers adapts to air traffic control the techniques of advertising, marketing and public relations applied elsewhere to influence public opinion and attitudes. Many of the techniques are orthodox and well proven. Surveys can establish what people see as attractive in the profession; test marketing can compare alternative forms of advertising and recruitment methods; previous successful recruitment campaigns can help to target future campaigns; readership surveys and surveys of the audiences for other advertising media can direct the recruitment literature towards the most appropriate and receptive target audiences. Even the most plentiful supply of applicants does not necessarily include those with the greatest potential to become successful controllers. If the best candidates never apply, the efficacy of all subsequent selection and training procedures is reduced. In any case, these procedures require periodic revision in response to actual and pending system changes. Recruitment procedures and methods may also require revision, to attract more suitable applicants, to improve the public image of air traffic control, to make applicants better informed, or to accommodate demographic and social changes that affect job requirements and opportunities.

In many countries, demographic trends predict a diminishing supply of able and willing applicants to become controllers. Air traffic control will face more competition in recruiting the people it needs. Recruitment procedures must become more effective simply to maintain the present quality of applicants, but the aims are to enhance their quality and to recruit more controllers.

8.2 Supply and demand

In the past some of the most contentious selection issues have arisen when an imbalance between supply and demand, particularly if coupled with a shortage of controllers, has led to pressures to relax selection criteria temporarily in order to satisfy urgent short-term increases in demand. The penalties of lowering standards depend on the quality and validity of the selection processes. If the processes are poor, the consequences of relaxing them may remain minor, but if they have high validity any relaxation of them may not yield the expected benefits and could be counterproductive by raising the costs of training without increasing the numbers who successfully complete it.

Selection is contingent upon successful recruitment. The more applicants there are, the more sophisticated and valid the selection procedures for controllers can in principle become. A gross surfeit of applicants usually enforces a two-stage selection procedure in which an initial paper sift based on application forms produces a smaller number of candidates who progress to the main selection procedure. Although the ratio between the number of participants in the main selection procedure and the number who will complete training successfully and become controllers depends greatly on the quality of applicants and the vacancies for controllers, nevertheless this ratio usually remains fairly stable in the short term, and it can provide practical guidance on how many candidates should be seen. The causes of any sudden gross changes in the ratio should be traced. A nearly constant ratio, whatever it actually is, can be administratively helpful because the numbers who will complete training become predictable. A continuous supply of fully trained controllers to meet envisaged demands can then be planned without major changes in selection criteria, without raising unrealistic expectations among candidates, and without the extra costs of administering the full selection procedures to too many applicants. Any gross change in the ratio of applicants to successes implies that there has been some change in the standards of applicants, the selection procedures, the training, the assessments or the requirements. Changes for any of these reasons may be present and acknowledged, may be tacit but implied, or may remain unrecognized. The explanation of any gross change in the ratio must be sought as reassurance that other unwanted factors have not intervened and that standards are not changing in an uncontrolled way. If they are, the outcome is inefficiency, associated with a surplus or shortage of candidates successfully completing their training.

A complicating factor is whether the selection procedure is intended solely to provide controllers or must meet further requirements using the same selection procedure or an additional one. An example of an extended application of the selection procedure itself would be the identification of candidates with potential managerial ability or computer skills, as well as the requirements for being a controller. An example of an additional selection procedure is the requirement for military controllers to possess officer qualities as well as meet controller requirements. Success as a controller may have little in common with success according to these further requirements. An example of a future complication for selection could be the recruitment of a few exceptionally able and evaluation-minded test controllers for air traffic control jobs parallel to those of test pilots (Westrum, 1994).

Several influences on supply and demand are particularly pertinent to air traffic control. The publicity that it receives as a profession does not generally give a realistic impression of it to prospective candidates. The recruitment literature and selection procedures have to make candidates better informed about air traffic control. Applicants are interested in the usual range of factors, including pay, career prospects, conditions of employment, level of autonomy, and the interest, satisfaction and challenge of the work. Air traffic control often entails shiftwork and unusual work schedules which can have occupational health effects on a few controllers or lead to domestic or social problems. Prospective controllers must be aware of this. Some applicants may have failed aircrew selection procedures and view air traffic control as the second-best option. If they cannot fly aircraft, at least they can control them. Such applicants do not necessarily become poor controllers but their initial disappointment must not result in protracted bitterness. Ultimately they have to be able to accept air traffic control as their profession rather than a second-best option that they had to settle for.

Air traffic control is still on the whole viewed as a career for life, but this view may not continue to prevail if the jobs in it seem liable to be automated out of existence despite the increasing numbers of aircraft requiring an air traffic control service. So far, increased automation of air traffic control has not significantly reduced the number of controllers, and indeed several nations are actively recruiting more controllers. However, the nature of air traffic control jobs will gradually evolve to take advantage of technological innovations, and human roles will change as more dialogues are conducted through human–machine interfaces. Many other jobs outside air traffic control will be evolving similarly, increasing the potential interchangeability of jobs and the commonality of skills between air traffic control and other professions. This implies greater mobility of labour because of enhanced transferability of knowledge and skills, and greater facility to leave air traffic control for other jobs if air traffic control becomes unsatisfying or other jobs look more attractive.

This greater interchangeability seems likely to affect supply and demand much more in the future. The winners will be those who provide professional jobs that are skilled and liked, with favourable working conditions that meet human needs and expectations. There is already evidence that many other professions are trying to recruit similar people to those sought by air traffic control. The supply of such people is declining as the demand for them is increasing. Those who can attract people with these scarce and valued skills and can keep them by providing jobs, careers, work, and conditions of employment which satisfy human needs and aspirations will win the tussle between supply and demand.

8.3 The identification of relevant human attributes

Extensive research has identified many relevant human attributes for inclusion in selection procedures (Smith and George, 1992). The first air traffic controllers often learned their skills within ground control systems that were either for air defence purposes or for handling air traffic (Hopkin, 1970). The requirements of air defence and of air traffic control seemed sufficiently similar for many of the skills and attributes that had proved beneficial in air defence to transfer to air traffic control, and there was an initial preference to recruit for air traffic control people with previous air defence experience. However, this was not always as successful as expected, and sometimes incurred high attrition rates and seemed wasteful. In an effort to lower attrition rates, more formalized psychological procedures for air traffic control selection were introduced, using the orthodox methodology of task and job analyses to identify relevant factors for which objective tests were then developed and validated. These procedures did reduce attrition rates (Cobb *et al.*, 1972), but there are still problems in recruiting civil controllers from military sources (Manning and Aul, 1992).

Various kinds of human attributes of potential relevance to air traffic control capability have been considered for inclusion in selection procedures.

One kind refers to measurable and relatively stable differences between individuals in human capabilities that have been deduced to be relevant from studies of tasks, skills and performance. General intelligence seems the most important of these, for intelligence well above average is necessary not only to be a controller but to learn to become one, although insistence on minimum educational qualifications for all applicants could partly cover this requirement.

Quite a large number of more specific capabilities have seemed advantageous. These include numerical ability, spatial ability, abstract reasoning, various aspects of memory, task scheduling, task sharing, verbal reasoning, mechanical reasoning, perceptual speed and accuracy, and linguistic fluency. The above attributes would generally be classed as cognitive. Psychomotor capabilities of relevance include manual dexterity, coordination, speed, accuracy and consistency in the performance of routine tasks, and effective usage of such input devices as a keyboard, joystick, trackball, mouse, puck, light pen, or touch-sensitive screen. Psychomotor skills are often the most readily quantifiable ones.

For some attributes which seem relevant, no suitable quantitative measures or tests have been successfully validated for the selection of controllers. Control over attention is one example; a propensity to misunderstand is another; differences between individuals in the kinds of human error that they typically make is a further example. The ability to listen must be relevant if tools to measure stable individual differences in listening ability can be devised. Abilities to make predictions, to take decisions, to solve problems and to understand and interpret situations correctly seem relevant cognitive attributes for which there are better prospects for developing appropriate measures applicable to air traffic control selection. Measures of the ability to match human thinking with automated cognitive aids could also be advantageous.

Another kind of human attribute concerns personality. The validity of personality measures in the selection of air traffic controllers is less certain than that of cognitive or behavioural ones. Although in some quarters controllers may cultivate a sociable and extrovert image, in most countries they conform quite closely with the general population in such personality attributes. Apparently desirable attributes for which personality measures might be helpful include high motivation, tolerance of stress, tolerance of workload extremes (gross overloading or underloading), acceptance of boredom, emotional stability, and possibly mild obsessionalism. The ability to reach and abide by decisions without undue prevarication and delay could be both desirable and testable, coupled with self-confidence but not the overconfidence of discounting contrary evidence. Excessive identification with one's own decisions can be hazardous if admission of one's own errors involves too much loss of face or loss of esteem. A good controller is not often wrong for he or she would not then be a good controller, but a good controller can admit to being wrong. Air traffic control requires the acceptance of responsibility, which occasionally will induce anxiety or worry. A safe controller is neither perennially over-anxious nor incapable of anxiety whatever happens. Part of the controller's job is to appraise potential hazards and act to resolve them before they develop too far, without intervening unnecessarily when circumstances can safely be left to run their course.

Air traffic control is an organized structure at several levels, with international, national and local policies and practices. Some of its management practices may conform with national stereotypes and be common to many professions. Some will reflect the priority accorded nationally to air traffic control. A few practices, such as the treatment of insufficiently collaborative workforces, may be politically influenced. If the policy is to select as controllers able and intelligent people who can be trained to act decisively, who are accustomed to being responsible for their own actions, who resist having decisions by others foisted upon them without proper explanation, and who value and identify with their professional procedures and practices, it should not be too surprising if these same controllers close ranks and resist any outside interference that seems to add to their burdens, to undermine their authority or to reveal ignorance of how air traffic is controlled. The selection procedures for controllers almost seem designed to recruit a workforce which will require management decisions to be explained, will suspect the motives of those who seek to make economies, and will need evidence rather than promises that proposed changes will be beneficial. Because some of the benefits claimed for earlier forms of computer assistance never materialized and some expected reductions in controller workload turned out to be increases, scepticism lingers, especially if the apparent interests of air safety, greater efficiency and cost reductions do not coincide.

Linguistic ability encompasses mastery of the language, vocabulary, formats and sequences of air traffic control messages, even under the most adverse conditions. These include understanding speech under stress by someone whose native language is not English, who mispronounces words, and who is using noisy or distorting channels of speech communication. Linguistic ability extends to the gleaning of further information from its nuances and subtleties, such as pauses, hesitancies, slight variations in phraseology, excessively pedantic or rigorously stereotyped message formats, acknowledgements that seem to lack understanding, minor flaws in repeated messages, and other signs of unsureness or lack of confidence in the speaker or the listener. Although major centres and airports may come to rely less on spoken messages in the future, nevertheless at less well-equipped airfields and even with data links it will remain an essential attribute of controllers that they can interpret the subtleties of spoken messages correctly.

Controllers must be able to work harmoniously as team members. Their behaviour must be sufficiently conformist not to seem eccentric or unpredictable to colleagues, yet retain some independence of thought and action. They should support their professional norms, standards and ethos. A relevant attribute is an ability to identify with a group and be loyal to it, and training for controllers includes team training. Effective teamwork includes supervision and assistance, and the controller has to be able to accept both. The controller needs to be conscious of the strengths and weaknesses of others: effective teamwork includes an ability to recognize when help should be offered, willingness to accept help when it is needed, and the forbearance not to intervene when help is neither warranted nor welcomed.

Biographical data are also pertinent. Many countries debar all applicants without minimum educational qualifications. Some express these as number of subjects studied and grades attained, whereas other prefer qualifications in subjects deemed to be most relevant to air traffic control. These are usually scientific ones although the supporting evidence for this is not strong. The possession of educational qualifications may also be construed favourably as evidence of a capacity for sustained study and motivation. While a good education is a desirable attribute in a controller, and the most highly qualified candidates may become able controllers, in the longer term those whose qualifications are of university entrance standard or include a university degree may be more liable to become bored by air traffic control, poorly motivated, and restive and disillusioned by a perceived lack of career prospects. Given that the objective is not only to recruit successful and safe controllers but also to keep them, the highest educational qualifications may not always be an advantage. Educational requirements can become a politically sensitive issue, because easing them can be a practical way to favour minorities or to increase the number of qualified applicants.

Recently biographical evidence has been gathered on attributes such as honesty and integrity and any history of minor criminality or delinquency, with the implication that this kind of evidence may have some predictive value in contexts where teams work closely together. Although more directly relevant to medical selection procedures, any history of drug or alcohol abuse may also be pertinent. This kind of evidence raises some difficult ethical questions about the reconciliation of individual rights with the requirements of aviation safety.

Biographical data normally include gender and age. Men and women appear to be about equal as air traffic controllers. In many countries but not everywhere, air traffic control is a predominantly male occupation because not many women apply to become controllers, and this is also true of several other jobs in aviation, including pilots and engineers. The unusual working hours, which usually include shifts and night work, may deter some women. The proportion of women applicants who pass the selection procedure is generally similar to the proportion of men.

In relation to age, the evidence is very different. For a long time, it has strongly favoured those recruited when young because older candidates are much less likely to complete training successfully. The factor of age even seems to outweigh experience. It was presumed that the greater experience of older former military air traffic controllers would outweigh the youthful inexperience of those for whom air traffic control would be their first job, but this presumption was wrong for both the failure rates during training and the attrition rates following training have been highest among the oldest applicants. The result is commonly a rigidly enforced maximum age of recruitment, usually about thirty, which is not relaxed because of experience or exceptional qualifications. The interactions between age and years of service depend on national policies on the retirement age for controllers and on whether retirement age depends on chronological age or on years of service.

A further kind of attribute of somewhat ambivalent relevance concerns the candidate's stated interests, particularly in aviation. The evidence about this is not consistent. A declared interest in aviation or air traffic control may possess more face validity than true validity in the selection procedure, and may be a better indication of motivation than of the ability to succeed as a controller.

A broad category of identified relevant attributes covers physical, physiological and medical differences between people. The most obvious of these concern the senses. Minimum eyesight standards, corrected or uncorrected, are strictly enforced because equipment is designed on the basis that every controller reaches this minimum. Controllers must have normal colour vision according to approved colour vision tests. Controllers must not be deaf, and their hearing is tested during the selection procedure. A serious stutter would preclude selection because it could be dangerous in air traffic control. There may be no space to adapt air traffic control towers or the steep means of access to them to accommodate a wheelchair.

Since air traffic control work is demanding, the hours are erratic and the training investment in each controller is large, standards of medical fitness similar to those for pilots are enforced. Controllers have to be basically fit and free from serious degenerative illnesses at the time of recruitment. They need to be emotionally stable and resilient, and should not have a history of mental illness. They should be tolerant of stress although there are considerable national differences in the incidence of stress-related illnesses among them, and in some countries it is below that of many other professions. Stress in air traffic control is never as severe or as prevalent as the popular press would imply. Selection procedures and subsequent training often combine to fail those who would be most prone to excessive anxiety or stress. Controllers have to pass an annual medical examination to retain their licence, and this strongly discourages certain conditions such as obesity.

Controllers must have reasonable strength and stamina. High blood pressure requires treatment, as does alcohol or drug abuse. They must not be irresponsible about self-medication. Many drugs, including some proprietary ones marketed without prescription, should only be taken under medical supervision by controllers because they could be claimed to impair job performance by inducing drowsiness, loss of concentration, inattention or excessive blitheness.

The range of human attributes identified as potentially relevant in air traffic control selection is large, though not untypical of other professions, but the penalty of wrong selection can be more extreme in air traffic control. Considerable resources have been devoted to the development and validation of appropriate selection procedures for controllers. Collins *et al.* (1980) and Rock *et al.* (1981) reviewed the earlier work, Sells *et al.* represented the state of the art in 1984, and Wing reviewed it in 1991. Manning and Broach (1992) look to future selection requirements.

The identification of relevant factors must be a continuing process as jobs, requirements and equipment change. Further changes suggesting an iterative approach would be in the perceived social status of air traffic control as a job, in expectations about what it has to offer, in the priorities of human needs and aspirations at work, or in public policy. Further human attributes relevant to selection can originate in three other kinds of factor.

The first concerns the impact of automation on air traffic control. One kind of desirable attribute refers to knowledge of computer functioning, familiarity with human–machine interfaces, and acceptance of working with computers as a congenial part of the job. It may be desirable to include in selection pro-cedures a test of ability to work with computers and to understand them. Any test would have to measure potential rather than actual ability or it would simply favour those familiar with computers rather than those with most potential talent to use them. Since it is often more economical and impartial to use computers to administer the selection procedures, any possible bias towards selection procedures that can most readily be computerized must be resisted in the interests of maximizing the validity of the selection process as a whole.

The second kind of factor concerns the relevance of identified human attrib-utes to all air traffic control jobs. Some air traffic control jobs seem very differ-ent from others. For example, how wide-ranging are the common attributes required for the jobs of the oceanic planner, the en route controller using radar, the ground movement controller of taxiing aircraft, and the military controller? How universal should the selection procedure be? Should the selection procedure be employed not only to select those to be trained but to allocate candidates to training for particular air traffic control jobs? This issue is gaining in relevance because it costs much more to train controllers for all jobs than to train them for a few.

The third kind of factor concerns the measurement of identifiable attributes. Many biographical and medical data do not change much, and measures of educational qualifications can be validated statistically. However, measures obtained by standard tests are vulnerable to individual familiarity with the tests, and those least familiar with tests and test procedures can be at a dis-advantage. This was not very important when most tests were not available to the public but in recent years coaching people in how to pass the selection procedure for controllers has burgeoned, using tests similar to the real ones. Performance on these tests may be boosted significantly by coaching and prac-tice with similar material. If those who have been coached are preferred, those who have not are disadvantaged, often because they could not afford the fees for coaching. The actual effects of coaching on the validity of the selection procedures and on the weighting of the items within them are still partly speculative but could be substantial. There is not yet sufficient evidence to tell whether those with the motivation or wealth to pay for coaching ultimately become better or worse controllers.

Meanwhile it is prudent, for candidates who initially fail the selection procedure and take the same tests again when they reapply, to use for selection purposes the scores obtained in their original application and not those at their second or subsequent attempts after coaching. This expedient may help to preserve the validity and integrity of the measuring devices while evidence on the effects of coaching is gathered. Extensive coaching complicates the identification and measurement of relevant human attributes. If those accepted after coaching are less able as controllers than those without coaching who would otherwise have been selected, or if coaching results in higher attrition rates, then coaching benefits no one except those who earn their living by it. However, it is essential at the present time to remain open-minded. If those who have been coached are not only more likely to be selected but more likely to complete their training and to succeed as controllers, this would argue for incorporating standardized coaching into the selection procedures for everyone, but if those selected after coaching are poorer than those selected without it this argues for using only the initial test scores or for deducting marks from those who have been coached or for discouraging rather than encouraging coaching. It is not satisfactory to allow the issue to drift as its consequences for selection can be substantial, and it introduces an additional source of unknown variance when all other efforts to improve the selection process seek to remove sources of unknown variance.

8.4 Interviews

Over the years evidence about the limited validity of selection interviews has accumulated, but they remain a standard part of most procedures. The best policy is not to disparage them, but to try to ensure that they are not more influential than they should be, that they are optimized within their limitations, and that they are correctly conducted.

There are some simple rules about interviews. The first, which is almost invariably resisted, is that interviewers need training. Since many believe that they are born interviewers and have high confidence in their untrained assessments, this idea of training is hard to accept, essential though it is. Once it has been conceded, interviewers have to be given the right training. An interview is not an intuitive and extemporary display of the interviewer's superiority for the benefit of the candidate. An interview should yield information which cannot be obtained as effectively in any other way and is valid for the selection of controllers; otherwise it has little point.

In selecting air traffic controllers the standard psychological principles for interviewing should be followed, in accordance with recent knowledge

(Keenan, 1989; Millar *et al.*, 1992). In particular, the interviewers should know what information they wish to obtain and to convey, and structure and standardize the interview accordingly. They should use agreed assessment methods that they are thoroughly familiar with, which should be as objective and quantifiable as possible and demonstrably independent of who actually conducts the interview. This independence must be provable. If necessary, interviewers should practise with mock candidates, make their assessments independently, identify where they disagree, establish why they disagree, resolve the causes of their disagreement, and repeat the whole process iteratively until they do agree because they are making the same assessments based on the same evidence. This procedure can establish the reliability of interviewing as a process that yields consistent results, but cannot establish its validity, which depends on the relationship between the assessments from the interview and capability as a controller.

In addition to the normal recommendations for interviewing, some attributes of specific importance in air traffic control may require assessment by interview. Clear articulation is essential for aviation safety and must not be compromised. It could be assessed from recordings but the interview provides an opportunity, which may not recur, to assess articulation under stress. Nervousness alone should not disqualify, but inarticulateness caused by nervousness may do so if it is extreme. The interview board may have to decide whether any deficiencies in articulation seem irremediable or would respond to training or to additional courses in air traffic control English. A controller has to be able to think while speaking and to remain articulate while thinking: either inability may point to training difficulties. Some problems can be awkward. Very strong national, regional or ethnic accents could be a handicap if they cannot be subdued sufficiently for all listeners to understand the spoken air traffic control language. This does not imply that the accent must vanish, but that it must not impair the intelligibility of messages transmitted through radio-telephones and heard through headsets. The language, vocabulary, pronunciation and emphases of air traffic control messages have to be unambiguous and standardized. The controller achieves this mainly by speaking plain English, but also by obeying specific air traffic control conventions such as pronouncing "five" as "fife" and "nine" as "ninah" to minimize confusion between them.

In an interview, it may be possible to separate two closely related factors with different significance. A sign of good motivation in a candidate is good preparation for the interview. This includes enquiring beforehand about the nature of the job, what it entails and what it offers. Serious candidates should have some rudimentary ideas about air traffic control from their preparation for the interview, which candidates may not have if they apply for air traffic control merely to gain employment. Knowledge as preparation for the interview should not be equated with prior air traffic control or aviation knowledge from longer-term interests, which may have greater face validity but less true validity. Whereas the interview permits previous knowledge to be explored, this should not occupy a significant part of the interview because of its suspect predictive value. A longstanding interest in air traffic control may have prompted a candidate to apply, but interest does not necessarily imply ability, and more modest knowledge may be a more valid predictor if it betokens good interview preparation.

Interviews must be extensively standardized. This is not just to make them more valid, though it probably does, but to ensure their uniformity as a procedure. The same psychological test can be administered in different places by different people with comparable results because of the standardization of the test and of the procedures in administering it. Comparable objectivity is more difficult to achieve in an interview if its outcome can be swayed by those who conduct it or by where it is conducted, yet the central purpose of the interview is the same as that of the test, to reveal differences between candidates. Thorough training of interviewers offers the best prospects for objectivity and standardization.

8.5 Psychological tests

The evolution of psychological tests applied to air traffic control selection has demonstrated steady rather than spectacular progress. The current position after some 30 years of toil is that a considerable range of testable dimensions apparently have predictive validity, mostly because scores on them correlate significantly with successful completion of training rather than with subsequent careers which have not usually been measured in this context. However, no single test, nor any optimally weighted combined battery of tests, has succeeded in accounting for more than a modest proportion of the total variance. This is a disappointment, following so much conscientious work. A belief seems to persist that psychological tests ought to be better predictors than they have hitherto proved to be. Job analysis and task analysis techniques, applied to air traffic control, can and do reveal human abilities, aptitudes and skills which seem relevant and for which standard tests exist or can readily be devised. Such tests, properly validated, should be helpful in selecting controllers. Therefore, efforts persist to analyse air traffic control, identify relevant testable dimensions within it, and devise, validate and apply appropriate tests. Much of the earlier research was collected by Sells *et al.* (1984).

There are good reasons for applying psychological tests in selection. In automated form, they can be self-administering, and even in paper-and-pencil form their scoring should be objective and free from any scorer bias. Most need not take much time, nor much of an instructor's or tester's time. Many can be administered to large groups concurrently or to many individuals independently for they are not location dependent. Scores are interpreted according to well-established norms, and the acceptable minimum, maximum or range is specified in advance. The highest score is not necessarily the best. For some dimensions it may be extreme scores that are not wanted. Psychological tests are inflexible if they have to be interpreted according to any criteria other than basic test scores. They cannot easily be adapted to support social policies in favour of any minority group, without seeming unfair. If they apparently show significant but politically unwelcome differences between population groups, the measurements may be blamed for the message. Psychological tests can raise difficult issues if their results do not conform with universal, national or legal guidelines about employee selection (Rock *et al.*, 1981; Wing and Manning, 1991).

Part of the rationale for using psychological tests is that they claim to measure fundamental human attributes which are relatively immune from improvement through familiarity. In practice the vulnerability of intelligence test scores to improvement through coaching or practice has been recognized for decades, and most of the psychological tests likely to be employed in air traffic control selection seem as vulnerable. Indeed there would be no point in including similar tests in coaching courses unless foreknowledge of similar material was an advantage. A potential disadvantage of tests and of some other selection tools is that coaching or practice could reduce their predictive power.

Quite a long list of testable dimensions has at one time or another been included or been proposed for inclusion in batteries of psychological tests for the selection of air traffic controllers. The list of dimensions is similar to that of identified relevant attributes. Tests include general intelligence, mathematical ability, numerical ability, arithmetical reasoning, familiarity or facility with small computers, verbal fluency, verbal clarity, abstract reasoning, spatial reasoning, directional judgement, task scheduling, task sharing, analogies, mental maturity, dial reading, manual dexterity, perceptual speed and accuracy, and language usage. There is no shortage of testable attributes positively correlated with success as a controller, although as more tests are included each tends to add less to the predictive value of the whole test battery, and intercorrelations between tests become more essential. Specific tests, nominally different, may be remeasuring the same attribute or may be redundant with a combination of other tests. The total variance that psychological tests can account for in the selection of controllers seems to be approaching its limit, and it may be worth trying to discover why.

The selection procedures for controllers cannot be validated by ideal long-term criteria such as successful completion of a lifetime career as a controller with an excellent safety record. This is impractical for it would take too long. Nor are the selection procedures normally validated by shorter-term data derived from controllers' performance, such as whether they stay as controllers or leave, whether any infringements of safety are attributed to them, whether and when they achieve promotion, and so on. Instead the most common validation of selection is by successful completion of training. How long this takes varies between nations. It can take several years and include considerable on-the-job experience, but the training can be much shorter but more intensive. The most serious problem is that neither the validity of the successful completion of training itself nor the validity of final training scores as a predictor of subsequent success as a controller is known, because whether those who are best in training become the best controllers has not been established. If the validity of this is no higher than that already achieved for selection procedures, efforts to improve the validity of the selection procedures themselves may always be vitiated until they no longer depend on training criteria. Perhaps some of the effort expended to raise the predictive value of the tests in the selection procedures should be redeployed to extend the criteria for validating those procedures into measurements of subsequent career progress, continuity, and safety.

It might be worth reconsidering general intelligence as a predictive factor. Controllers must be above average in intelligence to learn to be controllers, but the highest intelligence scores may not be of benefit in air traffic control. The extent to which general intelligence and educational qualifications overlap as predictive tools could be addressed, and the optimum intelligence for controllers should be sought. While those with maximum intelligence should not necessarily be selected, intelligence as a factor should not be ignored to the extent that it often appears to be.

The predictive value of psychological tests for air traffic control selection is lessened because they are administered to a far more homogeneous population than the general population. Therefore the individual differences in many test scores will be quite small. Coefficients between test scores and any criteria will be lowered by this restricted range. Any valid selection test must be highly sensitive within this restricted range rather than within the range applicable to the whole population. Some tuning of tests could be beneficial to enhance their sensitivity within the applicable range.

8.6 Personality

The image of the air traffic controller as calm, confident, unflappable, decisive, mature in judgement and able to cope with excessive task demands has always seemed to imply the relevance of some personality dimensions to capability as a controller so that the selection procedures for controllers should therefore include measures of personality. Attempts to convert this premise into a set of personality measures of proven validity and reliability have demonstrated faith in the quest but reaped few rewards.

The relevance to air traffic control of the most standardized measures of personality, with their scoring norms and interpretative manuals, is not self-evident. No personality measures have ever been devised specifically for air traffic control purposes. Some of the commonest personality tests were not originally devised for selection purposes or for applications within occupational psychology, but as clinical measurement tools. Most personality measures do not have the kind of theoretical basis that could provide a rationale for linking them to air traffic control.

The main approach has attempted to maximize the relevance and usefulness of existing personality measures for selecting controllers, rather than start from air traffic control, deduce attributes of personality that would be beneficial in controllers, and devise and validate tests of those attributes. The history of personality measures in air traffic control has been reviewed, particularly for the United States, by Convey (1984). Most countries employ some form of personality assessment of air traffic control applicants, either formally or informally, but no personality test has achieved widespread adoption. Research continues on the applicability of measures of personality to controller selection.

Nor is there agreement on how personality measures should be employed. When global measures yielding personality profiles have been administered to applicant controllers in the past, sometimes the assessments have been based on the whole profiles but more commonly only a few dimensions have been emphasized. Even the scores on these dimensions have not normally led directly to acceptance or rejection of a candidate. Extreme scores have been interpreted as a reason to seek fuller information about a candidate, perhaps through a clinical interview. This process may seem somewhat dubious if those without extreme scores but rejected for other reasons complain that their candidacy might have been successful had there been a reason for obtaining more information about them.

The personality dimensions which might appear relevant from studying air traffic control requirements do not seem to bear much resemblance to those measured in many of the ready-made personality tests that have been tried. Examples of attributes that might prove desirable include harmonious working with others, unselfishness, thoroughness, self-confidence, equability, and loyalty. Quite a long list of such attributes could be formulated, but there are no suitable existing tests of most of them. The available personality assessments vary greatly in their rigidity and structuring, their practicality, their acceptability, their face and actual validity, and in the reliability of their norms. In additional to standard personality tests in the form of self-reporting inventories of questions or choices, there can be several other kinds of substantiated or putative data on personality. Examples include clinical interviews, various projective tests, and further techniques, such as the use of graphology to select controllers in Switzerland or the employment of concepts such as accident proneness (Rodgers and Blanchard, 1993), whose scientific respectability is disputed.

Sometimes personality tests give puzzling results. The distinction between type A and type B personalities, that in many professions has successfully differentiated between the former with a greater propensity to develop symptoms of stress under pressure and the latter with less propensity (Strube, 1991), was not reproduced among controllers (Rose *et al.*, 1978), and a recent study has not found much relationship with biographical data or training progress (Nye *et al.*, 1993). Even the findings in relation to a single test in the same country about the relevance of personality dimensions to air traffic control have not always been replicated (Buckley *et al.*, 1969; Karson and O'Dell, 1974). Meanwhile the search for valid measures of personality in selecting controllers has not been abandoned. Della Rocco *et al.* (1991) have considered the appropriateness of current selection practices for future systems. Research relating personality to aptitude (Nye and Collins, 1991) and to training success (Schroeder *et al.*, 1993) continues to show relationships that are not large but not negligible either. Furnham (1994) has reviewed current knowledge on personality at work.

It is now abundantly clear that there is no ideal personality or personality profile for all controllers, as currently measurable by available personality tests. The efforts of the International Test Commission notwithstanding (Hambleton, 1991), the scoring norms for controllers cannot be extrapolated validly to other nations, and on this topic there are real limitations in what international collaboration can expect to achieve because findings and recommendations should never be transferred between nations without verification. Sooner or later the more contentious implications of employing personality tests to select controllers must be faced. The avoidance of personality clashes among controllers or between controllers and their supervisors is one practical objective. The full implications of such clashes for safety and efficiency are not known. One possible benefit from research on personality could be the allocation of individuals to teams with compatible personality characteristics. It is not even known how feasible this might be. Another factor is equal opportunity of employment. If applicants can be rejected because they lack aptitudes or intelligence or normal colour vision or are too old, could they be rejected because they have inappropriate personalities and seem incapable of learning to work productively and harmoniously as team members? What reasons could there be for not rejecting them on such grounds?

8.7 Air traffic control and aviation knowledge

All the selection measures considered so far, though of human attributes thought to be relevant to air traffic control, have been intended for other uses, and adapted if necessary for air traffic control needs. By contrast, measures of air traffic control or aviation knowledge are specific to air traffic control or aviation, and have been devised for air traffic control purposes and validated for the population of air traffic control applicants rather than for a different or more general population. Any scoring norms are not usually applicable to a wider population. There can be full control over the contents of tests of air traffic control knowledge, and every test item can be chosen to meet defined air traffic control objectives. Nevertheless, the outcome of employing tests of air traffic control knowledge or of air traffic control skills within a selection battery for controllers, or as predictors of the successful outcome of air traffic control training, has been similar to the outcome of employing other types of test, for tests of occupational knowledge of air traffic control do not usually account for large amounts of the variance but neither are they totally irrelevant. Typically the test scores have low but significant correlations with the criteria of success.

Such tests have the major advantage of high face validity. Candidates for jobs as controllers find it reasonable for the selection procedure to include tests that seem to measure their performance at rudimentary air traffic control tasks. This approach also seems reasonable psychologically, in terms of transfer of training and abilities in common, if tasks that actually need little prior air traffic control knowledge can actually tap some of the abilities required for it.

Dailey and Pickrel (1984) have described the development of the American Occupational Knowledge Test of air traffic control. One of the original purposes was to check the degree of agreement between actual air traffic control knowledge and experience and claimed knowledge and experience by using a simple test of knowledge. It was also believed that if potential air traffic control ability could be measured directly as part of the selection procedure this could reduce the attrition rates of those selected, which were causing concern. The inclusion of the Occupational Knowledge Test did apparently have this effect to a limited extent.

Of less historical interest but perhaps more significance is aviation or air traffic control knowledge as an indicant of interest or motivation. Some candidates may wish to become controllers because aviation or air traffic control fascinates them. Their air traffic control knowledge may have come from books, from talking to controllers or from paying for a coaching course on air traffic control. These forms of preparation may denote greater motivation and job interest which may encourage persistence throughout training and thereafter. What they cannot guarantee is sufficient ability to succeed as a controller.

A complication in ascertaining the validity of tests of occupational knowledge of air traffic control within a selection procedure is that foreknowledge of air traffic control may partly be confounded with age. If it is, this can reduce the predictive value of occupational knowledge in the selection procedure because age at the time of selection is a good predictor of the likelihood of becoming an air traffic controller but acts in the opposite direction to prior knowledge. The more the knowledge, the greater the probability of success, but the higher the age, the less the probability of success. Pitted against each other, higher age more than cancels any benefits of prior knowledge.

Occupational knowledge test scores might have value for allocating candidates to the air traffic control jobs for which they should be trained. This has been tried to a limited extent, using different cut-off scores for different air traffic control options (Mies *et al.*, 1977). Possible improvements in the selection process may be achieved by subtler weightings of different aspects of occupational knowledge tests or by administering different subdivisions of occupational knowledge tests for different air traffic control jobs.

8.8 Screening

The purpose of screening is to identify as early as possible individuals who seem unable or unfitted to become air traffic controllers or to remain in their present air traffic control jobs, for whatever reason. Screening is a continuing process throughout a controller's career but, as long as selection practices are validated by the successful completion of training, screening applies primarily to candidates who seem unlikely to complete their training. Several kinds of screening are associated with air traffic control selection. All are most concerned with the detection of potential failures. Screening as a process does not normally discriminate between degrees of success or failure unless job allocation is included in the screening programme. Pickrel (1984) has reviewed the historical development of screening.

Initial screening is based on the information supplied by candidates on application forms and may rely most on biographical data (VanDeventer *et al.*, 1983). There can be more than one sift, depending on the balance between supply and demand. All applicants must meet minimum requirements, such as age, education and medical history, and those who do not can be discarded at the first sift. Whether there is a second more stringent screening depends on practical administrative considerations, such as the desirable ratio between the numbers of applicants and vacancies according to historical precedents. The objectives are to recruit enough suitable applicants to fill all the available training places and all the envisaged vacant posts on completion of training. The disparity between the numbers who apply for selection as controllers and the numbers who actually begin training can be large but quite predictable. People drop out for many reasons. Some invited for selection never appear. Some fail during the selection procedures. Some conclude that air traffic control is not what they expected, and leave. Some abandon the idea of becoming a controller for personal, medical or financial reasons. Some have applied also for many other jobs, and opt for another job they are offered. The screening processes have to allow for all these factors and others.

If after the first sift the number of eligible candidates far exceeds the numbers sought, further successive sifts are implemented until the numbers become administratively manageable and of the right order of magnitude. They still have to rely on the completed application forms and must rely solely on objective criteria that are demonstrably fair, and it is difficult to devise a form that is adequate without being too long or intrusive. The application must contain sufficient information for screening purposes, given that neither the number of applicants nor the proportion who will pass any given screening criteria can be known precisely beforehand. Quantitative evidence receives most emphasis, and all criteria must have credence and human factors support and be legal. Factors such as gender, race, religious affiliation, postal address or family background must not be used. Complications can arise if national employment policies of positive discrimination have to be reconciled with the need to be fair.

The alternative to screening can become very expensive as it involves testing and interviewing very large numbers of candidates, an exercise which can become self-defeating if it merely yields more comprehensive evidence instead of more clear-cut evidence. The aim of screening is not to reject a predetermined proportion of candidates but the practical management of the subsequent stages of selection and training. For humanitarian reasons, it seems unfair to encourage unduly candidates with very slim chances of being accepted.

In some countries, specialist organizations charge a fee to complete application forms for jobs on behalf of candidates in ways which are claimed to present the candidates favourably. This apparently well-intentioned service can be counterproductive. The resulting immaculately presented application form may be so stereotyped as to be instantly recognizable for what it is and as not the unaided work of the applicant. This raises questions about how to treated applications with professional assistance fairly in relation to more amateurish unaided efforts. Perhaps resort to such professional help betokens a lack of ability or of confidence, or perhaps it is a sign of keenness and serious candidature. Evidence about this is urgently needed, in the interests of fairness.

Following this initial screening to reduce numbers, the main stages of selection also incorporate various screening processes. One of them is medical screening. This covers medical history, general health, eyesight, hearing, tolerance of shift work and of stress, any adverse conditions such as obesity, and any signs of substance abuse. Applicants with extreme body dimensions may also have to rejected if they are too tall to sit comfortably in the workspaces or too short to reach equipment items in frequent use. The interview also functions as a screening process, for candidates may be rejected because of behaviour during the interview or formal assessments based on it.

Psychological test scores and other quantitative data obtained during standard selection procedures are also employed to screen candidates. Opportunities during interviews and discussions are used to give candidates a better understanding of what is entailed by being a controller. There may be a tendency to be cautious towards applicants who view air traffic control as a second-best option. They may find air traffic control highly satisfying because it is a part of aviation, or never come to terms with it fully because they hanker after a flying career. A function of screening processes can be to attempt to make this distinction, and to select those who will become successful rather than disillusioned controllers.

Sometimes an inadvertent screening process is introduced by delays between phases of the selection process. Administratively the purpose of the delays may be to ensure a plentiful supply of applicants for any training places, but their consequences for screening deserve more study. The most dedicated candidates for training may indeed be the people who are most tolerant of the delays. Alternatively those who fail to obtain jobs elsewhere are most tolerant of delays, so that air traffic control does not get the best people. Such delays seem unlikely to improve the recruitment and selection processes but likely to reduce the validity of the selection and screening procedures.

Some countries have adopted a more intensive screening process which in effect has become an intermediate stage between selection and training. This additional screening process takes the form of teaching, familiarization and indoctrination in the principles and practices of air traffic control, to discover whether candidates have sufficient ability, enthusiasm, dedication and motivation to complete the full training procedures successfully. It has seemed prudent to begin with these screening procedures to save the expense of completing much of the training, only to lose students because they leave or fail. It seems preferable to establish their unsuitability by much shorter procedures which are not irrelevant but become the basis for subsequent training. Typically these additional formalized screening processes last several weeks. They teach rules and their application and assess the students' ability to learn, understand and apply what they have been taught. Performance measures of each candidate's progress are combined with other assessments into a composite score used as a pass/fail criterion for full training as a controller. This type of screening process generally succeeds in reducing attrition rates significantly during training. Screening has been described by Manning *et al.* (1989) and shows promise (Broach and Brecht-Clark, 1994).

8.9 The reliability and validity of selection procedures

The complex nature of air traffic control, the numerous factors which each make only a small but a significant predictive contribution to the selection procedures for controllers, the difficulties in specifying any small set of factors that are crucial for success as a controller, and the reliance on training for validation criteria, have combined to limit the reliability and validity of controller selection procedures. Although controllers and their supervisors usually agree about who the most able controllers are, the basis of these judgements has proved elusive despite their consensus. Factors mentioned include a tidy workspace, trustworthiness, steadiness, and a total lack of histrionics without actual placidity, but there are other factors that controllers cannot put into words. Attempts to quantify the basis of judgements to assist their validation have not made much progress. Yet a clear definition of the attributes of a good controller is necessary in order to tell if the selection procedures provide what is required.

Those who are selected form a more homogeneous group than the population from which they are drawn. Whether the rejected candidates could have become satisfactory controllers must remain unknown. The proportion of the population who could be trained to be satisfactory controllers is also unknown. If the selection procedures are elaborate this would seem to imply that this proportion must be quite small. The procedures seek individuals with many desirable but independent attributes, and must reject most applicants because there are so many ways in which they can be unsuitable. This philosophy presumes that the required attributes and potential abilities are essential prerequisites for successful training, and implies that selection is ultimately more influential than training because training can never compensate fully for inadequacies in the selection procedure.

The above represents one view but there can be another. It presumes that quite a large proportion of the population and of the applicants could be trained to be successful controllers because the number of attributes that are really essential for success in air traffic control, as distinct from being desirable, is quite small. Intelligence well above average, an age no more than about thirty, clear articulation, good health, emotional maturity, motivation, and the specified standards of eyesight and hearing may comprise the main requirements, and a lot of people can meet all of them. This argument is that the more elaborate selection measures do not contribute enough to justify their inclusion, and that success as a controller does not depend primarily on selection but on training. The issue of the respective contributions of selection and training has never been satisfactorily resolved, but it is necessary to ascertain their respective influences on the provision of able controllers in order to demonstrate the validity and reliability of both selection and training and in order to understand how to lower attrition rates. If selection has the greater influence, attempts to reduce attrition by improving training may achieve little, whereas if training has the greater influence more elaborate selection procedures may have little effect on the outcome. Although the existing evidence is not fully persuasive, it leans towards the view that quite a large proportion of the population could be trained to become adequate if not outstanding controllers.

Many criteria might contribute to the validation of an air traffic control selection procedure. One set of criteria uses the assessments, tests and judgements made during training and on its completion. Other criteria can relate to job allocation but these can become confounded to some extent with the criteria for validating the selection procedures if job allocation depends partly on measures during selection or training, or if the subsequent on-the-job training opportunities are not equivalent. Other criteria for validation depend on assessments of on-the-job performance by supervisors or peers, and can extend to career advancements, promotions and extra responsibilities, including any reduction in the quality of the workforce because the best controllers cease to practise on assuming managerial or other duties. More negative validation criteria concern the involvement of individual controllers in any incidents, safety infringements, investigations or official inquiries into their performance as controllers. Further validation criteria may be derived from performance in retraining or refresher courses and from the attainment of further professional qualifications. Attrition rates can provide further criteria if the reasons for leaving concern the job and not extraneous events, for a successful selection procedure chooses controllers who are not only competent but who stay. Unsatisfactory attrition rates during and at the end of training prompted a recent study that showed a need to reappraise the selection and screening procedures for controllers (Haglund, 1994). Another kind of validation criterion measures how rewarding and satisfying the jobs of controllers are, their contentment, their expectations, and their success in adapting to new equipment and facilities.

Although the above criteria may all contribute to validation procedures, there is no reason to believe that they are closely related. The only rationale by which they could be combined and optimally weighted is the empirical one of adjusting the factors and their weightings to give the best predictions. Many of these measures take years to derive, by which time system changes will undermine their validity. This problem cannot be circumvented by retrospective validation, whereby the original selection and training scores of controllers now in mid-career are examined to see if they could have predicted who would stay and who would leave, who would have an unblemished safety record, who would be most respected by peers, or who would earn career advancement. If the selection procedure is also expected to provide future management, planners, trainers and programmers, its validity becomes even more difficult to establish.

There are no adequate criteria to determine what the validity of a satisfactory selection procedure ought to be. The perfect procedure would select only controllers who remained safe and efficient throughout a lifetime career. Here, as elsewhere, perfection is unattainable. At present, a selection procedure has been improved if it reduces attrition rates and costs, but this is a real improvement only if those controllers who would formerly have been rejected but are now accepted are efficient and safe as controllers, and if those capable of becoming the best controllers are still selected. The quest to lower attrition rates can be furthered by better selection procedures or by keeping good controllers who contemplate leaving. Whenever a controller leaves, the reasons are often sought because the resources expended on training that controller have apparently been wasted. Perhaps better job allocation could place more controllers in jobs that suit them. Perhaps lower attrition rates can be attained without compromising efficiency and safety. The assumption that lower attrition rates must benefit air traffic control could be wrong. They should not be pursued so singlemindedly until there is better evidence that they are fully compatible with the highest safety standards.

If the present selection procedures for controllers are actually as successful as any selection procedures for controllers can ever be, this could not be proved, nor can it be predicted whether further work to improve the selection procedures will repay the effort involved. We could not recognize an optimum selection procedure for controllers even if we were using it. If the selection procedures in fact possess high validity and reliability, the criterion measures must have equivalent or greater reliability and validity in order to show this. Much statistical work has been done to optimize the predictive value of what we have, although the effects of coaching and practice on scores are becoming a widespread problem (Sackett *et al.*, 1989).

Critical re-examination of the common practice of using training scores and assessments to validate selection procedures is overdue. The purpose is not to select those who can complete training successfully but those who will give a lifetime of satisfactory service as efficient and safe controllers. The value of end-of-training assessments as predictors of subsequent progress as a controller is dubious at best. If they are poor predictors, they are unsuitable as validation criteria for selection and should be abandoned. In any event, selection procedures for air traffic controllers should be linked less closely to training and more closely to subsequent performance as controllers.

Some instructors have a propensity to judge quite early in a training course which students are going to succeed. These judgements need closer examination. If they are well founded, the instructors are tapping evidence which would allow those who will fail to be identified earlier during selection if the evidence could be converted into an objective form of measurement, but if these judgements are a self-fulfilling prophecy lacking more tangible support they should be firmly discouraged and replaced by more impartial assessments. In either case, practical action is needed, either to discourage such judgements because they are unwarranted or to incorporate more formally into the selection and early training procedures the evidence on which they are based if it is demonstrably valid.

8.10 Effects of automation on selection procedures

Automation affects the selection procedures for controllers in two totally different ways (Nyfield, 1991). The first effect originates in the progressive computer assistance of actual air traffic control tasks. Tests to select those most able to use such computer assistance effectively have to be devised and validated, using orthodox methods such as task and job analysis to identify the new required skills and abilities clearly enough for quantitative tests of them to be developed and added to any selection procedures that do not include them already (Della Rocco *et al.*, 1991; Manning and Broach, 1992). Appropriate tests would assess ability to use the proposed computer assistance, and to cope with such associated consequences as the reduced team roles, the increased self-sufficiency entailed in working mainly through a human–machine interface, and the need for greater self-generation of professional standards when others can no longer judge each controller's performance or read each controller's traffic so well. The full range of the effects of computer assistance on the controller should be considered for incorporation into the test batteries for selection wherever possible.

The second kind of effect of automation on the selection procedures for controllers is through the automation of the selection procedures themselves. This could mean diminished roles for the aspects of selection procedures that rely on human beings to administer them, such as interviews, and increased roles for those aspects which could be presented, scored and interpreted automatically. This applies particularly to tests which can be presented in paper-and-pencil or electronic forms and become self-administering. The tested candidate population can be expanded because automated testing is cheap to administer. The conduct of the selection procedure can be revised beneficially because the scores from automatically presented tests can be made available at once, for example to a selection board or interviewing panel. Test automation facilitates the administration of the same tests at many test centres, with standardized scoring that is demonstrably fair and impartial. Thus the automation of the selection procedures can represent a major impact of automation on selection.

A few potential disadvantages have to be allowed for. Those who have previous experience and ability at computer terminals may indeed be more suitable as future air traffic controllers and more likely to possess relevant skills, but it is not fair if the measures in a selection procedure depend less on innate ability than on whether an individual has access to a home computer or has been educated where computers are widely used. Candidates totally unfamiliar with computers, a diminishing group that is again confounded with age, may require an extra familiarization session just before the self-administration of an automated test battery. Such difficulties, once identified, can be overcome. Automation of at least some of the selection procedure will become the norm in future where it is not the norm already. The issue is not whether to apply it but how to apply it optimally (Bartram, 1994). There is current research on the predictive efficacy of automatically presented tests and procedures for air traffic control selection. The roles of rudimentary air traffic control tasks in selection are also being re-examined, since automation permits the presentation of these tasks in dynamic and interactive forms which have enhanced face validity and perhaps better true validity. This always has to be checked because face validity has never been a reliable indication of true validity in air traffic control selection.

8.11 The evolution of selection procedures to meet changing needs

Every selection procedure depends on accurate and comprehensive descriptions of the full range of controllers' tasks, and correct deductions of the knowledge, skills and abilities required. As air traffic control systems evolve to meet increased traffic demands, and as technological and navigational advances greatly add to the available information, there are consequent changes in the tasks and in the knowledge, skills and abilities needed for them. Logically, each significant change in tasks or in system functioning should be preceded by comparable changes in the selection procedures for controllers, and this need not be as impractical as it may seem. The timescale for planning and introducing major system changes is several years, and the main consequences of each change in terms of the required skills, knowledge and abilities of controllers become apparent quite early in this timescale. Tests and measures of these new requirements should be developed for selection purposes as they become known, because the processes of devising and validating them are quite lengthy. This linking of selection and system evolution processes, though feasible, seldom occurs. Instead, changes in the selection procedures often lag seriously behind changes in air traffic control systems, to the extent that the selection procedures for previous systems are sometimes still being followed when the revised system is controlling the air traffic.

The selection procedures must evolve in accordance with the full effects of changing needs, both the direct effects of computer assistance on tasks and its indirect effects on standards, team roles, supervision, levels of activity and responsibilities. Many of these implications for selection were identified more than ten years ago (Hopkin, 1982a), including the relevance of abilities to understand how software functions, to use computer assistance for problem solving, decision making and predictions, to relate effectively to adaptive machines, to work well within smaller and more fluid teams and to adapt to gross changes in demands and in levels of activity. Future selection procedures may also have to take account of some broader kinds of change. Demographic population changes may alter the size and nature of the population from which controllers are recruited. Social changes may alter expectations about work and work conditions. Attitudes towards jobs and acceptable forms of management may change. Expectations and attitudes about acceptable levels of activity and tolerance of boredom may alter. Changes seem to be occurring in the acceptable levels of stress induced by work itself or by working conditions. The extent to which a workforce is able or wishes to generate and conform with the ethos, norms and standards of a profession can change. An ability to share responsibility with other people or with machines in new ways may be required. A minimum level of computer literacy may become an essential attribute of a controller. The basic point is that the kinds of changing needs which may need to be identified and reflected by changes in the selection procedures can cover a very wide range and may include needs that differ from any which have had to be considered hitherto.

Changes in selection almost always involve a transition phase. Full validation of new selection procedures cannot be completed until data from on-the-job performance have been available for some time and have stabilized. Only then can the validity of the new selection procedure really be established.

Experience gained from the operational use of a revised system normally leads to some revision of the tests, of other measures or of their weightings, to improve the validity of the revised selection procedure. It is necessary to know the validity of existing selection procedures in order to judge when new selection procedures are warranted. While it is impractical to validate new procedures fully before they are introduced for it would take too long, it is prudent to be able to demonstrate that the revised procedures are an improvement over the former ones. The timing of the introduction of a change in selection procedures should not rely solely on empirical judgement and expediency but on the best available evidence about the optimum time to make the change. The adaptation of selection procedures to changing circumstances is a longstanding issue and remains a live one (Mies *et al.*, 1977; Sells *et al.*, 1984; Myers, 1992).

9

The training of controllers

9.1 *Training objectives*

The purpose of air traffic control training is to produce sufficient qualified controllers to meet staffing requirements, all of whom possess and can apply the requisite knowledge and skills to be safe and efficient and are motivated to remain as controllers. Air traffic control performance relies almost entirely on training because the typical air traffic control work environment contains very little direct information about the nature or conduct of air traffic control work. To an uninformed observer, most of air traffic control is not inherently meaningful, and it has only become meaningful to the controller because of training. An air traffic control human–machine interface is not at all like a computer terminal designed for public use, which presents its user with a series of simple alternatives until the desired objective has been reached.

Air traffic control is complex, more so than it seems at first. There is therefore much to learn in order to become a controller, and training takes quite a long time. How long varies according to the type and intensity of the instruction, the methods employed, the pace and continuity of the training programmes, and the sophistication, technical complexity and levels of traffic of the air traffic control system to which the training applies.

Although many of the standards of air traffic control are agreed internationally, the training requirements are not as uniform among nations as might be expected. Whereas some controllers in geographically large nations may handle internal air traffic almost exclusively, others in small nations closely surrounded by other nations may deal mostly with international overflights. National differences in training have to reflect to some extent geographical factors, the amount and distribution of air traffic, the types of equipment installed, and the feasibility of mutual support and supervision.

Because air traffic control is complex, training requires sophisticated support facilities such as real-time simulation. The total number of training establishments in the world is not large. It is impractical for most smaller nations to have their own. Some are within other organizations such as universities or training establishments that also cover other aspects of aviation, but many deal solely with air traffic control. They can have various names – school, college, institute or academy, for example. Most accept students from many nations, and are proud to publicize the high number of nations their students have come from. There is thus considerable international standardization of air traffic control training. English is the language used to control international air traffic. Colleges in non-English-speaking nations teach the aspects of the English language needed for controlling international air traffic, and colleges in English-speaking countries have appropriate supplementary English language courses for foreign students. The controller handling international air traffic must be able to speak air traffic control English clearly, unambiguously and reasonably fluently over radio and telephone channels so that all listeners can understand it. If training in this is needed, it must therefore precede most of the other air traffic control training. If the phonetic properties and structure of the student's native language are very different from those of English, some degradation in intelligibility when speaking in English may persist, but it must not be capable of inducing any misunderstandings or misinterpretations that could hazard safety. The language of air traffic control, such as the phonetic alphabet, attempts to minimize sources of confusion. Training in the content, phraseology, sequencing and nuances of meaning in spoken air traffic control English is essential for every prospective controller, who must not experience serious difficulty in speaking or understanding English spoken face-to-face or through communication channels.

In many nations, air traffic controllers are employees in the public sector and therefore policies on public sector employment apply to them. Ironically some of the most politically sensitive issues are the easiest to resolve in air traffic control where there is no significant evidence, for example, of gender differences in capability or trainability. However, equal opportunities begin with job applications, and any problem of too few applicants from a designated minority has to be corrected by recruitment policies for it cannot be corrected later during training, just as it cannot be corrected during selection.

A related issue, which could be equally contentious and more difficult to deal with, is beginning to emerge. It concerns cultural ergonomics (Kaplan, 1991). This refers to deeply imbued cultural differences between nations and groupings of nations, and affects attitudes and loyalties and responsibilities rather than abilities or performance. Some peoples may be able to learn all the knowledge and skills necessary for air traffic control but be unwilling or unable to subscribe wholeheartedly to the professional ethos, norms and standards of air traffic control, including the closing of ranks with controllers in other nations if the profession is challenged from outside. Many current controllers are loyal to air traffic control before everything else, but in future the culture of some controllers means that other loyalties may have their prime allegiance. At some point, the training implications of this issue will emerge, as they are already doing with regard to the inherently Westernized forms of command structure, team roles and shared responsibilities that are built into many air traffic control workspaces.

Two implicit training objectives can usefully be made explicit. One is to plan training to pre-empt gross swings in the demand for training. This is accomplished by combining long-term and short-term factors. Long-term planning of required future staffing levels, taking account of known attrition rates and retirement ages, can result in a near-constant requirement so that training facilities are fully and efficiently employed to produce a steady flow of newly qualified controllers. Shorter-term adjustments can avoid under-use of training facilities, sudden demands that exceed training capabilities, and undue delays for students between successive training courses or between completion of training and subsequent employment.

The other implicit training objective is that the air traffic service received by the user should not vary much according to which individual controllers happen to be providing that service, because all the air traffic control procedures, methods and practices are standardized and the competence of every controller can be guaranteed. Obviously the individual controller has some influence over the quality of the service that each pilot receives, but a function of training is to ensure that any differences among controllers remain safe and efficient and are never worrying. Safety cannot be compromised. It is a purpose of training to ensure that it never is.

During training, controllers encounter the range of equipment that they will use on the job, which should have been designed to meet training needs (Chobot and Chobot, 1991). Some training should therefore employ the latest equipment, not only refresher training on that equipment but initial training also, for otherwise some training will be inappropriate. Training establishments should therefore be among the first to receive new equipment about to go into service, but in fact they are often among the last. Training may even have to be done on old equipment that is obsolete elsewhere. This failure to appreciate the critical importance of proper training equipment and to fund it adequately is a better guide to the true priorities accorded to training than eloquent statements of its great importance accompanied by obsolete equipment for it.

Training relies on competent dedicated instructors, who have preferably asked to be instructors rather than been dragooned into it. For this to succeed, a period as a training instructor has to be perceived as a prized step towards career advancement and a token of acknowledged competence. If instructors are volunteers, success depends on satisfying two requirements. First, many controllers, including the most able ones, must want to be instructors, and secondly, there must be a validated selection procedure to choose the best instructors from the many applicants. All too often the opposite circumstances apply. Instructing, far from indicating career advancement, can be a move sideways that indicates the opposite. This is not the way to recruit good instructors.

Effective air traffic control instructors are not born but made; that is, they themselves are a product of training that includes acquiring teaching skills. If they are also required to make assessments as they are teaching, they need training in assessment skills also. The methods of assessment of students by instructors have to be agreed and be as impartial and objective as possible. In so far as assessments rely on subjective ratings they are open to charges of bias, and it is particularly important to be able to prove that the criteria for ratings are clear, well known, and highly reliable. Students should be confident that all ratings and assessments of them throughout training do not depend significantly on which instructors made them. This is not a matter of exhortation or reliance on each instructor's intentions of integrity and impartiality, but of firm evidence that the assessments are indeed reliable. Simple precautionary checks should be made frequently, such as independent assessments by different instructors of the same student doing the same task, to confirm the high level of agreement between them, and to trace the reasons for and resolve any discrepancies.

9.2 The content of training courses

For several reasons the content of many air traffic control training courses is currently under review. As the costs of training a student in all the main aspects of air traffic control rise, pressures increase to train each student to professional standards only for those air traffic control jobs which the student will actually do, and to supplement this with less detailed training in all other aspects of air traffic control that the student needs some knowledge of for such essential purposes as liaison, coordination, communication and the handover of responsibility, while stopping far short of full professional training in these ancillary aspects. Although the main driving force for this narrowing of training is usually to cut costs, the earlier completion of training may be a further reason if there is a shortage of controllers. Training content is also being reappraised in terms of the need for supplementary theoretical and academic knowledge, and of the respective roles of various forms of instruction, including on-the-job training. The need for new skills, knowledge and experience to complement new forms of computer assistance and new technologies can also induce reconsideration of training requirements because their correct use may depend on new kinds of controller knowledge, such as how to manage strategic resources (Herschler, 1991). An expansion of available training tools has prompted some revision of course contents and training methods (Smith, 1991; Baldwin, 1991). Stand-alone training packages offer options not formerly available and raise issues about how to use them and integrate them with more traditional methods. If their usage will cover assessment as well as training, their contents must be adaptable for this extended role. A further cause of reappraisal is if the objectives of training are extended: examples might be the instillation of professional norms and standards, and the periodic verification of professional competence through the solving of simulated problems.

Air traffic control training is expected to comply with broad international requirements that concentrate on the desirable outcome of training rather than on its detailed content and methods. A further concept that is highly relevant to training courses is teachability: it is not acceptable to devise, test and introduce a new form of computer assistance only to discover that controllers cannot learn to use it as it is intended to be used.

From the point of view of training, air traffic control is not a single entity. Some divisions within it, such as that between military and civil control, can be so fundamental that the training for them is totally separate. Other distinctions, such as those between traffic on and off airways, en route and terminal area traffic, and radar and procedural control, are different enough to require separate training courses and separate assessments. Air traffic control is not as homogeneous as superficial impressions of it suggest. Training courses commonly differ sufficiently for the correlations of the scores by the same students on different courses to be not very high. Successful completion of training on one course is not always a good predictor of the successful completion of subsequent courses, although it is considerably better than chance.

The policy on whether a licensed air traffic controller should be qualified on most control jobs or only on a few of them obviously has a major influence on training content. At one extreme the policy is to train individuals in most aspects of air traffic control in their initial training or as endorsements on their controller licences, so that they can be allocated to almost any air traffic control job, subject to any necessary refresher courses and updates. At the other extreme training is specific to a single type of air traffic control job at a single location, and a controller is qualified for a particular centre or tower but not for anywhere else. The latter training is much shorter and cheaper but far less flexible, and controllers cannot be reallocated at once if they are needed elsewhere. National policies and needs differ on the best balance between fully comprehensive and narrowly specialized air traffic control training, and no single balance would be ideal everywhere.

The taught content of most courses often starts with general principles of air traffic control, introduces procedural control without radar, and progresses to radar control, though this order is not sacrosanct. Training content is of several broad kinds. It includes fundamental knowledge about the principles of flight and navigation, the functioning of computers, radar and other equipment, and relevant topics such as meteorology, aviation law, and the international rules and regulations applicable to air traffic control. Basic principles of air traffic control are elucidated, including the rules of the air, vertical and geographical divisions of airspace, divisions of air traffic control responsibilities, relations between ground-based and airborne facilities, aircraft manoeuvrability and equipment, constraints such as separation standards, and objectives such as safety, expedition, the regulation of traffic flows, and acceding to users' requests whenever possible.

Training content refers to the practices as well as the principles of air traffic control. This encompasses procedures followed and instructions given and received, in terms of their substance, their sequences and their formats. The language of air traffic control and its correct usage are taught. The air traffic control workspace and its functionality have to be learned and understood. This includes the meaning and usage of every possible item of displayed information, of every means of communication, of every data input device, and of every task that employs them. The training content must include some understanding of the effects on system functioning of every controller action. Knowledge has to be gained about task sequencing, task sharing, mutual support, the detection and prevention of errors, dealing with emergencies, and the criteria for adopting normal or exceptional procedures. Controllers need to know about the sources of information in the system, how it is gathered, collated and presented, how reliable it is, how it is intended to be used, and how deficiencies or failures in the sources or presentation of information can be recognized. All instructional materials should comply with the human factors guidelines for their good design (Hartley, 1985).

All the above is essential knowledge. The content of training must also cover how to deploy and apply that knowledge. This includes the allocation of resources between tasks, the optimum timing and sequencing of tasks, the timescales within which tasks must be accomplished, the scheduling of work, the integration of strategic and tactical air traffic control, the optimum timescales for advance planning, the criteria for choosing among alternative procedures, and the respective roles of individual controllers, teams, assistants and supervisors. Training must provide practical experience of many specific tasks, such as arranging and maintaining the flow of en route traffic, the detection and resolution of potential conflicts, the sequencing of aircraft in and out of stacks and holding patterns if necessary, integration of converging flows of air traffic into a single uniformly separated sequence for final approach to a runway, and many other comparable tasks.

Having learned what to do and how to do it, the student must learn how well it has to be done to attain professional standards of proficiency. This is normally achieved through practical work using simulations or demonstrations rather than by classroom methods. The student learns by doing, often with some form of air traffic control simulation. There is emphasis on the integration of tasks, the correct usage and interpretation of the information presented, the requirements for and access to additional information, the acquisition of skill and dexterity in using all the input devices, the continuous planning of future work while doing current tasks, the acquisition of cognitive skills and fluency in speech, and the building of appropriate levels of trust and confidence.

In on-the-job training, all that has been taught is put into practice, initially under close supervision. Controllers then learn that real problems do not always fall into the neat categories under which they were taught, and that real pilots do not always act like simulated ones. They also gain a realistic impression of the stresses and responsibilities of the controller, from encountering real and not simulated incidents and realizing the emotional difference between them. The content of on-the-job training extends far beyond what is formally taught. Much that is learned is never formally taught. This includes the expected attitudes of controllers, to their profession, to management, to pilots, and to others they come into contact with. It includes the professional standards expected of controllers by their peers. It includes loyalty to the profession and willingness to defend it if it is challenged. It includes the establishment of harmonious and productive professional relationships with colleagues and supervisors so that the controller is an effective member of an integrated team. It includes learning how to gain and keep the trust and respect of colleagues. Many of these aspects of on-the-job training are vital although they may not appear in the training contents. If their significance is unrecognized, the curtailment of on-the-job training in order to cut costs can have some unexpected and unwelcome consequences.

All air traffic control training content should take full advantage of known principles of learning. It should be established beforehand that everything to be taught is in fact teachable. There should be clear guidelines on the best balance between the elucidation of principles and their practical demonstration and application. There should be opportunities for some overlearning and entrenchment of learning, particularly of the most important or difficult material. Immediate feedback and knowledge of results should be provided at every opportunity, for effective learning cannot occur without them. There should be sufficient flexibility in training content to accommodate the particular needs of individual students by building on strengths and compensating for weaknesses. There must be opportunities to develop, exercise and maintain skills because unused skills atrophy, and the value of skills should be demonstrated. Frequent active recall of knowledge should be encouraged in order to sustain it. There should be opportunities to minimize human error, and to learn from any errors made how to prevent their recurrence.

Training content should not only instil desirable habits but break undesirable habits when necessary. It is never advisable in air traffic control to attend exclusively to each problem until it has been resolved, but always necessary to continue to scan for other developments while doing the task in hand. Some training and knowledge should transfer between different aspects of air traffic control, and transfer of training principles should be utilized as much as possible. Training should aim to motivate and inspire realistic levels of confidence. Learning covers procedures and actions but also their timing, the reasons for them and their hierarchy of importance. The direction and switching of attention appropriately is essential in a controller, as is the ability to share tasks and collaborate with others without interfering unduly with their work or presuming that one's own tasks must be more important than theirs. All this has to be learned and so must form part of the training content.

9.3 *Methods of instruction*

Most training instruction in air traffic control uses several instructional techniques and tools. Baldwin (1991) distinguishes three broad training techniques:
institutional courses, on-the-job training and continuation training methods.
The relative emphasis on various methods has gradually changed, in response
to new technological options and cost pressures. For any new instructional
methods, the issues that arise are their applicability to air traffic control, the
functions for which they are fitted, and their integration with the training
methods in use. They also prompt a re-examination of opportunities for coup-
ling assessments with instruction and training.

The teaching of the fundamental principles of air traffic control still relies
heavily on classroom methods and on private study and committal to
memory. Modern visual teaching aids are coming into use but much teaching
employs exposition, explanation and demonstrated examples. A certain
amount of rote learning is unavoidable. The rudiments of air traffic control
have to be learned and understood before they can be consolidated and
applied. In some nations where the initial period of training also serves as the
final stage in the selection process, it is still necessary to teach within that
period some air traffic control fundamentals including some mention of its
context, rules, regulations, divisions of airspace, language, objectives, legal
framework, responsibilities, sources of information, communications, and
tasks.

Instruction by demonstration is sometimes used and can be very efficacious
although it can be a demanding method for instructors. Worked exercises that
illustrate principles in terms of procedures can identify and explain the most
crucial variables and trace their consequences. Demonstrations by instructors
and demonstrations by students may be related in two ways. In one way, the
instructor initially demonstrates that a solution is feasible and invites the
student to repeat the demonstration. With carefully chosen examples, the
process can have the salutary effects of revealing that what looks easy is in fact
difficult, that what is known to be possible can seem impossible, and that
current learning is still insufficient, particularly in identifying what is crucial in
terms of evidence and timing. In the other way, the student tries to solve the
problem first, and then the instructor demonstrates how it should be solved,
again revealing which actions and timings are crucial. Either kind of demonstration can bridge the gulf between theory and practice, reveal the important
interactions, show that scheduling and timing can be as vital as the correct
choice of actions, and help to build a mental checklist of most relevant factors.
Students can not only learn from their mistakes but learn to recognize circumstances when they are most likely to make them.

Another method of instruction is to perform functions and tasks in a simulated air traffic control environment. Training simulations vary greatly in complexity. The simplest may be little more than a communications channel or a
board of flight progress strips. The most complex simulations attempt to

include all the main features of real air traffic control workspaces. The appropriate level of sophistication of the simulation normally increases progressively as the training itself progresses. Some basic instruction on the ordering, sequencing, handling and interpretation of flight progress strips and on the language, concepts and message formats of air traffic control has to precede training in more sophisticated simulated environments. Often training employs a hypothetical region of airspace, designed to allow many air traffic control problems and knowledge requirements to be introduced in a controlled manner in order to aid and consolidate teaching and assessment.

Some aspects of air traffic control training must employ quite complex simulation because they are interactive and include team training. Previously, considerable training has been in a team setting, with each controller performing his or her tasks alongside other controllers performing theirs, but it is now more widely acknowledged that in the future teams must be trained as teams rather than as groups of individuals working together. Occasionally team training has been examined (Denson, 1981; Hopkin, 1994c).

Hitherto most training using simple simulations and all training using demonstrations has been by human instructors. Sometimes each instructor has taught many students in a classroom; sometimes each instructor has taught a single student or a few students in a simulated or real-life environment. A more recent method of instruction consists of training packages which the student loads into a computer terminal and works through independently. No instructor need be present. This form of instruction can be flexible, and can adapt to the individual student's needs by continuing to present examples of a particular type of problem until the student has demonstrated sufficient competence to progress to the next type of problem. Students can practise skills and acquire experience at their own pace in accordance with their own needs. Issues that arise are whether these packages should be a routine feature of training courses, whether they should merely supplement other forms of teaching, whether their main application is to support weaker students, whether their main role should be to teach or to assess, and whether such packages, if optional, could be construed as a form of favouritism (Hopkin, 1991a). Such training packages seem certain to be more widely used in future. The main question is not whether to use them but how to use them to best advantage.

As teaching packages for training flourish, the use of real aircraft for air traffic control training, controlled as traffic by students under close supervision of their instructors, is waning, mainly for reasons of cost. It is not the same as on-the-job training as the traffic flies exclusively for air traffic control training purposes, but its expense leads to perennial arguments over how its costs can be justified. Student controllers experience the responsibility for real aircraft in circumstances where every pilot can allow for their inexperience. A few students, often among the most able academically, resign because they find that they are unable to face having the permanent responsibility for the safety of aircraft and passengers. Attempts to detect these students during selection have failed, but it is both cost-effective and humanitarian to detect them as early as possible. Controlling real aircraft as distinct from simulated ones may be the first opportunity to do so.

On-the-job training is a further form of training that is comparatively expensive, and therefore its financial justification is queried regularly. The student is introduced gradually to the control of real air traffic under supervision, by performing ancillary tasks, by observing and learning from others, by exercising control initially when traffic is light, and by handling progressively higher traffic loads under close supervision. From on-the-job training the student learns how to practise and apply what has been learned, how real air traffic is prioritized, how workload is controlled and planned, how tasks are actually done, and how controllers function as a team. The student controller also acquires information on the many aspects of air traffic control that must be learned but are not formally taught (Hopkin, 1988), which can be learned only by controlling real traffic.

On-the-job training has some serious limitations: if a student makes a mistake, the supervisor or instructor may have to intervene and correct it by taking control from the student without warning. The interests of air safety must prevail over the student's ability to learn from mistakes. On-the-job training teaches the student when a situation generated by the student requires outside intervention because it is inherently unsafe, but does not allow the student to learn all the consequences of mistakes, as simulation can do. An effective combination is to replicate subsequently in simulation the on-the-job problem, explain why the instructor intervened in the particular way that was chosen, but allow the original situation to develop in simulation without the instructor's intervention and leave the student to try to recover from it and to appreciate its consequences. Recent developments in the reconstitution of air traffic control scenarios render this training method more feasible (Rodgers and Duke, 1994).

The choice of the methods that are actually most effective for different kinds of learning needs further research, plus the informed application of what is already known. Efficacy of learning cannot be the only criterion. All teaching and instruction has to be accommodated within a course of fixed duration, and the pace of learning can never become wholly flexible. There are constant pressures to cut training costs and total training time. In some countries these have been curtailed severely, partly by modernized training methods, partly by reducing the range of jobs for which the controller is trained, and partly to relieve a shortage of controllers because traffic demands have exceeded predictions. Future air traffic control training seems likely to retain multiple methods of instruction but to readjust their relative importance, the interactions between them, and their scheduling and sequencing.

9.4 *The role of simulation*

Real-time simulation in air traffic control training links theory with practice, knowledge with its application, and experience with skills. The student brings to simulation some knowledge of the rules, regulations, principles, procedures, instructions and objectives of air traffic control, and learns from simulation how this knowledge is applied through the workspaces, equipment, tasks, functions and responsibilities of air traffic control. To some extent, real-time simulation for air traffic control training is always dynamic in that it responds to the student's actions and presents situations to which the student must respond. From simulation, the student can gain experience of the following: what is easily accomplished and what is difficult; what is successful and what leads to further problems for which solutions have to be found; what has to be planned ahead; what the optimum timing, sequencing and scheduling of control actions is; what kinds of error can occur and how they can be detected, resolved or prevented; how, when and where information appears; how information is communicated; how information is updated, recorded and discarded; how all functions are exercised through input devices; what actions are initiated by controllers, including their timing and circumstances; and how much work there is in controlling air traffic.

Simulation can be selective. It can be used to teach and train students in separate aspects of the control of traffic, such as the interpretation of radar information, the use of paper flight progress strips, or the learning of communication, coordination and liaison procedures. Its use can then be extended to integrate these and other learned aspects into the total air traffic control job. Complex simulations seek to present the total job to the extent that the workspaces and human–machine interfaces, and the displays, controls and communications facilities, can look and function similarly to their real-life counterparts. However, the emotional climate of having the control responsibility for real aircraft cannot be simulated. Simulation can show the immediate consequences of actions in terms of their effects on traffic configurations and subsequent traffic handling, but does not cover their wider ramifications such as a controller error leading to the filing of an official incident report and some form of official inquiry. In principle there seems to be a strong case for including such ramifications in training, though this is neither done nor usually suggested.

The tools of simulation can promote training continuity through commonality of traffic samples and by building new learning on previous learning. Simulation enables more thorough practice of the more difficult aspects of air traffic control although, according to theories of learning, the training on each particular type of problem often ends prematurely when there has still been less experience of complex problems than of simple ones. In the later stages of simulation training, completeness of simulation can be more important than extreme realism or fidelity. Full fidelity and realism are unattainable in air traffic control simulation, and the utmost efforts to achieve them can lead to over-optimistic expectations about the range of training objectives that the simulation can achieve, but if a key aspect of air traffic control is not simulated at all this can limit severely the roles of simulation in training. A basis for simulation is transfer of training. Classroom training is transferred to the simulation, and simulation training to real life. The efficacy of simulation depends more on the conditions required for the transfer of training than on the technical or cosmetic properties of the simulation devices.

The technology of simulation has been in some respects revolutionized in recent years (Life *et al.*, 1990). Recent and pending technological advances applicable to air traffic control training have been extensively discussed (FAA, 1988; Wise *et al.*, 1991, 1993). Software packages can present simplified air traffic control problems on personal computers. More elaborate workstations, including suites for more than one person, can be employed to develop air traffic control skills. Networking encourages simulations for the concurrent training of several groups or several individuals on different aspects of air traffic control, with flexible integration of their activities. Displays such as videos, aids based on artificial intelligence or expert systems, and speech processing can be incorporated into simulated systems. Many training programmes using simulation are evolving from the standardized instruction of students in groups to the matching of training with individual needs. Simulation is a vital tool in the introduction of more flexible learning techniques. Another vital role of simulation is to provide immediate knowledge of results, not merely by indicating when the student has made an error, but by revealing all its actual and potential consequences. Such feedback is essential for training, and one of the strengths of simulation as a training tool.

9.5 *The assessment of progress*

Progress during training is assessed for several reasons, the most obvious being to measure the standards of skill, knowledge and proficiency of each student. This may occur at the end of a particular part of a course, if progress to the next phase of training depends on its successful completion. Another reason for assessment is to identify those who need extra training. Assessments of progress or the lack of it may confirm the termination of training of unsatisfactory students. Other reasons to assess training progress are to confirm the efficacy of any changes in training and to confirm the uniformity of the standards of success and proficiency where equivalent training is provided at several facilities and locations. Training may also be assessed to confirm its continued relevance to air traffic control. Sometimes the training progress of individual students is assessed to guide their allocation to suitable jobs at the end of training.

Assessments may be objective or subjective. Normally a combination of both is employed. Some aspects of air traffic control are well suited to standardized objective measures of progress and knowledge attained, especially with standardized air traffic control scenarios and problems. Scoring may simply record whether the student's solutions were correct or may evaluate them in more complex ways in terms of the choice, timing, sequence and appropriateness of actions, and the detection and resolution of implanted problems. Most simulation facilities can be used to assess as well as to train, by assessing performance automatically on a series of scenarios and problems devised specifically for assessment purposes. Each student's performance using any stand-alone training packages may also be assessed by such automated measures as the number of examples of each task that were attempted before a pre-set criterion was attained.

Objective assessments using standard material trace each student's progress and relate it to cumulative norms of performance, and therefore can measure either absolute or relative levels of achievement or performance. In the interests of actual and perceived fairness, much of the assessment of progress during training should rely on objective measures which can be scored automatically or at least impartially so that the outcome is unaffected by who does the scoring.

Subjective assessments inevitably may be criticized as unfair or biased even if they are not. Subjective ratings of performance are notoriously unreliable, and have been known to be so for some time (Henry *et al.*, 1975). Thorough training of all those who make subjective assessments can help to ensure that they all use the same information, criteria and standards, but even such training may not suffice to free subjective judgements entirely from suspicions of bias, which may or may not have some substance. The difficulty in defining the crucial attributes of a good controller is inevitably reflected in subjective judgements of progress and in initial disparities between assessors in what they consider most important. Subjective judgements can be valuable and some kinds of evidence can be obtained in no other way, but it is unrealistic to expect their levels of validity or reliability to equate with those of the best objective measures. Subjective judgements can also be influenced by self-fulfilling prophecies originating from instructors' beliefs about individual students' abilities, in ways in which objective measures cannot.

Independent assessments by several instructors in order to identify and resolve sources of disagreement between them, or assessments by others who make no direct contribution to the teaching, may help to increase and stabilize reliability. Continuous or frequent assessments are likely to be fairer than infrequent ones, which may be biased by a few trivial but memorable events. One kind of potentially useful assessment has not really been tried, namely, the ability of the student to perform unaided a difficult task that the instructor has just demonstrated. Such an assessment might test the student's understanding of the solutions and of the reasons for them.

In future, assessments are expected to become more standardized and more independent of the teaching. The combination of a range of simulations and of stand-alone training packages can make it more practical to centralize assessments by objective tests administered to students at different training facilities and scored automatically. Some careful interpretation of scores might still be needed: if performance at some facilities was consistently poorer than at others, this might be the fault of the students but could also be ascribed to poorer teaching, to fewer opportunities for extra practice, or to less encouragement. A possible trend in subjective assessments is for controllers to assess students in terms of their similarity to themselves. The instructors have beliefs about the most desirable attributes and look for them in their students. This source of bias does not seem as rife in air traffic control as it is in some other contexts, but it is still necessary to be on the alert for it. If instructing becomes a high-status job and instructors perceive themselves as an elite, there may be some validity in preferring students like themselves, but if instructing is a low-status job fulfilled by some less able rather than more able controllers judgements based in part on perceived similarity between instructors and students could be both invalid and undesirable.

9.6 On-the-job training

On-the-job training is frequently scrutinized on cost grounds as it seems expensive compared with most other kinds of training. There is widespread agreement among trainers that on-the-job training is beneficial in air traffic control, but less agreement on the aspects of air traffic control training for which on-the-job training is indispensable, or on the aspects for which on-the-job training is most effective and is therefore recommended even though learning could occur in other ways. The objectives of on-the-job training may extend beyond training to conclude with validation procedures that certify an individual's professional competence as a controller.

It is in on-the-job training that the role and influence of the instructor can be most critical. The instructing controller needs to be thoroughly competent, professional and knowledgeable in air traffic control, capable of building comparable skill, knowledge and experience in others through instruction, and able to apply that skill, knowledge and experience to assess the progress of each student. The controller who is not a very good controller can scarcely be expected to teach a student to be a good controller, for that would entail the handing on of skills that the instructor does not have. The highest standards of on-the-job training require the best controllers as instructors, which in turn requires instructing to be a high status job.

A crucial aspect of on-the-job training is judgement. For how long can students be allowed to discover and resolve the consequences of their own mistakes before potential hazards to air traffic control efficiency and safety compel the instructor to intervene? If the student generates a problem that forces the instructor to intervene, the student has to understand the reasons for the sudden intervention and how the problem originated so that similar problems do not recur, but the student's self-confidence should not be undermined. On-the-job instructors must combine the functions of instructing, assessing, monitoring progress, and assuming control if necessary. The instructor should combine the skills of a good controller, a good teacher and a good assessor, and this combination of skills should be suitably rewarded. In order to teach self-confidence and assurance by example, the instructor must possess these qualities. On-the-job training usually presents the first opportunity to imbue professional standards, norms and attitudes. Instructors may themselves be learning new skills, particularly familiarity with new forms of computer assistance for instruction which must be integrated with traditional teaching methods.

Instructors need training so that their assessments are consistent and concordant. Their training has to be quite formalized so that their assessments rely on criteria that are uniform in their content, stringency, and the weighting of factors. During on-the-job training, the instructing controller usually retains the legal responsibility for the safety of the air traffic. It is important that instructors do not mete out to their students the treatment that they may have received as students. An instructor who was unfairly treated as a student must not treat his or her students unfairly in turn, for this would perpetuate a potential shortcoming of on-the-job training and weaken the case for its retention.

For the following kinds of learning, some on-the-job training seems mandatory or highly desirable, since simulation, classroom instruction, demonstration, or other kinds of tuition seem insufficient. From on-the-job training the student begins to learn professional norms and standards, the expectations of other controllers, and the criteria for judging a good controller. The instructor assesses how effective each student's learning has been, how well the student can choose and implement procedures, rules, instructions and standards, and in what respects the student's ability to apply knowledge is approaching professional standards. The stress and responsibilities of real air traffic control may be glimpsed by students sufficiently for others to judge whether they will be able to cope with the responsibilities. On-the-job training reveals that real problems do not always conform with the neat categories in which they are taught, and their solutions can make some of the initial training seem simplistic.

It is not usually possible to spend much time during on-the-job training in discussing how an actual incident arose, but it may be essential to consider it afterwards in some depth through debriefings, recordings or replays. A vital lesson is to identify what all the options were, whether they were recognized, whether the reasons for rejecting those not adopted were correct, and whether in retrospect another solution should have been preferred. In simulation, teachable examples are chosen but in real air traffic control some problems and their solutions may be difficult to teach. This aspect of on-the-job training can be difficult to integrate with the simulation training that has preceded it.

Another difficult decision in on-the-job training can be when to stop. Normally simulations cover a fixed predetermined programme of work. On-the-job training has to tackle problems as they arise. The problems that actually arise at any particular job will not be fully representative of all the problems that could arise elsewhere. A decision to end on-the-job training may be taken when the students have gained most of the experience that the particular job can offer. If the training objectives are vague, neither the instructor nor the student may know when progress warrants discontinuance of training.

During on-the-job training, the distinction between what is learned and what is taught becomes most acute. What is taught is professional knowledge, experiences and practice in dealing with real aircraft. What is learned includes the attitudes of controllers to their management, to their equipment, to their condition of employment, to the system, and to their profession; notions of professional ethos and loyalties; the expectations of professional colleagues; the pride and satisfaction in being a controller; and the extent to which controllers support each other as team members. Most of this kind of learning is never taught, and cannot be learned from simulation. Team training is an important feature of on-the-job training. Even where there is one instructor per student, training must include working harmoniously and effectively as a team member and fulfilling all the expected air traffic control roles as a team member. Some on-the-job training must teach groups of controllers to work as teams.

It is easy to treat on-the-job training exclusively as learning, and to forget that some of it may be unlearning. Some practices taught in simulation may not be adopted at the facility where the student's on-the-job training is done. Without constant efforts to relate the training to the work, disparities between training and real life practices continue to sow confusion in students, which has to be resolved during on-the-job training.

9.7 Training for new tasks

For the foreseeable future, air traffic controllers will be faced with a succession of new or greatly modified tasks, as an inevitable consequence of combined requirements to control more traffic, using new forms of sensed data and of data processing, and using computer assistance. Training for these new tasks must be devised, validated and implemented.

New tasks, equipment or computer assistance are often tried out and evaluated first in a real-time simulation, for which procedures and instructions have to be devised and participating controllers trained. These evaluations present the first opportunity to evaluate the associated training, and there is much to learn about training from them. The participants in the first evaluations may misunderstand what they are supposed to do, or fail to use the equipment as it was intended to be used. It may prove difficult to teach them what is required of them. If professional instructors took part in the preparation of these evaluations they could assess whether the envisaged tasks would be teachable, recommend task changes to improve teachability, advise on how training for the evaluation should be given, compile the instructions for the participants in collaboration with those in charge of the evaluation, and note aspects of tasks that caused learning difficulties or misunderstandings, with a view to changing them before training for the real operational tasks began. The extent to which the findings of an evaluation can be an artefact of the adequacy of the training is often underestimated.

However, instructors seldom have the opportunity to fulfil this useful role. Instead, much of the training for evaluations is based on expediency: the participants learn as quickly as possible what is deemed strictly necessary for the evaluation and no more, since the expense of having a prototype system on-line for training purposes rather than for the evaluation itself always seems to be resented by those who do not really understand the main determinants of the cost-effectiveness of an evaluation. Normally alternative training techniques are not examined as a controlled variable in evaluations, and the evaluation cannot therefore yield direct evidence on whether the training for it could have been improved. It tends to be presumed that new procedures will be teachable until there is evidence that they are not. New tasks are not planned from the outset to be teachable, though that would be a more cost-effective approach.

Human factors contributions towards training for new tasks include the specification and validation of appropriate training techniques that build on what the controllers already know and on transfer of training principles. They cover advice on how to derive guidelines for training from the findings of evaluations, and on how to make the training contributions and initial evaluations more mutually supportive. An objective is to verify the teachability of new tasks beforehand and to improve teachability through adjustments to tasks or teaching methods. The human factors specialist can deduce the attainable level of proficiency, specify the best measures of progress towards proficiency, and suggest criteria to gauge when further training is unlikely to yield further significant improvements in performance. Human factors guidelines on the preferred order and sequence of learning usually require the basic principles to be taught first, and then recommend the preferred duration, scheduling and pace of further training. Human factors advice also covers appropriate training materials and examples to elucidate principles and general procedures beyond the specific examples themselves, and the choice of examples to increase task difficulty in appropriate incremental steps for the progress of learning.

Many of the possible errors, delays, confusions and misunderstandings which can arise during training may be predictable in advance. They can thus be avoided during the first stages of learning, and demonstrated by suitable examples in the later stages. Further examples explore the consequences of any remaining errors, delays, confusions and misunderstandings so that students can learn to recognize the initial signs of them and hence to prevent any serious consequences that are preventable by training. Efficacious extra training can be suggested for individual students encountering specific difficulties that can be resolved by further intensive practice. Stand-alone training devices may be particularly suitable for this application.

Two aspects of air traffic control training that always have to be addressed are the extent to which it can be an isolated activity or must take place within a team setting, and the extent to which a realistic air traffic control environment is essential for successful training and appropriate feedback. Training for new tasks cannot be divorced from previous familiar tasks but the training problems differ depending on whether existing knowledge can be transferred or the familiar tasks must be forgotten because they are no longer appropriate. Better guidance than is currently available is needed on the problem of training people to forget what was once familiar (Hopkin, 1988). An integral aspect of training for new tasks is the derivation of procedures to assess the success or failure of training objectively and to provide clear guidance on when training has achieved its purposes to the extent possible and should be ended.

9.8 Retraining

Retraining usually becomes necessary for one of three reasons. Firstly, if a controller moves to another air traffic control job, much knowledge that is specific to the new job must be acquired through formal or informal retraining. Secondly, if the performance of an individual controller has for any reason become unacceptable, retraining may restore performance to an acceptable level in some circumstances. Thirdly, any major changes to an air traffic control system which entail significant changes in the controller's knowledge, skills, tasks, procedures or instructions will also entail appropriate retraining in those changes.

The extent and nature of retraining depend on its objectives and on several other factors. For the individual controller, these include the time that has elapsed since the controller's previous training, the extent to which tasks have changed in the interim, and the extent to which the controller's abilities, experience and knowledge have changed. Sometimes older controllers appear to have more difficulty than their younger colleagues in learning new tasks, but the more difficult practical problem can be that familiar actions, procedures and skills that are no longer appropriate have become more overlearned by older controllers and correspondingly more difficult to unlearn, discard and forget.

Some particular retraining problems arise when revised tasks are an uneasy amalgam of familiar actions and new ones, because some existing associations are then retained but not others. If tasks in their new form remain quite similar to the original tasks, this may encourage the extension of existing habits into inappropriate conditions if it remains possible to perform the new tasks quite well in nearly the same way as the original tasks were performed. If tasks are totally new, much less interference from the previous tasks and much less unwanted transfer of learning from the old to the new can be expected, but if the revised tasks are a mixture of the old and the new it may be the inappropriate rather than the appropriate aspects of the original tasks that persist. This suggests that the preferred aims should be either to transfer existing skills almost in their entirety or devise tasks that require new skills to be taught and learned. If new tasks permit gradual reversion to old inapplicable habits, the task designs are flawed. No feasible actions in the new tasks which were present in the old ones must ever become hazardous in their new setting, and a function of retraining is to ensure that such circumstances can never arise. Tools such as task analysis can help to illustrate how similar the new and the old tasks are and where they differ, and hence can provide guidelines on the most appropriate forms of retraining that retain the previous skills that still apply, teach the new skills that are needed, and discourage the persistence of old skills that are now irrelevant.

A different aspect of retraining applies when a new job is grossly different from the previous one, as when an active air traffic controller occupies a full-time managerial position. It is partly on the basis of former skills that this new appointment has been made. Erstwhile colleagues of the new manager expect their former colleague to understand and sympathize fully with their problems, and can become embittered when the new manager apparently forgets his or her roots so quickly and seems to turn traitor. Retraining new managers to learn the managerial skills that they will need is clearly important, but if former controllers who become managers can also keep their skills as controllers alive they might remain more conscious of controllers' needs and aspirations and combine more effectively their knowledge of users' requirements with their management responsibilities.

9.9 *Criteria of training success*

Administratively, a fixed training schedule and the completion of each stage of training before embarking on the next are normally essential. Subsequent stages of training must build on the presumed success of completed previous stages. Two features of training have to be agreed before it begins. The first concerns the training content, and its validation in relation to the training objectives. The second concerns the forms of assessment that establish if each student has reached an acceptable standard.

With regard to the first, the objectives of each training course must be stated clearly and unambiguously beforehand. Only then can the actual training be validated, but validation is an essential preliminary to training. The validation of training establishes that it is apposite and relevant, and that the training content can satisfy the training objectives. Numerous aspects of training are covered. They include the training content and methods, the correctness of the assumptions about the knowledge, experience and skills already possessed, the timing and pertinence of new training material, the appropriateness of the progressive increments in the difficulty of the training material, the adequate coverage of the methods of assessment during and on completion of the course, the specification of milestones or subgoals to be achieved during each course as a condition of further progress, the level of difficulty of the course material in relation to the course objectives, and the teachability of the material. The validation of training must also take account of such practical matters as the teaching and assessment burden on instructors, the training costs entailed, the effective deployment of training resources, the scheduling and usage of various equipment items and facilities, and the adequacy of the administrative support.

The second aspect of training to be settled in advance refers to the available means to confirm that the student has met all the training objectives to an acceptable standard. The relevance of the material to the attainment of the training objectives should therefore be established first, because the extent to which the student can meet the objectives must depend in part on the quality of the training materials. Whether the material should be validated and the students' mastery of it assessed are not the main issues, for both must be accomplished. The main issue is how to accomplish them in advance. A practical balance must be struck; if the standards are too high or the material is not teachable, few will complete the course. The attainable standard must be realistic, in terms of the objectives, timescale, facilities, and teaching materials. Instructors should know how to use the training material optimally, and it should be suitable for demonstration purposes. The training material must have been checked to ensure that it contains no unknown errors, and be known to be satisfactory because it has been tried out in advance. If it is necessary to devise parallel forms of material of equivalent difficulty, it is also necessary to prove that they are of equivalent difficulty.

Assessments of the success of training must rely on taught material, plus any ancillary material such as manuals and self-teaching aids which are mandatory parts of the course. It may be helpful to use a training management model to describe the operation of the training system (Gibson, 1993). Students should have had equal opportunities to acquire experience of all the tasks and facilities that can affect the outcome of the assessments that determine training success. The assessment criteria must be as objective as possible. If the scoring can be automated, so much the better, since everyone is then seen to receive equal treatment. If the training includes functioning as a member of a team, then the assessments of individual task performance within such teams must reflect fairly the respective contributions of each student to the success of the team, and measures of team performance must therefore permit some individual assessments. Students in a team which performed poorly as a team because of the inadequate contributions of other team members must not be penalized if their own contributions were satisfactory.

9.10 *The reliability and validity of training procedures*

The concept of reliability, when applied to air traffic control training, refers to the consistency of the results achieved. Courses that purport to be similar should attain similar standards, and the causes of failure should be attributable to the individual student. Ideally, each student could take the same course elsewhere, with the same result. No extraneous factors should have gross effects on what is taught, what is learned or what is practised, for this would signify poor training reliability. Most expedients to improve success rates and reduce attrition rates during training incidentally increase reliability by correcting training weaknesses and by ensuring the attainment of minimum standards of knowledge, skill and practical experience. Numerous procedures and forms of assessment during and at the end of training can also promote reliability. Among the procedures are extra tuition, recoursing, additional demonstrations, extra opportunities to practise, knowledge reinforcement through specialist computer training packages, and specific tuition, explanation and demonstration by instructors to improve each individual student's understanding. Among the forms of assessment are standardization of the training syllabus, defined milestones during training, and impartial objective testing conducted by an agency that is independent of the actual tuition. But there are practical constraints on what can be attempted in training; it costs too much and takes too long to develop prototypes only to discard them, for example (Baldwin, 1993).

If ostensibly parallel courses at different facilities do not share the same training methods, philosophies and priorities, independent testing may become essential to guarantee that the basic knowledge, skills, understanding and ability to control air traffic are learned by every controller, no matter where and how the training has been given. Such independent testing is intended for assessment, but it also measures training reliability. Training reliability may suffer if there are gross changes in demand. Any inclination to lower standards in order to compensate for an urgent, albeit temporary, shortage of controllers should be resisted, for any resultant short-term benefits are not in the long-term interests of the safety and efficiency of air traffic control unless the training standards were incorrect anyway, which would be a serious but quite separate problem.

Another aspect of the reliability of training deserves consideration. This is the level of agreement between the scores achieved by the same student for different courses or phases during training, both in terms of the absolute scores by each student converted to a common form such as percentages, and in terms of the relative scores of the students within a group expressed as their orders of merit according to different assessments. If the assessments and scores from separate parts of training do not correlate highly with each other, so that there cannot be a consensus on who is a good student from the absolute scores of an individual or on who are the best students from the orders of merit, this limits the attainable reliability of the course as a whole, and raises the issue of whether air traffic control can legitimately be treated as a single entity for the purposes of determining training reliability. If students are not consistent, with some better at theoretical and others at practical work, or with some better as radar controllers and others as procedural controllers, the reliability of overall assessments is curtailed.

Whereas the reliability of training procedures has not been a major issue (although perhaps it should have been), their validity has often been a frequent subject of study and contention, usually in relation to specific training (Boone, 1983; Broach and Manning, 1994). Many have succumbed to the temptation of treating the completion of training as an end in itself for the purposes of establishing its validity. The validity of assessments early in a course for predicting its successful completion has been examined assiduously, for costs could be saved if potential failures could be detected sooner.

The validity of training assessments as predictors of subsequent career success seems ripe for reappraisal. It is far from certain that those who will give a lifetime of satisfactory service as safe and efficient controllers are necessarily those who complete their training most successfully, yet they should be if the training itself has high validity. Not only is there considerable uncertainty over the predictive value of training assessments for subsequent career progression, but also uncertainty about the levels of validity that it is reasonable to aim for. At what points are further efforts to increase the validity of training no longer repaid by commensurate improvements?

Many questions about the validity of training dispute whether what is taught is what is needed. Much training emphasizes knowledge of the practices and procedures of air traffic control, but emphasis on the successful completion of practical examples may not imply complete understanding of the underlying principles. The value of basic theoretical knowledge may be disparaged during later simulations and practical on-the-job training. The validity of training procedures is impaired by any gulf between what is taught and what is claimed to be needed. It is seldom easy to decide what must be included in training. Among the more contentious issues are the need for basic knowledge of associated disciplines such as meteorology and navigation, the level of understanding of system functionality that is essential in order to use it properly, the appropriate understanding of system software, and the level of exposure to emergencies or rarely used procedures that should be included in training.

Training may be invalid for different reasons. It may teach information, skills or knowledge which are irrelevant or never used. The teaching may be at the wrong level of detail or in the wrong order. Understanding and insight may be sacrificed in favour of the rote learning of standardized practical examples. Training may lack validity if it is on out-of-date or inefficient equipment or if the training environment or training tasks differ too much from real life for learning to transfer.

There are several criteria for testing the validity of training. The long-term one is that the training produces controllers who are highly efficient, who are safe and who stay, but in the meantime the training will have changed. Another criterion considers what still has to be learned after training has been completed. There may be subsequent entrenchment and the acquisition of further skill through experience, but the training itself should have covered all the fundamentals. The validity of training is reduced if the quest for uniformity, consistency and lower attrition rates results in the successful completion of training by some who are incapable of remaining as highly efficient controllers in the longer term. The validity of training is also impaired if certain factors crucial for air traffic control success are not covered at all during training. Examples include acquiring realistic career expectations, subscribing willingly to the norms and standards of the air traffic control profession, and not being overstressed by the burden of having the sole responsibility for the air traffic under control.

It is a mistake to check validity solely through attrition rates, or to assume a close connection between different indices of validity. If the attrition rates have been reduced it does not necessarily follow that the accepted controllers are of better quality, or that those rejected could never have become equally successful as controllers. The supporting evidence for such conclusions is currently lacking. If those students who were most successful at the end of training do not prove to be safer, more efficient or in any measurable sense better than those who were borderline, then the validity of training is low, and perhaps some of those who failed would have been as good as some of those who passed. Both the reliability and the validity of training should be high. Further work to develop better quantitative measures of reliability and validity and then to apply those measures iteratively to increase the reliability and validity of training is needed.

9.11 Automation and computer assistance during training

A useful distinction between automation and computer assistance in relation to training is that the automated processes with no provision for human intervention apply mostly to scoring and assessment, whereas the computer assistance in support of the human applies mainly to the teaching processes. The application of computer assistance to air traffic control training processes represents a revolutionary step that can no longer be postponed (Wise *et al.*, 1991). Several reasons combine to make the more widespread computer assistance of training imperative, and Smith (1991) has discussed many of them. There seems to be the political will to spend more on training. Reliance on existing training procedures and on classroom tuition is weakening. The preoccupation with attrition rates and the prevention of failure is yielding to increased concern about the quality of those who do succeed. Many training methods have seemed too inflexible, with insufficient encouragement of those who have difficulty with them. Individual differences have been treated as an encumbrance instead of a potential asset. The possible applicability of many recent theoretical and technical developments in training and education to air traffic control has not been considered seriously enough, although they are beginning to be considered for flight (Gopher *et al.*, 1994; Williams, 1994).

Several factors that can seem remote from automation and computer assistance in training are in fact very pertinent to it, such as recruitment policies for candidates, the status of air traffic control instructors, and the share of air traffic control funding that should be applied to training. The widespread introduction of automation, mainly as computer assistance, would involve the curtailment or abandonment of many hallowed practices, and would therefore be resisted if it seemed to represent change for the sake of change. Any computer assistance in training must prove its worth; it cannot afford to seem like a fad, a palliative or a cost-cutting exercise.

Smith's (1991) call to implement the radical changes in air traffic control training envisaged in various policy documents included a list of changes in training practices that would accord with current educational concepts, involve some automation and improve learning. He listed over thirty changes, which indicates how comprehensive and far reaching the impact of automation and computer assistance could be. Many forms of computer assistance are available (Fabry and Lupinetti, 1989; Galotti, 1993). They differ greatly in complexity and level of detail, but most share a single revolutionary feature—selective computer assistance permits training to be flexible and adapted to the needs of each student. This can be done now, with or without the student's knowledge that it is occurring. As a result, a much wider range of policies on the scheduling, administration, assessment and direction of training could be adopted now if they were agreed.

In principle, some kinds of automation and computer assistance in training can allow all students to learn at their own pace and in their own way and can permit continuous comprehensive appraisal of the learning processes and progress. It becomes easy to identify problems that are causing difficulties for individual students, to provide appropriate extra tuition and practice, and to test the efficacy of such additional training. Self-administering separate training packages can supplement the main training programme, at the behest of the automation, the instructor, the student or an independent assessor, these not being mutually exclusive. A problem is to direct extra training to those most in need of it. The most enthusiastic and motivated students who may be keenest on automatically presented extra tuition may need it least. The packages adjust the level of difficulty to individual progress so that they are not identical for every student, and give immediate feedback of results. The student, the instructor, an assessor or the package itself can determine when training with the package is discontinued.

Computer-assisted training can help the student to develop a mental checklist of the important factors in solving any given kind of air traffic control problem. One technique is to present a series of superficially similar problems that in fact have different solutions. The controller has to learn how to tell the difference, and what the crucial factors are. This kind of learning lends itself well to a sequence of demonstrations within a stand-alone training package. Computer assistance in training can expand the other kinds of teaching available. For example, a problem can be developed and then frozen while the full consequences of alternative solutions of it are explored in turn. This can also be done iteratively by setting up the same problem and working through or demonstrating a variety of solutions.

Radical technical innovations may suggest radical forms of training. For example, students can learn not only by resolving problems but by devising them. This can be a salutary and memorable way to discover what is difficult but seems easy and what is easy but seems difficult, particularly when a student can devise a problem but cannot solve it. Computer assistance may lead to more flexible forms of feedback. Early in training, the computer should be a model of patience with the student and always provide appropriate feedback and never procrastinate, but later in training perhaps a careless error or a potentially dangerous action should elicit a sharp rebuke by the computer.

Signs of complacency warrant an automated challenge. Training devices might not be receptive to over-complex solutions of simple problems, to unnecessary aircraft manoeuvring, to dithering when there is a lot to do, or to carelessness in routine tasks. There is no place for these in real air traffic control and perhaps there should not be much machine tolerance of them in the later stages of training. This is an aspect of training on which some initial research would be required. The emphasis has been on using self-sufficient packages to provide opportunities to practise in order to acquire knowledge and develop skills, but they could also be intolerant of human incompetence and perhaps they should be in the later stages of training.

A further more speculative application of computer assistance during training relates to functions fulfilled in operational systems now that may have to be fulfilled in future as part of training. How far can computer-assisted training help to establish professional norms, standards and attitudes? What are the respective roles of computer-assisted training and on-the-job training? If future systems incorporate means to convey the controller's intentions to the system in order to enlist its collaboration in achieving them, suitable training to optimize this new kind of human–machine relationship will have to be devised. The best forms of computer assistance for team training in air traffic control will have to be proved.

A future application of automation to training will be for controller assessment. One of the best safeguards of impartiality and fairness in assessment is to use computers as much as possible to devise and present problems, and to assess and score individual controllers' performance in dealing with them. This process could become widespread and standardized, but evolve so that assessments that were independent of the training could be superseded by more continuous evidence of progress obtained directly from the training itself.

The pace of computer assistance for air traffic control should determine the pace of the corresponding changes in training to prepare controllers for work in more automated systems. This statement seems obvious to the point of triteness, but it represents a departure from most previous policies. Sometimes it has been necessary to adapt evaluation equipment for training, when suitable training equipment has not been commissioned in time. Those concerned with training should have been actively involved in the specification, development and planning of the computer assistance to ensure that the implied human tasks, procedures, skills and instructions will be teachable. These issues are at last being addressed (Chobot and Chobot, 1991), but the actual scheduling is still usually far from optimum.

9.12 Effects of air traffic control automation on training

Before a new form of computer assistance can be used by controllers, they have to be trained in its use and capabilities. Before the controllers can be trained, their instructors have to be trained. Before the instructors can be trained, the tasks, procedures, instructions, scenarios and examples have to be specified, verified and validated, and the teachability of the controller's tasks proved. Methods of performance assessment and tests of acceptable levels of proficiency have to be devised, and adjusted if necessary. All this requires extensive time and considerable resources. The timescales allowed for training seldom seem sufficient. It is not unknown for the operational date of a new air traffic control system to be announced in advance, usually for political reasons, when everyone knows that the date is premature, and resources have to be diverted to cobble together a temporary system to meet an unrealistic deadline. It is a tribute to the professionalism of controllers that the outcome has merely been inefficiency and not loss of safety.

The more complex the forms of computer assistance, the less the available evidence on the optimum human–machine match and on the best forms of training. Expensive and potentially beneficial forms of assistance may be provided but not used unless they are fully understood and trusted. When computer assistance is optional, training has to cover not only how to use it but the conditions that do and do not justify its use. If the computer assistance presents options for the controller to make a choice, training has to include sufficiently detailed understanding of the computer assistance to guide the choice correctly. Both the nature and content of training courses are changed by the computer assistance of air traffic control (Baldwin, 1991).

Forms of training that are best suited to the human tasks in complex systems are still being developed. Cognitive issues are generally most crucial (Hopkin, 1988). The controller relies on training to know the tasks, the options, and the criteria for choosing among options. Decisions on how self-evident the functioning of the human–machine interface should be have major implications for training and should be influenced more by these implications. Unrealistic deadlines discourage thoroughness in ensuring that the computer assistance is fully teachable.

Proposals for extensive computer assistance in air traffic control with a change of emphasis from tactical towards strategic control place additional demands on training beyond those strictly required for the actual use of the computer assistance. Future controllers must work effectively as managers of substantially automated systems, and the best ways of doing so are probably not intuitively obvious but require training. Experience from several contexts suggests some of the training needs. New roles for teams will require modified forms of team training. The feasible kinds of leadership and the skills required for them will change. The roles of supervision and the practical means of supervision will change. The forms of communication between pilots and controllers must be adapted to automated data links (Hopkin, 1994d).

Generally automation adds to training. The trained controller must still be able to control air traffic without computer assistance, yet be skilled too in the use of all the forms of computer assistance available. The ability to use data links effectively must be combined with the ability to communicate directly with pilots. The kinds of training issues raised by Herschler (1991) have to be resolved, and the resultant changes to training devised, introduced, optimized iteratively and verified.

9.13 The allocation of controllers to jobs

Several factors influence policies and practices in the allocation of controllers to jobs, including the following:

1. The job vacancies in air traffic control which have arisen or will arise, for which new controllers have to be selected and trained.
2. The extent to which the selection procedures are general and intended for all air traffic control jobs, or are specific to particular air traffic control job categories.
3. The training objectives, particularly whether training is for all or only for some air traffic control jobs.
4. The specificity of training courses.
5. The stage at which job allocation is introduced; practices vary, and range from allocation only on completion of all training to a brief initial general training followed by job allocation before the main training that is specific to a particular location begins.
6. Policies about the weighting in allocation decisions of various influences such as stated preferences, air traffic control needs and vacancies, the capabilities and limitations of individual controllers, and the predictability of longer-term air traffic control requirements.

Nations differ in their attrition rates during air traffic control training, mainly because of different job allocation policies. The highest attrition rates are associated with a requirement for every controller to reach the minimum training standards or be rejected. The lowest attrition rates are associated with a different policy whereby acceptance for training as a controller almost guarantees subsequent employment as a controller because the ablest controllers are allocated to the most demanding jobs and the weakest to the least demanding. Low attrition rates are achieved at the cost of subsequent reduced mobility and flexibility in the deployment of manpower resources.

Considering that the processes of the allocation of individual controllers to jobs have the potential to determine the efficacy and validity of all preceding recruitment, selection and training processes, it is remarkable how little research has been conducted on allocation, to the extent that even its significance for success as a controller compared with the other processes of recruitment, selection and training has apparently not been seriously examined. But if allocation seems arbitrary or is done solely in relation to vacancies rather than as part of a deliberate policy of matching job requirements with demonstrated individual strengths and weaknesses, then this must limit the validity of any processes intended to compare the efficacy of recruitment, selection and training procedures with ultimate long-term success as a controller. Research is needed to determine how important the processes of allocation are in relation to subsequent efficiency and career development, as a preliminary to trying to optimize those processes. The actual allocation of a specific job to a controller may have little effect on that controller's subsequent efficiency, safety or career, or it may be the most important single influence. It would seem prudent to find out.

10

The work environment

10.1 Traffic flow management

Where the air traffic demands regularly outpace the handling capacity of the system, or where the normal conduct of air traffic control is subject to widespread disruption by severe weather, broad solutions of the resulting problems are devised and implemented by traffic flow management, although the workspaces currently being introduced for traffic flow management are the first to be designed specifically for it, as distinct from dedicated to it. Traffic flow management applies primarily not to single aircraft or to current traffic but to patterns and flows of future traffic, from at least half-an-hour up to a few hours ahead, From data about planned flights and long distance flights already airborne, future demands are estimated, and matched with capacities and constraints so that the traffic can be handled, if necessary by re-routeing, by re-planned flows, or by delays to departures. Traffic flow management may require the re-routeing of many aircraft away from a popular route temporarily affected by bad weather onto specified alternative routes, but the air traffic control centres and controllers convert such traffic flow management directives into instructions to specific aircraft or allocated slot times. The involvement of traffic flow management in air traffic control must become more common as increases in traffic demands render system saturation more frequent.

Because traffic flow management has evolved to meet demands, the ideal information for its purposes may not have been gathered or may not be available in a suitable form. Current and forecast meteorological information can be the cause of traffic flow management interventions and the main constraint on practical solutions. Strategic traffic flow management tasks concerned with the planning of traffic flows may have to use tactical information about single aircraft, some of which is of uncertain provenance. Too much detailed information of suspect value can result in cluttered displays that obscure flow patterns, especially where suitable codings to represent flow patterns have never been devised and proved.

Traffic flow management applies to very large airspaces that can contain thousands of aircraft. Some partitioning of the airspace into regions and consequent division of the responsibilities can become a practical necessity, yet traffic flow management must apply to the whole airspace, and problems must not merely be transferred to another region. Significant human factors problems that arise include the integration of the activities of different regions, the definition of team functions, the compatibility of information throughout the system, and the selective retrieval and coding of uncluttered data for traffic flow management. There have also been some difficulties in providing knowledge of results in forms that facilitate learning from experience. Harwood (1993) mentions some relevant work in the United States, and Duytschaever (1993) describes the European system.

10.2 Air traffic control centres and control rooms

Although the public image of air traffic control associates it primarily with towers from which controllers can see some of the aircraft under their control in daylight and fair weather, most air traffic controllers actually work in centres away from the air traffic. Far from providing views of aircraft, these workspaces often lack windows, but in principle this permits ergonomic optimization of some visual aspects of the workspaces. The system depends on the selective presentation to controllers of remotely sensed and transmitted information, and on messages through speech channels. Since all the information utilized within the air traffic control centre has to be sent there, its location need not depend much on the air traffic but can be influenced by other factors such as national policies for employment and regional development, subsidies, site accessibility, local transportation, the local infrastructure and amenities, and the attractiveness and expensiveness of the area as a place to live. A centre that controls heavy traffic over a large region is a major installation and source of local employment (Figure 5).

The physical aspects of an air traffic control centre, being controllable, can in principle be simulated very exactly. Sometimes it is possible to deceive a casual onlooker, or even a more informed one, that a simulated air traffic control centre is real. This verisimilitude has beguiled investigators into believing that real-time simulations of centres can address and answer validly any question about them, but this is not the case. Physical duplication is not all that is required for functional similarity. To the air traffic controller, a simulated centre may look like a centre but not function like one.

Within a centre, air traffic control is actually conducted in an operations room, a large purpose-built room containing numerous suites of air traffic control furniture linked by elaborate communications facilities in an accessible and carefully specified physical environment (Figures 6 and 7). Each centre is largely self-sufficient in its support services, with considerable delegated

Figure 5.   A new air traffic control centre

Figure 6.   A centre for controlling civil air traffic

Figure 7.   A large control centre

authority for the safe and efficient control of the air traffic within its region of responsibility. It therefore contains a great deal of equipment for supporting functions, including the gathering, storage, manipulation and presentation of data, communications, serviceability and maintenance, quality control, recording, and standby and backup facilities. Much of the equipment in centres is standardized because it is cheaper and more efficient to procure, install, maintain and replace the same equipment everywhere and to train controllers to perform standardized tasks on standardized equipment. Standardization makes centres more compatible with each other, and assists the transmission of data between them and the consolidation of all their data into a single database of all the air traffic within a much larger region. Such data can then be applied to plan routes and flows, to control traffic capacities and flows strategically, to trace the full effects of local events on the larger system, to provide a record of all traffic for such purposes as costing and long-term planning, and to identify trends in the amount, mix and distribution of air traffic. In these respects, centres may form a network.

   The detailed contents of displays and the specific communications facilities differ between centres and within centres and workspaces, depending mostly on the geographical region, the traffic and the equipment. Nevertheless, these differences should not violate any general human factors principles for the display of information, the choice of input devices, the specification of communications facilities, and the physical environmental requirements. As a result, different centres can look quite similar, but their superficial similarity can conceal substantial differences. Figures 8 and 9 show suites at different centres. The most striking differences are likely to be in traffic flows and traffic demands. Less obvious differences relate to shift work and rostering practices, conditions of employment that can be negotiated locally, management styles, the loyalty of the workforce to the centre, and its reputation within the air traffic control profession.

The adequacy and acceptability of an air traffic control centre as a work environment is influenced by several mundane practical matters. The availability and accessibility of canteen and refreshment facilities of good quality and modest cost is a practical way to demonstrate interest in the well-being of the workforce. The convenient location of well-maintained toilets and of well-furnished places to relax for a short time should improve on the statutory minimum requirements. Controllers often arrive at work and leave at unusual times, and the incoming controllers must arrive before the outgoing ones can leave, which should be reflected in ample car parking facilities. From time to time, a centre must accede to requests and requirements beyond its normal functions. A suite may become unserviceable; investigative procedures may be undertaken; there may be additional training commitments; the needs of many who must enter the operations room must be met, for such purposes as work measurement, quality assurance, research and development, evaluations, incident investigations, and audits. To allow for these extra requirements, a

Figure 8. Suites in a control centre operations room

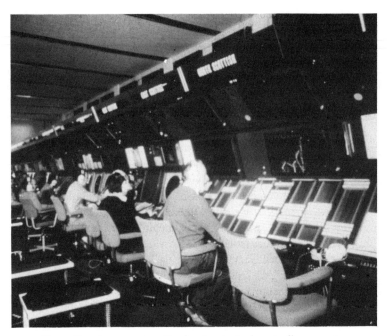

Figure 9.   An en route suite showing typical controller grouping

centre may have more suites than those required for its maximum traffic-handling capacity. A centre housed in a well-designed and pleasant building conveys a favourable impression to those who work there and to visitors. Initial impressions strongly influence subsequent attitudes.

Air traffic control has always attracted more than its fair share of visitors. On some pretext or other, many people like to visit air traffic control work-places and to watch and listen to controllers working. It is prudent to design each centre to permit visits but avoid friction. Arrangements which are con-genial and welcoming to visitors while being minimally disruptive to the con-trollers themselves constitute good public relations through reassuring impressions of orderliness, good planning and quiet efficiency. Viewing bal-conies sound-proofed from the operations room, separate suites staffed to explain to visitors what is happening but without direct access to control pro-cedures, professionally made video demonstrations, and off-line workspaces that provide some hands-on experience of air traffic control for visitors, are all effective and acceptable means to promote air traffic control and inform people about it without disturbing it. The few occasions when visitors may enter the operations room should be minimized, and no visitor, no matter how exalted, should ever be free to interrupt operational controllers without their prior agreement.

Many of the most influential visitors stay for only a short time. If their understanding of air traffic control is limited, they rely, perhaps too much, on immediate fleeting impressions, and it is sensible to plan their visits accordingly. Although it is inherently unsatisfactory that visits on which key decisions depend are often too short, and it would be better if those who make such decisions spent more time in real air traffic control environments, this seems unlikely to happen. In the best interests of air traffic control, all possible public relations opportunities, including press visits, should leave as favourable an impression as possible, while not hiding the truth where changes are needed. It is important that any genuine issues of concern are raised and clarified during relevant visits and that balanced appraisals are not jeopardized by eyecatching trivia. Given the importance of brief impressions, it is worth some effort to make them favourable.

10.3 Air traffic control towers

Air traffic control towers are called visual control rooms in some contexts, though the former term is employed here. Many of the above points about centres as work environments apply to towers. However, at least five significant differences between towers and centres are important for human factors. First, the physical environment, particularly the lighting, is not constant in towers but variable. Ambient lighting levels can vary from bright sunlight to those provided within the tower during the night. The direction of the light source varies with the time of day: in certain positions sunlight may shine directly onto, and be reflected from, operational displays. Secondly, the tower controller integrates into the air traffic control tasks the view from the tower of aircraft on or near the airport, and thus combines direct and indirect sources of information. Thirdly, the controller has a panoramic view through a large visual arc and must be free to scan in many directions, with the minimum interference between the height of any consoles or other equipment installed in the tower and the controller's view over them looking down at aircraft manoeuvring. The tower controller's postural and viewing positions cannot be wholly standardized. Fourthly, the controller must see and perhaps hear aircraft movements against a wide range of visual backgrounds including sky, cloud, darkness, restricted visibility, and light patterns on the ground, none of which must induce misperceptions. Fifthly, towers have to be sited in relation to the runway length and orientation, the surrounding terrain, other buildings and facilities within the airport, and the positions of the sun. Many local conditions should influence their siting, to minimize interference with the controller's view of circuits, approach and departure paths, runways, taxiways and aprons. The controller's view must not be impeded more than absolutely necessary by stanchions or other structural features of the tower, or by other controllers at their workstations in the tower. The designs of towers must overcome these limitations and satisfy their requirements (Figure 10).

Most of the work of tower controllers involves communications with centres rather than with other towers. As a result of all these factors, the designs of

Figure 10.   An air traffic control tower

towers are much less standardized than those of centres although greater standardization is envisaged (Ammerman *et al.*, 1986). At first, each tower was a
one-off design. Nowadays the standardization of equipment and of viewing
conditions has resulted in more features that are common to many towers, but
in principle towers must differ more from each other than centres do. Tower
designs are difficult to specify completely in general human factors terms, and
several of the following aspects of each tower design usually have to be verified
specifically for it. If the handover of control responsibility for an aircraft is
represented by the physical handover of its flight progress strip from one controller to another, it must be impossible to misplace or lose the flight strip.
Blinds protecting one controller from the sun's glare must not obscure the
view of another. All relevant forms of supervision must be possible without
impeding the view of other controllers: this applies both to the formal duties
of the supervisor and to controllers supervising ancillary tasks such as those of
the lighting assistant who sets the lighting on taxiways and aprons that directs
the aircraft along the correct routes between the stand and the runway, at the
behest of the controller. The layouts of the equipment and workspace within
the tower must permit the controller to exercise proper supervision.

Tower workspaces can be studied by simple mock-ups to check basic functions, lines of sight, broad divisions of responsibility, layouts, and information
flows. It is more difficult to use simulation to study the dynamic aspects of
tower operations or for training, because some of the tower controllers' know-

ledge comes from the direct sight of aircraft which can be difficult to replicate adequately by mock-ups, although there have been many attempts at it. Surprisingly often, model aircraft have been moved over a model of the airport and its environs. The most recent simulations have extended this principle and resolved the simulation problem by generating very realistic aircraft and aircraft movements electronically, to provide a dynamic simulation in which aircraft are seen to respond to controllers' instructions with great realism. This equipment is currently used for training, but it could be a powerful research and evaluation tool.

Although human factors recommendations about the design of air traffic control centres can often be based directly on the information in ergonomic handbooks subject to some verification, this approach may be less satisfactory with towers. It is often necessary to evaluate and solve any practical human factors and workspace problems of existing towers on site, and to employ mock-ups and simulations to confirm ergonomic recommendations about planned towers because the applicable constraints on the validity of the evidence may be specific to each tower. In towers, space is more limited than in centres. Towers and centres therefore tend to pose different human factors problems. Centres are a more controllable environment that encourages attempts to find optimum solutions. The human factors constraints in towers are more numerous and more specific, and encourage acceptable human factors compromises that will suffice for all conditions but probably be optimum for none. Because of external environmental influences, few general principles of workspace design can be applied directly to towers unmodified, but most require adaptation to characteristics that are specific to each tower.

10.4 Principles of workspace design

A fundamental point about air traffic control workspaces is that they are unique to air traffic control. They are designed for air traffic control tasks and have no other use. Yet they may not look very similar, as examples of suites for en route control (Figure 11), for approach control (Figure 12) and for oceanic control (Figure 13) illustrate. The design principles that apply to them are not those for simple tasks performed by unskilled people but those for complex tasks performed by a carefully selected and quite homogeneous skilled workforce, and they must take account of workforce characteristics. The planning, design and specification of air traffic control workspaces are the responsibility of system planners and designers, but must accommodate the requirements of many other disciplines, including accountancy, architecture, engineering, computer sciences, medicine, human factors and air traffic control itself. At the workspace design stage, the prime concern of human factors is to ensure that the practical implementation of the planned jobs, tasks and equipment already specified meets all identifiable human factors requirements

Figure 11.   A suite for controlling en route traffic

so that the envisaged roles and functions for humans can all be achieved to operationally acceptable levels of safety and efficiency, without impairing well-being in any way.

The practical human factors contributions to air traffic control workspace design often rely on compromises between human factors requirements and those of other disciplines. Any human factors ideal is usually unattainable, even if it can be recognized. Many standard human factors recommendations can be adopted (Ivergard, 1989), but air traffic control tasks and jobs often impose serious constraints that preclude some standard recommendations and require effective compromises to be sought. There may be so much displayed information that it cannot all be presented within recommended viewing angles, yet it is necessary at different times for various jobs and must appear

Figure 12. An approach control suite

Figure 13. Oceanic control positions

somewhere. There may be too many input devices to house them all within recommended reach distances, yet sometimes all of them are needed. Workspace designs may have to be compromised to allow jobs to be split or amalgamated. The workspace must provide space for an incoming and an outgoing controller during the handover of responsibility. Its design must encourage the satisfactory fulfilment of a large range of ancillary functions, including supervision, on-the-job training and assessment, and some maintenance and cleaning. The requirement that the air traffic control environment must not harm those who work within it conjures up images of excessive stress and workload, but postural and visual deficiencies in the workspace design are more probable causes of harm, and must be rigorously excluded at the design stage.

Workspace design cannot proceed until most of the system planning and task analyses have been completed. Its three levels of detail cover the general workspace environment, the workstation of a single controller, and single items of equipment within each workstation. The planning of the workspace design has to follow this order, with some provision to correct specific inadequacies by revising more general requirements. Tasks, jobs, instructions and procedures provide guidelines for stating what the workspace design must achieve. Less obvious, but more critical, are the indirect effects of workspace design in terms of functions that it hinders or prevents. Many aspects of policy, for example on responsibilities, task sharing, staff levels, flexibility, handovers, training maintenance, assessment, skill development, and especially on safety, can be positively promoted through appropriate workspace designs if their needs have been recognized, but they can also be thwarted either by deliberate decisions that the workspace cannot accommodate them, or by default if their requirements remained unrecognized throughout the workspace design.

The objectives of air traffic control workspace design were described in detail some years ago (Hopkin, 1982a) and have not changed much since. They include:

1. To achieve the highest standards of operational efficiency and safety by avoiding all unnecessary constraints on task performance and on its accuracy, reliability and pace.
2. To design workspaces that promote system policies and objectives regarding the division of tasks, the allocation of responsibilities and the balance between individual and team decision making.
3. To ensure that the workspace permits the whole envisaged range of staffing flexibility, so that staffing can always be matched with traffic demands.
4. To fulfil within the workspace all its ancillary functions, including handover, training, assessment, supervision, assistance, and maintenance.
5. To meet all the envisaged requirements for liaison, coordination and communications between air and ground, between human and machine, and between humans within the system.
6. To provide every controller with all the information needed and with

means to access it readily, and to provide all the equipment needed for the whole range of tasks in forms that are easily identifiable and promote the efficient performance of each task and smooth transitions between tasks.

7. To reflect task structures and the relationships between tasks through the equipment, facilities and layouts.

8. To check that all items within the workspace meet the applicable ergonomic recommendations.

9. To incorporate in the workspace design all practical means to prevent foreseeable critical errors and omissions and to recover from any that are not prevented.

10. To provide physical work conditions that are not obtrusive or distracting but pleasant and harmonious.

11. To check that all the information necessary for jobs and tasks is present in forms which are intelligible and that it can be manipulated to meet operational requirements and to harmonize with the needs of other tasks.

12. To ensure that no features of the workspace could have adverse effects on the well-being of those employed within it.

13. To foster interest, job satisfaction, morale and self-esteem within the workspace.

From system specifications and job and task designs, many essential features of the workspace can be described. For example, all the displays and input devices used by the same controller must accord with the recommended viewing and reach distances for a single workstation, and where work is shared among the controllers in a suite the displays and input devices must meet the viewing and reach distance recommendations for all who may use them. Work by teams requires communications appropriate for individuals who must function as team members. If information used by more than one person cannot be displayed so that all who need to see it have a good view of it, it must be duplicated or the requirements for it must be revised. The specific workspace and human factors problems depend on the form of the shared information, whether within consoles or on wall-mounted displays. If the latter are chosen, the suites and consoles have to be positioned within the room so that all who need to see them while seated at their workstations can view them comfortably. This is achieved by matching the known viewing distances and angles with the detailed design of the contents of the wall-mounted displays, allowing for other intervening factors such luminous flux, ambient lighting, and minimum eyesight standards. The workspace design must accommodate other functions, like general briefings and demonstrations for visitors. The number, location and content of any well-mounted displays illustrate the influence of workspace design on the feasibility of intended human functions, and also illustrate the essentially interacting nature of most human factors recommendations and requirements.

A further example may reinforce this important point. It seems quite a simple matter to decide whether the main radar displays for controllers should be mounted horizontally as they often have been in the past, or vertically as they generally are now and will be in the future. A historical reason for horizontal displays is that each aircraft track on a radar display was originally labelled by writing its identity on a plastic plaque that rested on the flat radar display and was moved to maintain track identity. The decision on horizontal or vertical radar displays seems primarily a display decision but is also a communications decision, because on it depend the kinds of communication required, the feasible team roles, and the nature and degree of consultation and liaison. The problems of lighting horizontal and vertical displays are very different, and other differences concern methods of inter-console marking, the relationships between displays and input devices, and ergonomic and occupational health problems associated with vision and posture.

In workspace design, it is essential to reach agreement on the general requirements before tackling the details. Issues to be settled include maximum staffing levels, the flexibility of staffing levels and the methods of changing them, the space needed for handovers, interrelationships between jobs, the integration of supervision and training requirements, team coordination and communications within the control room, the space requirements for accessibility and maintenance, and the functions and locations of any general wall-mounted displays. Such broad issues should be decided before the specification of each suite and workstation is tackled, and long before the fine details of input devices and display contents and formats are finalized. The facilities have to be checked to ensure that envisaged responsibilities can be exercised and procedures followed.

Workspace design can employ a variety of practical tools. The commonest in air traffic control have been life-size physical mock-ups and models, to check lines of sight, flows of information, the layout of suites in relation to each other and the layout of displays and input devices within suites. This technique can be particularly effective in evolving acceptable compromises whenever the preferred solution of one ergonomic problem such as the layout of displays in relation to each other, incurs other ergonomic problems such as poor relationships between displays and input devices or restricted viewing angles for important information. For more limited purposes, miniatures of the control room may be constructed. A more recent option is to use software tools for workspace development, as computer-aided design becomes more widely employed to specify air traffic control workspaces and to verify their ergonomic acceptability. Ideally, only designs that do meet ergonomic requirements can be generated electronically. Some of the special needs of air traffic control, notably in staffing flexibility, can mean that computer-aided design methods developed for other contexts may not transfer to air traffic control without extensive modification. However, they can be used to formulate alternative layouts that would meet all the normal requirements, and can reveal what the residual problems associated with various computer-generated alternatives would be. In the earlier stages of workspace design, this can be a very effective way to identify problems that must be resolved before it is worth

Figure 14. Furniture mock-up of a future tower layout

Figure 15.   Alternative mock-up configurations for workstations with two radar displays

building mock-ups, although the success of the method is very sensitive to the correctness and completeness of the air traffic control data and the defined requirements. The role of small-scale or actual-size physical mock-ups may then become one of verifying proposed computer-aided design solutions and gathering controller opinion about them. Figure 14 illustrates a physical mock-up of a proposed tower layout, and Figure 15 shows mock-ups of alternative console configurations for workstations requiring two radar displays each. In the past, mock-ups have often contributed to the planning of air traffic control workspaces but their future roles may veer towards verification.

10.5 Suites and consoles

Before suites and consoles are specified in detail, their positioning within the control room or tower and in relation to each other should be fixed, taking into account all the workspace requirements which collectively provide guidelines for deriving optimum compromises among the many influences on the positioning of suites. From these requirements, and the task and job descriptions, the displays, input devices, and communications facilities that will be needed can be gauged. It is necessary, before suites and consoles are designed, to start with a complete list of everything to be fitted within them, together with essential relationships between listed items. Basic ergonomic considerations include reach and viewing distances and angles, anthropometric data on the controller population, the envisaged sharing of facilities, and the provision of means of mutual assistance. The suite and console designs must make suitable provision for all the other workspace requirements, such as stowage for job aids and manuals and temporary instructions, the needs for controller comfort and support, and even places where drinks can be placed if they are permitted so that accidental spillage will not damage equipment. A further factor is the strong trend towards standardized or modular air traffic control furniture. In its most extreme form, every workstation becomes physically identical and the differences between jobs depend entirely on software. A more flexible variant of this policy has a standard basic console or suite, with flexibility through a range of optional extras.

The first design decision about the layout of suites and consoles is whether they should be in a continuous row, in small groups or island suites, or in crescents or other configurations within the centre or tower. This decision is linked to what each controller needs to see: for example, the occupants of all workstations requiring a view of wall-mounted displays must face towards them. The next design decision is whether seated controllers must be able to see over their consoles. This decision has many consequences for design constraints, and accounts for much of the difference between the air traffic control furniture in towers and in centres. When these key decisions have been taken, more specific designs of suites and consoles can begin to evolve. From this stage onwards, all decisions about the suites and consoles should also consider the verification and validation procedures applicable to them.

The appropriate flexibility in the design of air traffic control console profiles has not yet been settled. There are no universally adopted practices but two different approaches. One is to fix the profile of the suite or console and build it rigidly so that it is not adjustable. Anthropometric differences are compensated for externally by such means as seat adjustments, platforms and footrests. The other approach is to cater for anthropometric differences by adjusting aspects of the console itself such as the shelf height, and to meet some of the requirements for flexible staffing of the suite by pivoting vertical displays so that each of them remains at an approximate right angle to the controller's line of sight. A consequence of this latter solution can be to render much of the information on the displays in one suite invisible from adjacent suites, but this may be acceptable. The applicable ergonomic principles are not the same for both these basic approaches. Each incorporates some kinds of flexibility but sacrifices others. Muddle can result from attempts to combine both approaches, or from a failure to appreciate the full ergonomic consequences of either.

A related problem in the design of consoles has been to make them adjustable in accordance with users' needs, without teaching users how to ascertain what would be the optimum adjustment of the console for them. It seems to have been assumed, wrongly, that everyone knows intuitively his or her own optimum seat height, posture, viewing distance and viewing angles. However, none are intuitively obvious. They all have to be taught, but seldom are. In some simple workspaces, operators who have received appropriate training can be left to set up the system themselves: a word processor with its input devices may be moved about on a desk to suit each operator. Air traffic control workspaces are much more complex, and the permissible flexibility in adjustment has to be carefully planned in relation to the environment.

Other major influences on the design of suites and consoles are the evolution from multiple small displays to fewer larger displays and even to a single very large radar display with windows containing tabular information, and the evolution from conventional keyboards to electronic input devices with touch-sensitive surfaces or to input devices that can be moved across the shelf. This development also has a balance of advantages and disadvantages ergonomically: compactness and fewer reflecting surfaces, yet more cluttered tabular displays in heavy traffic and more elaborate but less self-evident input devices.

Air traffic control suites and consoles have to accommodate a large range of body sizes. Candidates would have to be exceptionally tall or short to be rejected because of their height, and obese people are usually rejected on medical rather than anthropometric grounds. Air traffic control suites and consoles should aim to be adjustable for 99% of the population (three standard deviations) rather than the more customary 95% (two standard deviations), using the appropriate anthropometric data for the controller population. For example, Western and oriental peoples have significant anthropometric differences, and the air traffic control suites and consoles should not have the same dimensions for both.

The surfaces of a console interact with each other so much ergonomically that it is not obvious which to specify first, but it is usually best to start from the height above the floor of the front of the upper surface of the horizontal or nearly horizontal shelf. This immediately introduces the alternative approaches, for in one it is fixed and in the other it is adjustable. If fixed it should be about 700 mm above the floor, and if adjustable it should have a range of height adjustment of more than 100 mm, centered on 700 mm. (These and subsequent data are for Western peoples.) The maximum thickness of the front of the shelf is 80 mm, and the recommended thickness is much less. The shelf can be horizontal particularly if hard copy is extensively used, or can slope by up to 10 degrees from the horizontal if it will house small displays or input devices which themselves slope. This decision about slope can have major consequences for the placing of light fitments within the suites and within the room to prevent glare or reflections. There must be space on the shelf for other functions, such as writing, or resting hands, wrists or elbows. A maximum shelf depth of 300 mm is recommended. At any greater depth, input devices not on the shelf begin to be beyond the comfortable reach of shorter controllers. Sockets to house plugs and other setting-up devices should be recessed within the shelf thickness and offset to one side rather than on the top or underneath surface of the shelf. Cables for telephones or headsets should not trail across the shelf where they can catch on or obscure other devices, but should be led through clips or grooves or be plugged in at the front of the shelf.

After shelf height, the next step is to determine eye position. Again the two approaches must not be mixed for their principles and practices are not the same. Small people will want to lower an adjustable shelf and sit lower, whereas tall people will want to raise it and sit high. The resulting large range in eye heights, which approximates to the sum of the range in seated body heights and the range in seat height adjustment, can make it more difficult to avoid all sources of glare and reflections on the displays. In the other approach, the idea is for everyone to have the same eye height above the floor. Short people are encouraged to sit high and may need a foot rest, and tall people should sit low and must have plenty of room to stretch their legs under the console. In practice, a single eye height is not normally achievable, but a range of eye heights less than the range of sitting heights is. A single eye height can provide a viewing position from which the line of sight is approximately at a right angle to every display, again more easily achieved in theory than in practice.

The recommended eye height above the floor of about 1150 mm includes a correction for slump because the controller in the normal work position leans forward so that the eye is not directly above the front edge of the shelf but nearer to the displays by between 50 and 100 mm. This eye position and the associated posture should normally be used to calculate viewing and reach distances. However, a very large main display can force controllers to sit back to view it as a whole. If they do so regularly, it is prudent to design all the visually displayed information content for a viewing distance of about 750–800 mm, even if from this viewing distance they have to stretch forward to reach some input devices. Sitting back should not force controllers to peer at the displays.

Eye position having been established, the layout and viewing angles of displays and reach distances can be settled. The controller's line of sight should be approximately perpendicular to a point slightly above the centre of the display, and the whole of frequently viewed displays should be below the controller's eye level except for very large displays. For a single display, a downward viewing angle between 10 and 15 degrees below the horizontal is considered optimum. If several air traffic control displays must be viewed frequently, looking from one to another should generally require horizontal head and eye movements which are less tiring than vertical ones. Gross differences in the viewing distances should be eschewed because they involve tiring refocusing. When the eyesight requirements have been met, the input devices can be positioned in relation to the displays and to each other, to reflect their respective operational significance.

A complication in air traffic control concerns staffing flexibility or the sharing of suites and consoles. It may be necessary to place near the middle of the suite any input devices to be shared by controllers, or to duplicate them if two controllers may need to operate them independently at the same time. Different types of input device are not interchangeable ergonomically. Some, such as a rolling ball, are much superior to others, such as a light pen, when used by the non-preferred hand.

For controllers alongside each other, the seat centre to seat centre separation should be at least 650 mm for temporary or intermittent occupation or for suites with a maximum of two occupants, and at least 750 mm for permanent occupation or for larger suites. The seating separations must permit handovers without disturbing others. Seats should move readily on castors. If they have armrests, increased separations between them may be required. The seat centre to seat centre separation influences suite design and suite width, but the reach distances are the overriding factors as small controllers must not have to stretch to reach input devices.

Certain factors can aggravate the common air traffic control problem of too many displays and input devices for the workspace. One is large boards of flight progress strips, all of which the controller must be able to reach and annotate. Another factor is the out-of-date horizontal radar displays in some older but busy air traffic control environments. The ergonomic problems in matching paper flight progress strips with a horizontal or with a vertical radar display are different, but there are no ideal solutions for either. Certain touch-sensitive surfaces that are both visual displays and input devices try to meet the ergonomic requirements of both through uneasy compromises which are optimum for neither.

A hallmark of well-designed suites and consoles is that some of their extra functions have been thought of. Job aids and instructions are not merely housed but are housed in such a way that they are readily identified and easily accessible, suitably indexed to be retrieved quickly when needed. There should be somewhere to stow headsets permanently if the policy is to leave them at the workstation rather than take them away, and somewhere to place them temporarily if the controller has to move about in the workspace. The cabling is often attached to the headset although it seems better not to wrap the cable round the headset but to detach the headset from the cabling and coil the latter in the suite. Headsets are personal equipment for most controllers, who become used to their own. This is also a hygienic preference, and the earmuffs of headsets can be equipped with hygiene covers and other personal additions.

10.6 Staffing levels

The commonly required flexibility of air traffic control staffing levels poses some problems of workspace design that seem to be unique to it, so that little relevant information can be gleaned from other contexts. The solutions of all the consequent ergonomic problems must satisfy the basic requirements of good workspace design. The envisaged range in task demands, and the extent to which jobs and tasks can be amalgamated and split to cover the range in demand, determine the range of staffing levels within the centre or tower that the workspace design must allow for. The more practical method to amalgamate or split jobs is usually to change the region of airspace for which each controller is responsible. The alternative of combining several jobs or roles so that they can all be conducted by one controller within the same workstation is more difficult to organize satisfactorily in ergonomic and operational terms.

The main practical human factors consequences of a large range in staffing levels are the following:

1. The communications channels available at each workstation must be variable and switchable so that those of several workstations in heavy traffic are all at the same workstation in light traffic, and their availability there is indicated. The utilization of each will be much diminished in the amalgamated state.
2. If the workstations to be amalgamated have different data entry functions when separated, it must be feasible to combine them in a single workstation, but this must not interfere seriously with optimizing the workstation for efficient functioning in heavy traffic, since the greatest task demands on the controller will not normally occur in the amalgamated state.
3. All the information required for the amalgamated jobs must be displayed or available for display within a single workstation whenever the jobs are combined. Implied effects can include changing the scale of a radar display to cover a larger region, expanding the number of reporting points although there will be fewer aircraft reporting to each, and ensuring that

the conventions for classifying aircraft on displays remain appropriate so that, for example, the continued ordering of aircraft by flight level or route does not impose excessive search tasks.

4. Procedures at handovers and for the transfer of control responsibility between controllers may have to change when sectors are combined or split. It must always be obvious where the control responsibility lies. Handovers normally must be completed without any break in the control service and without disturbing adjacent controllers. They may take only a few seconds in light traffic but considerably longer in heavy traffic while an incoming controller builds the traffic picture or is briefed by the busy outgoing controller. There must be no ambiguities at handover concerning how jobs have been split or amalgamated. The workspace design must provide space for efficient handovers.

5. In some air traffic control environments, the supervisor remains remote, but in others the supervision is active and even interventionist, and the workspace must make this kind of supervision possible if policies do not preclude it, and must allow the efficient rematching of supervisory and controller responsibilities whenever staffing levels change.

6. The workspace design must make provision for the assistant to deliver to the workstation paper flight progress strips prepared remotely, and to place them conveniently for the controller or put them in their correct order at the top of the flight strip board if the controller requests this. There must always be space for the new strips whatever the staffing levels are, and whatever the airspace region controlled at each workstation.

7. There must be room for both a student and an instructor at workstations where on-the-job training facilities are a requirement, or separate training suites within the workspace designed to be compatible with all other activities. Effective on-the-job training, supervision and assessment are contingent on workspace designs that make suitable provision for them, and may include the training requirements of different staffing levels. An ancillary training requirement may be for different recording facilities within the suite.

8. Staffing levels change what controllers can know about colleagues' activities. Several controllers in a single suite with a good view of each other can work closely as a team, but if a suite is staffed by one controller or if the team of controllers in the suite has responsibility for a much larger region of airspace because the traffic is light, the relationships between them and the range of team activities on which they collaborate may change substantially, and the workspace design must allow for this.

A crucial influence on the practical range of staffing levels is whether a single optimum workspace design is sought. It is inherently unlikely that a workspace design that is optimum for one staffing level will also be optimum for any other staffing level. The overriding requirement is for the job at any workstation to be safe and remain practical and reasonably efficient at every envisaged staffing level; otherwise inadequacies of workspace design will curtail the range of staffing levels. Several kinds of compromise can be made. One is to determine the most typical staffing level, to attempt to optimize the workspace design primarily for it, and to adapt this design for alternative higher or lower staffing levels so that they are feasible though not optimized. Another compromise starts from the premise that a highly efficient workspace design is most important when the traffic demands and the staffing are at their maximum, and the workspace design is optimized accordingly; the resulting procedures may be more cumbersome in light traffic, but this may be acceptable because there is more time to perform them. Perhaps the commonest and most practical solution is to concede that an optimum workspace design might not be recognizable or provable even if it was achieved, so that the aim should be a workspace that is usable at every staffing level though optimum probably for none. The fact that a single workspace design cannot be optimum for all staffing levels is not sufficiently acknowledged in air traffic control. More effort could profitably be spent on trying to discover whether the workspace designs are already nearly optimum or could be improved significantly.

10.7 Seating

The seating and the suite and console design should be specified together. The seating requirements and adjustability depend on the console profile characteristics and on the planned reach and viewing distances. Figures 16 and 17 illustrate some recommended seating adjustments in relation to two alternative but fixed console profiles. Any requirement for the controller to be able to move along the suite while seated depends on suitable seating. The achievement of eye heights within the planned range relies on adjustable seating and its manner of adjustment. A further influence on the seat adjustment is the thickness of the front of the shelf: it should be as thin as possible, preferably no more than 20 mm. The recommended minimum vertical separation between the seat and the underneath of the shelf is 200 mm, to ensure adequate thigh clearance.

The optimum seat height is about 420 mm, with at least 100 mm of adjustment if the shelf height is fixed, and more if it is also adjustable. Thigh clearance requirements set the upper limit of seat adjustment. The lower limit is set by the reach distances for input devices, and by comfort considerations. Controllers require training to adjust the seat height optimally as it is not intuitively obvious, and they should always be encouraged to adjust the seating and not to sit in the seat as they find it. Suitable chairs provide adequate back support in the normal work posture and support for the lower spine in more relaxed postures. Chairs designed for offices are often disappointing for air traffic control as they are not robust and durable enough for continuous occupation, and their padding becomes compressed and their surfaces grubby and shiny. Ten years in an office may equal two years or less in an air traffic

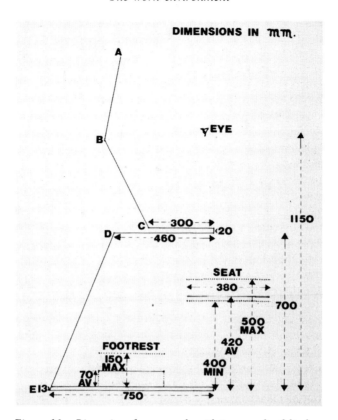

Figure 16. Dimensions for a console with two panels of displays

control environment. However, standard office chairs are cheaper and may be preferred if their seat height, back angle, arm rests and tilt can be adjusted.

The seat depth should be about 380 mm, a compromise between support for the thighs of tall people and unwanted pressure on the back of the knees of short people that forces them to sit forward without adequate back support. The seat itself should be horizontal or tilted at a maximum of five degrees, with the front higher than the back. To accommodate the longest thighs, the underneath of the shelf should extend back without obstructions for at least 460 mm wherever a controller may sit. The space at ground level should extend back at least 750 mm from the front of the shelf so that the tallest controllers can stretch their legs without encountering any part of the suite.

Chairs often have armrests. The arguments for or against them are finely balanced. Correctly adjusted armrests can add to comfort and provide support whenever controllers are not leaning forward, but armrests normally have to be quite short to clear the underneath of the shelf when the seat is high, and they must not dictate or impede seat height adjustment. Armrests can be an encumbrance when members of a control team are sharing a single workspace

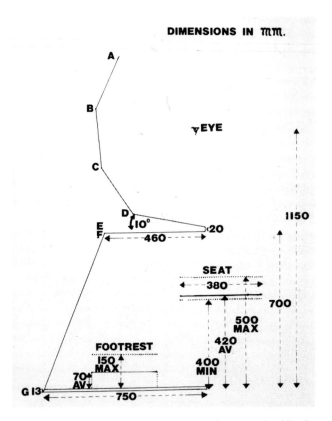

Figure 17. Dimensions for a console with three panels of displays

or during on-the-job training. The centre-to-centre separations between adjacent seats may have to be increased so that armrests do not get in the way when controllers push chairs back or rotate them at handover or for other reasons. If horizontal radar displays are still in use, the problem of sitting comfortably round them cannot normally be resolved by adjustments to the seating since it is physically impossible to sit comfortably with knees underneath a large horizontal display, and controllers have to sit partly sideways to it, thus inducing potential postural problems.

10.8 Anthropometry

There are now sufficient anthropometric data for the purposes of specifying air traffic control workspaces for most nations. Anthropometric data are used to specify preferred console profiles and ranges of adjustability to accommodate the body sizes of a defined user population. Such data determine the required range of eye positions, seat adjustments and reach distances. The specification of viewing distances is an optimum, but the specification of reach distances is not a range or average or optimum but a maximum, so that those with the shortest arms can reach all the input devices. For air traffic control workspaces, adjustability to accommodate the 99th rather than the 95th percentiles of the population is recommended, with compromises only if the more stringent requirement cannot be met.

From anthropometric data on body sizes and reach distances, the range within comfortable reach of each controller can be determined (Hopkin, 1982a), and hence the parts of a console within which any input devices to be shared by more than one controller must be placed can be specified. Anthropometric data are employed to compute the conditions under which reflections or glare will be perceived, usually by either the shortest or tallest controllers depending on the range of possible eye heights.

The anthropometric data for the whole population are often used for air traffic control, since the main anthropometric differences relate to gender and both men and women are employed as controllers. Mock-ups, simulations or other techniques are needed only to verify, as the standardized data should suffice if correctly applied. This is one topic that does not normally justify research because existing knowledge provides an adequate basis for planning and implementation (Pheasant, 1986). The resulting operational requirements should be satisfactory. Because of this, any instances of ergonomic problems because of anthropometric criteria should be foreseeable at the planning stages of the system. The commonest example is a surfeit of input devices within the recommended reach distances. This problem can usually be detected while appropriate compromises in the form of revised layouts can still be struck. The adequacy of these compromises can be verified by mock-ups and simulation or computer-aided design techniques. It is vital to avoid such anthropometric problems. Failure to take them into account can result in serious occupational health problems for a minority of controllers, which can be difficult and costly to deal with. Recent concerns with issues of cultural ergonomics have revived interest in the adequacy of anthropometric data worldwide.

10.9 The physical environment

10.9.1 Lighting

Ambient lighting has been a frequent source of human factors problems in air traffic control work environments. Controllers have described the lighting issues that are most important to them (Bashinski *et al.*, 1990). In centres and towers the lighting problems are very different.

Technological limitations of radar displays when they were first introduced into air traffic control demanded low ambient lighting levels. Extra lighting often had to be provided for other tasks such as reading hard copy, but the information on electronically generated displays could not be depicted with sufficient brightness contrast unless the lighting was low. Sometimes the spectrum of the ambient lighting was adapted to optimize the visibility of the displayed information by minimizing ambient lighting of the same wavelength as the predominant hue of the monochrome radar display. Horizontal radar displays, which are still in use in some busy centres, pose particular problems wherever reflections of light ceilings in the displays reduce the visual contrast of the information on them. Originally a plastic plaque showing the identity of each aircraft was moved across the horizontal display as each new blip appeared. This was retained for a time as a reversionary mode in the event of failure.

The problems of ambient lighting were not helped by subsequent descriptions of more advanced displays as "daylight" displays. This proved to be a misnomer, especially in air traffic control towers where these displays were unusable in sunlight. Even the normally recommended lighting levels for offices were too high for electronically generated displays until quite recently. Now there are displays which can remain usable under very high levels of illumination. Typically the ambient lighting in an air traffic control centre should be at about the minimum recommended level for offices, since there is often a requirement to read hard copy. Another influence is the need for reasonable visual uniformity within the environment, without pools of light or of darkness created by localized lighting. Variations in ambient lighting should not be large enough to induce significant changes in pupil size as the eye scans slowly round the work environment, because such changes lead to complaints of visual fatigue although the eye is robust.

It is ironic that just when advances in display technology have produced displays with adequate brightness contrast in high ambient visual lighting, air traffic control is adopting even larger displays which lack this capability, but which can accommodate a radar display and tubular displays in windows within the same screen. It must be hoped that pending technological advances can result in the combined benefits of large displays and adequate brightness contrast in high ambient lighting.

Ambient lighting is a product of the sources of luminance and the surfaces within the work environment. The specification of the spectrum, intensity and means of provision of the ambient lighting should be part of the workspace design. It is never fully satisfactory to design the workspace first and then try to light it adequately. The specification of the workspace should include the reflectance characteristics of all the surfaces. The surfaces of the furniture, walls, floor and ceiling should always be matt when new and not become shiny with use. The display filters and the coatings of display surfaces should minimize glare and reflections, whether they are from light sources, light surfaces or clothing. Rapid progress has resulted in many satisfactory displays, but new requirements, such as touch-sensitive surfaces, sometimes reintroduce familiar problems, such as reflections, glare or obtrusive finger marks on the surface. Such problems must be prevented, as it is very difficult to cure them retrospectively without major costs. Severe reflections seriously degrade information on displays, and the controller should not have to wear dark clothing to reduce reflection problems. It is a sign of poor design if the controller has to make otherwise unnecessary head movements to avoid glare from the display.

Uplighting reflected from the ceiling can often provide reasonable ambient lighting levels for the whole workspace of an air traffic control centre. It can result in uniformity of lighting, freedom from shadows and an easily controlled and specified visual environment. It works best in high rooms where the ceiling is less likely to be reflected in the controllers' displays. Downlighting is more difficult to control evenly but it may be necessary if the ceiling has to be dark to avoid reflections. The spectrum and intensity of the lighting within the air traffic control environment should not differ grossly from the lighting in its immediate approaches such as corridors.

Normally the lighting in air traffic control environments should be nearly white. Slightly bluish ambient light can suggest coolness and slightly pinkish light can suggest warmth, but this should not become a palliative to avoid adjusting the temperature. The use of colour in air traffic control environments for coding and structuring information is becoming more widespread, and clearly will be the norm in a few years' time. The spectrum of the ambient lighting must not distort the appearance of colour codings unless it has been deliberately planned to do so. The lighting environment must remain satisfactory for the whole range of operating conditions. Glare and reflections must not become a problem if all positions are manned because the traffic is heavy, or if few positions are manned and those on duty have to view some displays at relatively unusual angles. The lighting of each workstation should be from diffuse or cowled sources so that no light source is directly visible from the normal working position.

10.9.2 Acoustics

Air traffic control towers must have good sound insulation to ensure that aircraft noise does not degrade speech intelligibility. Temporary air traffic control environments for events such as air displays can suffer severely from the noise of departing and arriving aircraft, although the noise can act as feedback that an unseen aircraft has in fact departed or arrived.

Controllers talk to colleagues in the same environment, to controllers in other environments, and to pilots. Speech may be direct, or transmitted through RT channels, telephones or loudspeakers. The controller wears a headset and hears messages through it, but may also have to use a telephone and hear what colleagues in the same room are saying. There is, therefore a problem of striking the correct balance between various sources of spoken messages, including loudspeakers and normal speech in the environment, messages through the RT and messages through the telephone. Headsets are designed in relation to the acoustic properties of the RT channels to which the user has access, and are intended to extenuate external sounds significantly, including speech. A controller may listen to the RT through one ear but offset the headset in order to hear speech from other sources in the other ear.

If the noise levels in an air traffic environment start to rise, the consequences are cumulative as everyone has to speak louder to be heard. Low background noise levels are essential for efficiency, ideally as low as about 55 dB. Extraneous sources of continuous noise, such as fans and pressurized ducting to remove heat generated within consoles, have to be suppressed. All fans in control rooms should be large and slow to keep the noise level low.

Sometimes speech on RT channels, telephones and even within the room is quite difficult to hear, but contains vital information. It should always be spoken against the minimum achievable background noise. Air traffic control environments should deaden noise. Sound-absorbent plasters, acoustic tiles, silently operating doors, silent compressed air attachments to doors, carpeting with good-quality underlays on floors, sound-absorbing curtaining over acoustically reflecting surfaces and all other practical expedients should be adopted to reduce noise levels. A noisy environment has no advantages at all. Everyone working in a quiet environment can talk quietly, and the distraction of raised voices is avoided. Extraneous background sounds are not transmitted through the RT or telephones. Speech that cannot be amplified much, for example direct speech or speech over the telephone, never has to be loud.

10.9.3 Heating, ventilation and airflow

The recommended thermal environment in air traffic control is in accord with that for other sedentary occupations. Usually clothing is quite light. In some countries the tradition is to work in shirt sleeves whereas in others a sweater or jacket is customary. The thermal environment should be controllable. The temperature should be about 21°C in European environments and slightly higher in the shirt-sleeved environments common in the United States. The heat generated by equipment should be ducted away. If direct sunlight can enter the workspace, it may be necessary to use blinds, not only to prevent glare and other lighting problems but also to maintain comfortable temperatures. Various expedients to reduce noise in towers may also provide extra insulation, and render the thermal environment more controllable.

Thermal comfort depends not only on temperature but on humidity. The preferred humidity is around 40%. High levels above 70% seem hot and stuffy, and low levels lead to dryness in the throat and coughing, particularly if people have to speak a lot.

The third main determinant of thermal comfort is airflow. If the temperature and humidity are satisfactory, a flow rate of about 10 metres per minute is recommended. Faster air movements can be pleasant if the air is warm or humid, but can be felt as a draught in cold air. The aim is to maintain a satisfactory and unobtrusive thermal environment by combining recommended temperatures, humidities and airflow rates. Any need for localized fans suggests inadequate planning of the airflow. Properly planned air intakes and air movements must take account of the positions occupied by controllers in relation to the furniture, and their combined effects on the airflow. The designated airflow is not that for an empty room, but for the workspace with all the furniture installed and all the workstations operational.

10.9.4 Radiation

Controllers sometimes express concern about possibly harmful radiation emissions from the visual display units that they use. This problem is real because people worry about it rather than because of actual radiation hazards. National and international standards based on medical evidence set low permissible maximum radiation emissions from marketed displays, and the compliance of air traffic control displays with these standards is checked independently. Specialist equipment is needed to detect such low levels, and several radiation bands are checked. The extensive literature on radiation from displays is applied.

A recent anxiety has been possible effects of radiation on pregnant women. Air traffic control regulations comply with any general recommendations based on medical evidence, since this issue is not specific to air traffic control. However, some women might experience difficulties in air traffic control environments during the later stages of pregnancy if they have to stretch to use input devices which at other times are comfortably within reach. It is therefore important to discover the true origins of any problems that do arise, as remedies for misdiagnosed problems will be ineffective.

In general, radiation from displays is not a significant problem for air traffic controllers. Any problems are more likely to affect maintenance staff than controllers, and the remedy is always to comply with safety procedures for maintenance. Many studies have found higher radiation levels in home and leisure environments than in air traffic control workspaces. In air traffic control the levels are checked with a thoroughness seldom applied elsewhere.

10.9.5 Decor

First impressions of air traffic control workspaces are important. Decor contributes substantially to them. Control rooms should have pleasant visual proportions, and be high enough to house comfortably any wall-mounted displays, to permit reasonably uniform reflected lighting, and to avoid ceiling reflections in the displays. The workspace should convey an impression of spaciousness and absence of clutter. It should be and look clean. Even where the air traffic control is of an exceptional standard, the opposite initial impression may be conveyed if the environment looks a mess. A dedicated and committed workforce may rise above disarray, cramped workspaces, poor accessibility and ugly decor, but these indicate that the quality of the work is taken too much for granted and not appreciated, and lead to complaints and disillusionment. A poorly designed and unattractive work environment does not encourage its occupants to be tidy and orderly, whereas a neat and attractive environment can encourage everyone to keep it that way. To some, tidy environments and workspaces are signs of professionalism.

Decor can promote several aims. To some extent it can influence the visual impression of the room dimensions, particularly its height in relation to its breadth and width. Its apparent height can be slightly exaggerated by decor that emphasizes vertical room features rather than horizontal ones. Impressions of size can be altered by choosing plain, small textured or larger textured carpeting, or by carpet tiles in different colours that blend to form a discriminable but not obtrusive pattern. In general, all surfaces should be light and there should be no saturated colours in the decor at all, except for a few small features of operational significance. The visual environment should be designed as a whole, and not give the impression that different features of it have been designed independently. The colours and textures of everything in it, floor, walls, ceilings and furniture, should blend with each other and with the displays, and neutral or pastel colours are more suitable for air traffic control decors. The colours of the furniture and of any equipment items housed within it should never contrast grossly. Design consultants experienced in domestic or other industrial environments do not always possess appropriate expertise to specify the decor for air traffic control environments. They may try to blend the colours on the displays with the decor, but the opposite objective may apply to air traffic control. For example, if highly saturated green displays tend to induce slightly pink after-images, almost any colour is preferable to green for the decor. Some design requirements in relation to decor may therefore be quite specific to air traffic control.

10.10 *Standardized ergonomic data applied to air traffic control*

The ergonomic data from standard sources can often be applied to air traffic control, suitably modified to accommodate air traffic control conditions or requirements (Helander, 1991; Shneiderman, 1992). The data cover effects on health and well-being as well as effects on safety, efficiency and performance (Evans *et al.*, 1994). These sources can serve as a checklist so that no significant ergonomic factor is overlooked (Boff and Lincoln, 1988). Specific features of air traffic control generate numerous additional ergonomic considerations, including those now listed.

1. Handbook data on viewing distances and legibility are normally for viewers with twenty-twenty vision, corrected if necessary. The minimum visual requirement for controllers is often less stringent. The sizes of all display symbology, including alphanumeric characters, may therefore have to exceed the handbook minima.

2. Handbook data are normally for a viewing distance of 450–500 mm. In some air traffic control environments it can be 750 mm, and 900 mm is not unknown. Recommendations have to be adjusted for these larger viewing distances.

3. Normally the recommended line of sight is about at a right angle to the display, and it is for most controllers using their own display. However, if they have to consult colleagues' displays for any reason, their viewing angles may be up to 45 degrees or even more, and the displayed information must be readable from this angle.

4. Depending on the design principles followed, eye heights may be nearly uniform or extend over a large range. The consequences may require modifications of the standard data, either to assist viewing or to prevent glare or reflections.

5. The lighting environment may have to be specified precisely, particularly if the lighting levels on the radar must be much lower than those on the shelf to preserve its visibility.

6. In some air traffic control environments, highly saturated monochrome displays have tended to induce enduring after-images in the complementary hue, and it is prudent to avoid highly saturated colours and to choose other colours for the decor.

7. Many air traffic control jobs have to be performed satisfactorily by a single controller at a workstation, by one controller using two or more adjacent workstations, or by two or more controllers sharing a suite. Major questions arise about facilities that should be duplicated within the suite or must be placed centrally because they may have to be shared. Standard recommendations affecting the layout of the whole suite may have to be compromised so that it can function satisfactorily at every envisaged staffing level.

8. The ergonomic design of the workstations and suites must allow ancillary functions such as supervision, assessment and on-the-job training to be fulfilled, even with maximum staffing of the suite.

9. The design of the workstation should facilitate the handover of responsibility for it. All relevant information should be apparent to an incoming controller who can build a picture of the traffic without interrupting the working controller, and there must be room for a smooth transfer of responsibility without disturbing other adjacent controllers.

10. Air traffic control suites may be in an operational environment which never closes. It may be necessary to design the workspace so that some maintenance, cleaning and hygiene tasks can be done while it remains operational.

11. All air traffic control environments should be quiet, with sound-absorbent materials for the walls, floor, ceiling, furniture and equipment, so that telephones, RT and other speech can be heard clearly by controllers and pilots against a quiet background, with no need for voices to be raised.

12. Air traffic control equipment must be robust if it could be in continuous

or extensive use. This applies in particular to chairs. They require the standard ergonomic adjustments for seat height, tilt, back support and armrests, but in addition must be resilient enough to withstand continuous occupation.

13. Air traffic control centres require more stringently specified visual environments, in terms of the spectrum and intensity of ambient lighting and the prevention of glare and reflections, than most other environments such as offices.

14. In an air traffic control tower which must remain fully operational in extremes of direct sunlight or exterior darkness, the ambient lighting has to be more flexible and adaptable than in most workspaces.

15. Suites are positioned in the workspace to allow access for maintenance, handovers, supervisors, and assistants delivering flight progress strips. In towers, the controller's line of sight of aprons, runway approaches, runways and taxiways must not be impeded by any other console or controller, or by the tower structure.

16. Decor suitable for air traffic control may be unsuitable for other work environments and some specific standard ergonomic recommendations about decor may not apply, but the underlying principles of good decor should be followed.

This considerable list generally consists of modifications rather than new ergonomic principles. A human factors specialist with experience of the particular requirements of air traffic control should know when to apply standard ergonomic recommendations unmodified, when they can be followed subject to certain modifications, and when circumstances specific to air traffic control and not commonly encountered elsewhere render them invalid.

11

Air traffic control displays

11.1 Location within the workstation

The general principles of workspace design, and the specification and layout of suites and consoles, should determine broadly where various air traffic control displays are located within the workspace. In this section, the main influences on the arrangement of the displays within the workstation and in relation to each other are considered, with particular regard to their effects on the conduct of air traffic control.

The main influences on the positioning of displays are the following:

1. The identification, by task and job analyses or other techniques, of all the essential information at each workstation for all the tasks performed there.
2. The application of ergonomic evidence and human factors principles to specify which essential information should be presented visually on displays.
3. The further application of this evidence and these principles to the relationships between tasks, to determine whether the information is sufficiently specific to the tasks of individual controllers to be presented at each workstation or sufficiently general to appear on wall-mounted displays that can be seen from many workstations.
4. The identification of visual information that is common to several workstations and can be the basis of communications between them.
5. The identification of visual information that has to be divided, synthesized or otherwise manipulated whenever jobs are split or amalgamated to achieve flexible staffing.
6. Line-of-sight constraints, the main ones being wall-mounted displays beyond the workstation, a view of runways and taxiways unimpeded by displays within the tower, observability of others' displays for such purposes as supervision and training, and suitable provision for visitors and others entitled to be in the workspace to see the displays without affecting the work.

7. The provision of lighting of appropriate spectrum and intensity for the room and for each display without any perceived glare or reflections from any viewing position, which normally requires the lighting and the display locations to be specified concurrently and in conjunction.
8. Consideration of other attributes of the displays or the workstation, such as whether the main radar display is horizontal or vertical and whether there are electronic flight progress strips or paper strips in holders on flight strip boards.
9. The dimensions of each workstation, especially its height and width.
10. The intended configurations of workstations, their grouping into suites, their sharing, and the required staffing flexibility, particularly if this is associated with pivoting displays in the suite according to the staffing needs, since pivoting a display to aid one controller can impede the view of it by others.
11. The decision on whether, during low staffing or at other times, the controller should remain at a single position within the suite or should be able to move along the suite.

Broad layouts of the displays within suites can evolve and be evaluated with respect to the above influences, but progress remains limited until further characteristics of the displays have been agreed.

11.2 Physical characteristics

Two main physical dimensions of a display that have human factors implications are its size and its shape. Although the size and shape of the displays should obviously be sufficient to present all the information required at any given moment, attempts to follow this simple precept are frequently thwarted in air traffic control because more information is needed concurrently than can be portrayed in full accordance with ergonomic recommendations. Technical advances have made larger screens available. The optimum visual diameter of a display with a viewing distance of about 500 mm is probably no more than 350 mm. Much larger displays have introduced complications and further options. One is to replace two or three displays with a single large one within which different kinds of information are treated as windows or tiles. Although the information may be more compact on the large display, part of this benefit is lost if the controller habitually has to sit further from it to view it as an entity, especially if the size of alphanumerics and symbols has to be increased to compensate for the greater viewing distance. This problem is compounded if the size of coded information has to be further increased to retain its visibility because the largest displays provide lower maximum intensity and hence lower brightness contrast, and if the ambient lighting also has to be reduced to maintain display readability. Although some advances in display technology allow the ambient lighting of air traffic control centres to approach the levels in offices, some of the most recent advances have resulted in larger displays that are so dim as to reintroduce ambient lighting constraints that seemed to have been removed at last.

The largest displays have the potential advantages of increased flexibility and a closer association between the tabular information about each aircraft and its displayed geographical location. Tabular information can be moved around and repositioned within the display if it threatens to obscure other information. The largest displays have the disadvantage of being more difficult to use in association with other displays, because only rarely used displays may be placed above them and any displays placed alongside them must be viewed by the controller at a considerable angle. Therefore some of the potential advantages of large displays are lost unless all the information required for tasks can appear on them without other supplementary displays. A mixture of display sizes thus poses ergonomic problems, and it can also be difficult to specify ambient lighting that is satisfactory for all of them.

The physical dimensions of displays must reflect the planned capacity of the system. A flight progress strip board must hold all the strips for the maximum planned traffic-handling capacity. A system functioning at maximum capacity has problems enough, without adding an inability to display all the essential information. Electronic flight progress strip displays seem particularly prone to this problem, to the extent that condensed flight strips, known as mini strips and containing less information, may replace the normal strips in heavy traffic. An associated problem, not yet completely solved, is the need to expand the whole flight strip or part of it while amending it, in order to facilitate the positioning of the marker and see the changes well enough to verify them.

Physical display dimensions influence more factors than are immediately apparent. A common problem in air traffic control is that the information needed for all the tasks at a workstation cannot all be presented without violating some recommended viewing angles, viewing distances, codings or contrasts for displayed information. Advance signs of such mismatches include displays with insufficient lines of data or characters per line, coding difficulties or overlapping because too much information has been crammed onto displays, or characters that are too small for the viewing distances. Variables that could be changed might include the number of displays, their sizes, the formatting of information, its permanence, its level of detail, feasible more compact recodings, task reallocation, size reductions through improved legibility or reduced viewing distances, and revised minimum eyesight standards for controllers. Not all of these theoretically feasible options are practical, and many interact with each other and must be considered together and not in isolation. The severity of the problem influences the solutions that are considered seriously.

Some jobs pose particular display dimension and layout problems. A few current directing tasks employ two radar displays. Some future tasks envisage tabular information, current radar information and predicted information on three different displays. Controllers often prefer larger displays because they believe they can obtain a clearer view by looking closely at them. An advantage of larger displays is that they may show more pending information and thus reveal more of the air traffic control context of the controller's actions. A less favourable implication is that a large display may give the false impression that the location information on it has become more accurate. It is vital in the interests of safety that advances in display technology, particularly in clarity, are not misinterpreted as enhanced accuracy. This problem may be transient, as future advances in navigational sensing will genuinely improve accuracy, but many current displays provide no direct information whatsoever about the accuracy of the information depicted on them, with the risk that positional accuracy is perceived as an attribute of display size.

In human factors terms, the respective benefits and disadvantages of larger displays are thus quite finely balanced. Among their potential benefits are their technical availability, their popularity among controllers, their flexibility, the need for fewer displays, the greater area in which to manipulate and lay out information, reduced clutter and label overlap, and more direct relationships between different kinds of data such as positional and tabular information. Among their potential disadvantages are greater viewing distances, reduced maximum intensity and poorer brightness contrast, spurious impressions of enhanced locational accuracy, greater weight which has to be housed in stronger and heavier frames, higher initial costs, and more costly and elaborate means to adjust the angles and positioning of the display surfaces. Because larger displays are often darker they may be more difficult to light satisfactorily, and because of their size it may be more difficult to see over the top of them. Time will tell how widely adopted very large displays become in air traffic control workstations.

Coatings, filters, and surface textures are further significant display characteristics. At any display exhibition it is very apparent that some displays reflect clearly and distractingly both the viewer and the light sources within the room, whereas others do not. The surface texture should not be so smooth that it shines and reflects but should scatter the reflected light by very fine grain irregularities. Suitable coatings of display surfaces can also curb problems of reflectance and help to prevent glare. Filters may confine the display emission to a narrow part of the spectrum, and lead to high saturation of a monochrome hue if little light from other parts of the visual spectrum is mixed with it. Such technical virtuosity is a human factors bane, inviting persistent after-images in the complementary colour. The highest colour saturations have no significant human factors benefits, and filters should not aim specifically to achieve them.

11.3 The relative positioning of displays

The main objectives that can be furthered by the optimum relative positioning of displays within the workspace and workstation are listed below. Sources of the evidence utilized in attaining these objectives include task and job analyses and the procedures and instructions followed by controllers. Modifications of these, in addition to ergonomic adjustments of display positions, represent the most practical means to effect positional changes since they provide the rationale for any changes, the explanation for their effectiveness, and insight into the limiting constraints on them.

Among the objectives are to minimize the need to obtain essential information during the performance of a single task by cross-referral, whether between displays within the same workstation, between a workstation display and a wall-mounted one, between displays at sufficiently different viewing distances as to entail refocusing or reaccommodation, between displays presenting different information that has to be collated, between displays with gross differences in luminous flux, between displays that differ greatly in readability or legibility, or between displays that impose significant head and eye movements particularly in the vertical plane (Stein, 1992). Further objectives in positioning displays are to ensure that the display layouts facilitate task performance by presenting the requisite information in a logical order related to the sequence of its usage, that they facilitate smooth transitions between information sources, that adjacent displays have compatible ambient lighting requirements, that the relationships between displays and input devices are satisfactory, and that any unusual ergonomic requirements of particular displays are met (for example, that both the reach distances and the viewing distances of touch-sensitive surfaces are acceptable). Another set of objectives includes positioning displays to permit the splitting and amalgamation of jobs and the sharing of work between team members where this is applicable, to allow the displays to be properly physically supported within the furniture, to meet all requirements for accessibility and maintenance, and to contribute towards an attractive and unitary workspace design.

A further influence is the move towards greater standardization of the air traffic control furniture and the location of displays within it, culminating in workstations that all look alike when switched off and fulfil their different air traffic control functions through software rather than hardware differences. Thus some display configurations that are satisfactory ergonomically may be discouraged if they are non-standard. The extent to which differences among tasks and jobs can be accommodated by software rather than hardware has already influenced display positioning significantly, and this influence is expected to increase in the future.

All displays that have to be consulted frequently should be positioned so that they can be viewed comfortably by the whole population of controllers in the normal work position at recommended viewing angles and distances. Displays positioned higher in the furniture should never contain information that must often be referred to in the course of normal duties. If supervisors or others must be able to view displays, they should never disrupt controllers' lines of sight of essential information while doing so. The relative positioning of displays should reflect their operational significance, with the most important information directly in front of controllers under normal conditions, although some modifications may have to be introduced for non-standard staffing levels or non-standard conditions of use such as en suite training.

Some traditional air traffic control practices affect the positioning of displays. New flight progress strips are added to the top of strip boards and old strips discarded from the bottom. Redesigned flight progress strip boards and electronic replacements for them may perpetuate this well-established principle, although the latter provide an opportunity to re-examine its virtues. A major constraint on air traffic control workstation design has always been the need to tip the angle of the flight strip board back so that all the strips do not tumble out whenever a strip is discarded from the bottom of the heap. Because the radar provides a plan view, this limits the flexibility of the associated display layouts, particularly where windowing or other stratagems attempt to link tabular and plan information. Cross-coding problems between plan and tabular information are intrinsic to air traffic control. Both types of information are essential, but the layout constraints of each tend to rule out any compromise that is optimum for both.

It is confusing for users if more than one layout principle is followed when positioning displays. The layout may have to promote other objectives such as distinguishing between predicted and actual information, and the layout principles should follow a single common logic to the extent that if a further display has to be added to those already present it should be clear from that logic where it should be positioned.

11.4 Information content

The available information content of air traffic control displays must cover all the visual information required for every task at each workstation, in forms matched to controller needs. More detailed characteristics of particular codings are discussed in the next section; some broader issues are addressed here.

Originally in air traffic control the visual information that could be made available was limited. Its main forms were paper flight progress strips, visual labels for facilities such as communications channels, clocks, and maps including weather maps. Although the strips themselves were visual, their contents were updated mostly from auditory information obtained through speech with pilots or other controllers. The advent of radar provided a plan view which has been successively elaborated and processed to its current state of synthesized information based on secondary radar and supported by labels containing alphanumerics and symbols that provide fuller details about each aircraft and its flight. The information content can be extended to present the results of various computations and other forms of assistance, such as conflict alerts and predictions of future traffic scenarios. Electronic displays of tabular information may replace paper flight progress strips but substantially retain their information content. Various tabular or graphical information displays may be added separately, or via manipulable windows in large displays.

The main information content on most air traffic control displays still presents aspects of the current air traffic scenario, but often expanded or supplemented by predictive information, planning information, warning information, directing information, background information about prevailing conditions, evolving information such as weather, or more detailed information about specific aircraft. There is usually too much information to present all of it, and therefore the display contents may have to include some guidance on what else is available for access and on items that can be selected or are mutually exclusive.

Despite these developments, one kind of information is conspicuously absent from all air traffic control displays. Typically the displays contain no information at all about what they are, about the nature or purpose of the portrayed information, about the tasks for which it is intended, or about their timing or sequencing. In other words, there is nothing self-evident about air traffic control display contents, which are well nigh meaningless to the naive. The contents are designed for trained users with professional knowledge of air traffic control who must rely on their training and knowledge to make sense of the information content and to use it as intended. The information content provides what they need but is not instructionally supportive. This point is of crucial importance. The meaning of the contents of air traffic control displays is not contained within the displays themselves, but is added by the user. Controllers with different knowledge, training and experience may not interpret the same information contents in exactly the same way. If system changes have been introduced that alter the information content of displays, it is not normally sufficient to let the information content change without ensuring that corresponding changes are also made in the controller's knowledge and training so that the meaning of the changes is known. The information content of almost all air traffic control visual displays is quantitative. The display itself contains little or no information about the sources of the information, its validity and reliability, how large a change must be or how frequently it must occur before the display is updated, or how far the displayed information should be trusted. It also contains little information about failure modes, about the signs of failure, or about the extent and ramifications of any failures that may occur.

The required information content of air traffic control displays should ideally be derived from comprehensive job descriptions and task analysis. The intention is not just to identify all the information that must be provided but also to show which information is independent, which requires other information for its correct interpretation, when the sequence of presentation is important, and what level of detail is most appropriate for the tasks and their timescales. Because the amount of information available far exceeds the task needs and the presentation capabilities, various principles of selectivity have to be applied. One considers permanent display versus temporary display versus display on demand, allowing for any additional tasks incurred if the information content has to be manipulated. Task analysis can suggest how important various kinds of information are, how pervasive their usage in different tasks is, and therefore how visually prominent each kind of information should be. There are many different air traffic control tasks requiring different combinations of information content. Any given kind of information is likely to be used in conjunction with several other different kinds of information to serve different purposes in different tasks. Some display requirements are nearly universal, such as a capability to find required information quickly without undue searching, the prevention of visual clutter, and the need for visual stability. A balance has to be struck between selectivity of information content in order to avoid clutter and excessive detail, and controller preferences for all the available and relevant evidence to be visible, or at least very easily accessed.

The information content of displays should match human capabilities and limitations. It should help rather than hinder controllers' learning and understanding. To promote learning, the consequences of actions need to be clearly apparent. Understanding is a product of the match between the information content and the controller's knowledge and training. Display contents should map directly onto the controller's picture of the air traffic, and be immediately compatible with it. In practice this implies that the same information codings should be employed for the displays and the mental picture, so that displayed information does not have to be recoded by the controller before its meaning can be interpreted correctly and understood. It is necessary to be able to demonstrate that the meaning of the information content is clear and unambiguous and also that the content is teachable.

A contentious issue is how much the controller needs to know about the state and functioning of the system, as distinct from the air traffic. At the simplest level, some information about the serviceability of the displays must obviously be included in the display contents. A display must not look the same whether it is functioning correctly or not, but the controller is less concerned with the origins of failures or faults than with their consequences. It is essential to know what remains functional and can still be used. A good display should therefore indicate what can still be relied on despite impaired system functionality. Few display contents include this kind of information.

Air traffic control is a continuous process. The information content must therefore be capable of showing continuous changes. It must indicate states and changes of state, and distinguish between changing and unchanging states. It must show relationships between data, particularly where certain kinds of information are partly related, but not entirely or exactly related, as when height information on radar and on tabular displays does not quite agree because the data have been derived from different sources.

Changes in air traffic control requirements such as the preference for strategic over tactical air traffic control, and advances in the technology available to make these changes, have major implications for information content, not only because they seem to imply different tasks with different information requirements but also because they may make some of the most cherished traditional forms of air traffic control information redundant or inappropriate. It can be very difficult to convince experienced users that information which has always been essential in the past will never be needed in the future. Some new kinds of information are perceived as helpful and are welcomed, but it is still necessary to check that they could not mislead or be misinterpreted. The categorization of different kinds of information must not become a source of potential confusion between them.

11.5 Visual information coding

11.5.1 Dimensions

Visual information can use one, two, three or four dimensions, depending on whether it takes the form of points, lines, areas, volumes, or movements in space over time. A point can denote location, and several points can define relative location but a context requires linear or area information in addition to points. Locations on radars can be treated as point information in some respects, although in practice other kinds of information have to be added to make it meaningful.

Lines are used quite extensively for information coding in air traffic control and to segment information categories. They can depict boundaries, linear features such as coastlines, and distances by range rings. Pecked or calibrated lines, which thereby become two-dimensional, can form an electronic ruler or scale to assess distances or separations, for example on final approach. Trailing lines behind points or a row of points may depict track history. Projected lines ahead of points may show predicted routes. It is vital to code historical and predictive information very differently so that they can never be confused with each other. Linear and other codings are often combined to provide continuous or segmented lines, thick or thin lines, or lines of different colour.

Areas can be depicted by lines to show their boundaries, or by washes or infills with or without boundary lines to show their extent. Air traffic control area information includes airways or air corridors, control zones, danger zones, regions of restricted flying, and regions affected by particular weather conditions. Different kinds of area information cannot normally be superimposed without causing confusion or misinterpretation, although sometimes principles such as visual transparency may circumvent this coding limitation.

Volumes can actually be three-dimensional or be attempts to depict three dimensions on a two-dimensional surface. Since aircraft fly in three-dimensional space, volume information can seem ideal for air traffic control. In practice the numerous attempts to match real three-dimensional air traffic control displays with human visual requirements have all failed. A two-dimensional plan view of the traffic with flight level information shown digitally and not pictorially has always been more successful. Attempts to depict three dimensions by two dimensions often founder because of problems in the depiction of perspective, but nevertheless these attempts continue.

The depiction of moving traffic in real three-dimensional space by electronic means is not currently being widely pursued, although the idea has not been abandoned. It has tended to be superseded by two kinds of technical development. One is the projection of aircraft onto models of airfields, which has led to very convincing simulated depictions of aircraft traffic as seen from a tower that can be used as a high-fidelity tool for controller training. The other is the emergence of virtual reality, which clearly has considerable unrealized potential for the future representation of air traffic in four dimensions.

11.5.2 Shape and alphanumerics

There is an extensive psychological literature describing perceptual experiments in which psychophysical parameters of shapes are systematically varied. Most of this work sought to build theories of form or pattern perception and yielded surprisingly little data applicable to shape as a visual information coding. In retrospect, the excessive theoretical partitioning of the subject matter seems unproductive, though it lingers on. Studies which apparently have much in common remain unrelated because they share no constructs or theories, and the will to collaborate is lacking. Examples abound. The vast database on the Rorschach projective test has contributed little to our psychophysical knowledge about shape and pattern perception. Studies of discrimination in order to build or test psychophysical or neural theories have often employed variables of no practical value, used animals as subjects, and increased the incidence of errors of discrimination near the visual threshold through laboratory artifices such as very transient or dim or meaningless stimuli, which undermine both the practical utility of the findings and the generalizability of the theory to real-life conditions. The ignorance of relevant work in other disciplines is so widespread as to seem wilful: for example, ergonomic recommendations pay scant heed to the cartographic literature on alphanumeric coding (Hopkin and Taylor, 1979), or to the typographical literature on the legibility of typefaces and alphanumerics. These other disciplines have been equally culpable. Some of the earlier alphanumerics generated electronically or by dot matrix printers were so poor that some users tended to shun shape and alphanumerics as efficacious codings. Fortunately technological advances have brought improvements.

Shapes have several parameters. They may be open, such as a cross, or closed, such as a circle. The latter may be hollow or filled. They may or may not have a familiar meaning and name. Most shapes, even familiar ones, do not have any self-evident air traffic control meaning, and therefore any meaning assigned to them has to be learned: an exception is an upward or downward arrow on tabular displays to denote that an aircraft is climbing or descending. Shape coding should generally be used quite sparingly in air traffic control.

Recently shape codings have been developed in the form of icons that denote states, instructions or information categories within tabular displays and human–machine interfaces (Garcia *et al.*, 1994). Because their meanings must be learned and each must be easily identifiable, icons should not be allowed to proliferate but be employed conservatively. Shape coding and colour coding share a major source of ambiguity for neither shapes nor colours can usually be combined successfully by superimposition. Just as colours when superimposed may form a different colour, the constituent colours of which cannot be recognized, so superimposed shapes may form a new shape which conceals its constituent parts. It is tempting to try to represent different states or progressions through an air traffic control procedure by a series of shapes or by changing icons. It is seldom possible to assign one shape to one state, another to another, and to represent both states by superimposing the shapes, and this should never be attempted without independent confirmation of the principle. Almost any coding alternatives are likely to be better than the superimposition of shapes to denote combined or coexisting states or conditions.

Sets of shapes or symbols are chosen to maximize their visual discriminability, but not solely for that reason. Often it is essential to be able to refer to a shape verbally in messages to others. This confines the choice of shapes to those with universally known names. Shapes which are highly discriminable from others visually but lack any obvious name may therefore be of no practical use. Naming is not synonymous with the imposition of meaning. The latter may be conveyed through pictorial resemblance to a real object or by other means, whereas the name, such as a square or circle, refers to the shape itself.

Alphanumerics can be treated as shapes with preassigned familiar names and meanings. In air traffic control, their shapes must be visually discriminable and their names audibly discriminable. Collectively, a set of alphanumeric characters may acquire a unique meaning as an entity, such as the group of letters and numerals that constitute the callsign of an aircraft and thus denote its identity. The standard recommendations for discriminability apply to shapes and alphanumerics used in air traffic control as labels, on radar displays or on tabular information displays. For viewing distances around 450 mm, the minimum character height should be about 3 mm, minimum width about 2 mm, minimum horizontal separation between characters about 1 mm, and minimum vertical separation between rows of characters about 50% of character height. Visual design features that can help to differentiate between characters most likely to be confused with each other, such as S and 5 and 8, capital I and 1, and O and zero, should be exaggerated, or tasks and displays should be designed to exclude their possible occurrence at the same time in the same place. On the same principle, auditory differences are exaggerated by distortions of pronunciation, for example of "five" and "nine".

Modern sets of alphanumerics now generally attain high standards of legibility, and none that fails to do so should ever be adopted in air traffic control. Nevertheless, different alphanumeric fonts have not been used for different categories of air traffic control information, although there could be advantages, for example, in tabulating headings and heights in different and distinctive fonts. The legibility of characters also depends on other factors including stroke thickness, brightness contrast, incremental changes, ambient lighting and eyesight standards, and on the phosphors and pixels in electronic displays. Although the discriminability of shapes may be increased by slightly exaggerating their most salient features, this must not result in gawkiness or caricature. However, the salient features of small characters can often be made more prominent in the interests of discrimination without impairing their visual appearance and attractiveness significantly if this is done circumspectly. The cross-referencing of air traffic control information in different media may be enhanced by employing several fonts or typefaces and using each of them in the same way on all electronic displays and hard copy. In this sense, alphanumeric fonts may be treated as a separate coding dimension, suitable for sub-categorization within other dimensions such as intensity, size or colour.

11.5.3 Colour and monochrome

The human factors implications of colour coding on electronic displays constitute a live issue in air traffic control and other applications (Hopkin, 1991b). Several textbooks provide evidence and offer guidelines (Durrett, 1987; Travis, 1991; Widdel *et al.*, 1991; Van Laar and Flavell, 1993; Jackson *et al.*, 1994). Colour has been a useful coding in air traffic control for a long time. When the controller's primary tool was paper flight progress strips in holders on a board, the strips were coloured differently, most commonly to differentiate between inbound, outbound and crossing aircraft. Automated flight strip printing resulted in adaptation of the principle so that the strip holders rather than the strips themselves were colour coded. Radar displays remained monochromatic throughout the technological advances of secondary radar, synthesized displays, alphanumeric labelling, and brighter displays in lighter rooms. Safe, efficient and practical monochromatic codings for all the information categories on radar displays that were vital for air traffic control tasks evolved. When reliable colour coding became available within budgets, strenuous efforts were made to use it on electronically generated air traffic control displays, for three distinct reasons. One was simply its availability, which meant that the proposed timing of its introduction depended on technology rather than on air traffic control needs or human factors requirements. Another reason was that it seemed self-evident that such a powerful and useful coding must benefit efficiency and safety. The third reason was controllers' strongly expressed preferences for colour coding because it could be visually attractive, seemed efficacious, and symbolized advanced and up-to-date equipment.

There have been dozens of proposed applications of colour coding on air traffic control displays. In retrospect, many were so garish as to belie its claimed attractiveness. Slowly the evidence accumulated in air traffic control, as in many other contexts, that colour coding did not always improve performance and that any benefits were often smaller than expected and accompanied by unpredicted disadvantages, despite the universal and adamant contentions of users that colour did indeed improve their performance. Colour aided search tasks only if the colour of the sought item was already known. Colour did not always resolve clutter. Comparisons within the same colour might be aided, whereas cross-colour comparisons were disadvantaged. Perhaps the main benefits were on memory or understanding or data structuring or other attributes not adequately covered by direct measures of performance, so that the subjective impressions were correct but unsupported by objective data because the measurements were inappropriate or insensitive. Perhaps the objective measures were correct, and the attractiveness of colour beguiled its users into believing that their performance was better than it was. The problem was not that colour failed as a coding, for performance with colour was typically similar to that with monochrome, but the benefits of colour had to be more tangible than subjective preferences to justify the extra costs of providing it. Any direct comparisons were generally between monochromatic codings which refinement by long usage had rendered near optimum for air traffic control, and colour codings which were undeveloped technically and unoptimized for air traffic control applications. Because colour is a dominant coding that is difficult to ignore, its usage should be confined to distinctions of major operational significance, yet these are the very distinctions for which there are practical and proven monochromatic codings.

Colour is almost invariably redundant as a coding in air traffic control, whether it is intended to be or not. It is almost impossible, and never recommended, to devise a set of colours which do not also differ in brightness and in brightness contrast with the background. In air traffic control, monochromatic codings are not necessarily grey but often green, which is not only the most sensitive part of the visual spectrum but avoids the potentially excessive contrast of yellow or white and the potentially inadequate contrast of saturated reds or blues. Green maintains good brightness contrast throughout most of its saturation range.

The coded information on any air traffic control display is used for several tasks. Colour codings devised and optimized for any single task may fail to aid other tasks, or even hinder them. Before any colour coding of air traffic control information is adopted, the optimum balance of advantage for the whole range of tasks should be ascertained, allowing for their relative operational significance and frequency, and for any circumstances when a usage of colour could mislead or induce errors in a particular task. Information in the same colour tends to be perceived as related whether it is or not, and relations between information in different colours may not be perceived.

From time to time, the numerous studies on colour coding of air traffic control information have been reviewed (Hopkin, 1970; Kinney and Culhane, 1978; Hopkin, 1982a; Narborough-Hall, 1985; Reynolds, 1994). The findings of many further studies have never been widely disseminated. Nevertheless there seems to be an emerging consensus about the principles to follow. It is recognized that the effectiveness of colour as a coding does not depend primarily on hue but on brightness contrast, so that light blue on dark blue can be quite effective for example (Figure 18: see colour section after page 238). The three colour dimensions, hue, saturation and brightness, cannot usually be manipulated independently, particularly on electronic displays. Colour contrast without brightness contrast is insufficient for information that has to be read (Figure 19: see colour section after page 238). Adequate but not excessive contrast ratios for coloured alphanumeric and symbolic information are of overriding importance. Colour is useless for coding point information and of limited value for lines. It can be effective for area information especially when colour can be combined with transparency so that several colours can be seen as superimposed. To achieve this effect is complicated, because the transparent colours must be distorted differentially according to their backgrounds in order to appear to be the same colour even when they are superimposed on others or others are superimposed on them, but the results can be very worthwhile when, for example, severe weather over an airway straddling a coastline beside a danger area can be depicted successfully by transparent superimposed unsaturated dim area colours that collectively employ very little brightness contrast (Hopkin, 1994e; Reynolds, 1994). An application of the principle of transparency is illustrated by Figures 20 and 21, which, despite initial impressions, contain the same information (see colour section after page 238).

Several principles should be followed. Highly saturated colour information should normally be avoided entirely. Extreme saturation has serious disadvantages in colour coding and no advantages. Few colours can be presented electronically in saturated form without inadequate brightness contrasts. A related principle in the effective use of colour is that for most of the time the display should be substantially monochrome. Colour should draw attention to exceptional rather than normal data, to dynamic information that the controller must heed, or to emergency or safety-critical information (Figure 22: see colour section after page 238). If all is normal, if nothing is urgent, and if immediate action is not required, then monochromatic codings can often suffice. Colour as an information coding should generally be used sparingly, not necessarily by having only a few colours but by quite low saturations and by denoting special circumstances.

Given these constraints, many standard recommendations for colour coding can be applied to air traffic control displays. Background colours, if extensive, can distort foreground colours, and saturated background colours can have significant and unwanted colour induction effects (Walraven, 1985). Colour can code information in categories, can emphasize information, and can be used to warn. Although colour may reduce clutter if applied with care, its success depends on the successful manipulation of a series of visual layers with the most important information in the top layer, and it is imprudent to rely entirely on colour to separate items in this way: presenting different labels in different colours seldom makes overlapping labels readable.

Colour can effectively impose visual structuring, designate closely related but physically separated items, or induce the perception of visual entities through the principles of similarity and closure. However, the power of colour to structure has limitations, and linear colour coding cannot wholly overcome significant physical separations between items (Figure 23: see colour section after page 238). A standard notation must be employed to recommend colours or colour palettes, so that colours intended to be the same but in different locations or generated differently are perceptually very similar and not excessively equipment dependent. When recommendations for air traffic control take the form of general perceptual principles of colour coding to be satisfied rather than specific hues to be used (Reynolds, 1994), these requirements should be met by codings based on standard ergonomic recommendations for the application of colour.

Generally each colour should have only one meaning. If there is no possible interchange of personnel, the same colour may sometimes acquire different meanings in different parts of an air traffic control system, so that a colour has different connotations for controllers and for maintenance staff, for example, but even this invites misinterpretations and is not to be encouraged. Colour may be introduced to draw a distinction not hitherto made, although any new distinction may be too trivial to be suitable for colour coding if it has not been needed before. Colour can reinforce or render more important an existing distinction. Colour generated by different technologies, intended to appear the same and given the same name, may not be perceived as the same. It is vital to specify colours by a standard notation and to be aware of its multidimensionality (Hunt, 1991). Quite small and unobtrusive differences in some dimensions such as brightness contrast can nevertheless be reliably discriminable (Figure 24: see colour section after page 238). Precipitate applications of colour that are an artefact of our current technological state should not preclude potentially better future applications of colour that require further technical advances.

Colour coding has been proposed to show predicted routeings, control responsibility, height banding, future versus current information, emergency versus routine data, dynamic versus static information, and many other distinctions in air traffic control. The fact that these numerous proposals have all been entertained seriously and many tested empirically is a practical reminder that there is no self-evident, intuitively obvious application of colour in air traffic control. Whatever the convention adopted, it will have to be learned and is therefore potentially vulnerable to human error because of inadequate learning. Colour coding must not add significant sources of human error or decrease the system safety. It is not easy with a multiplicity of tasks to rule out such potential consequences, particularly if evaluations cannot measure them, but it behoves everyone concerned to try to do so.

11.5.4 Size

Size as a coding dimension for air traffic control information has a few limited but important applications. Most air traffic control displays have incipient problems of excessive clutter, especially with heavy traffic. To use size extensively as a coding dimension can aggravate problems of clutter because in practice it must employ increases in the normal size since the substantial reductions in size required to achieve reliably perceptible differences would bring the data too near to a visual threshold to maintain readability. In practice only two sizes are advisable for any frequently used application, with more than two confined to rare usages and not recommended even then.

Size as a coding can make categories or items of information stand out without becoming too obtrusive. It is better as a relative than an absolute coding, so that both sizes should therefore be present concurrently whenever it is used. Judgements that rely on recognizing an absolute size without error require size differences so gross as to be impractical on most displays. Differences in size should reflect differences in importance or urgency. Large-sized data should be active, significant and preferably short-lived. A common effective application is to increase substantially the size of a data block while it is being amended, which makes it easier to read, commands attention, simplifies the checking of amendments, and means that the positioning of markers or cursors within the data block to designate what is being changed may be done more quickly. Size is often recommended as a coding to denote data being changed by the controller, as distinct from data being changed or updated automatically.

A further use of size coding, in which it is applied not to the information itself but to its structuring, should not be forgotten. Differences in the size of blocks of data or of the gaps between them can be used to distinguish between them or to show differences in the degree of relationship between them or in the nature of the association between them.

11.5.5 Brightness contrast

Contrast refers to the relationship, or the difference along a dimension, between stimuli treated as foreground and stimuli treated as background. Brightness refers to the perceived light intensity of a stimulus, which is not necessarily the same as its physical intensity because stimuli of equal physical intensity need not seem equally bright subjectively. Brightness can be distinguished from lightness, which refers primarily to the location of a stimulus within a dimension from black to white. However, brightness can have a similar connotation to lightness in the context of colour, where brightness is used, in conjunction with hue and saturation, or refer to the black-to-white dimension. Brightness contrast refers to the difference between foreground and background in terms of contrast, which is most crucial for legibility and readability. Colour contrast compares the foreground and background in terms of colour. Those who apply colour coding often start by believing that colour contrast is sufficient to ensure legibility and learn from hard practical experience that it is not, because legibility depends primarily on brightness contrast and not on colour contrast. The notation of brightness contrast can be confusing. Most commonly it is treated as relative brightness expressed as the ratio of the brightnesses of the foreground and of the background, and this notation is adopted here. An alternative formulation, common in theoretical studies though rare in applied work, expresses this ratio in relation to the absolute background brightness and thus takes account of the luminous flux of the display background.

Brightness contrast is considered primarily within each display. Ratios of three-to-one or less are enough for information which should be present but not obtrusive and does not need to be read. Examples include airways and non-airways, range rings, coastlines, danger areas, restricted flying areas and some weather information. Very low ratios indeed can and should often be employed for such information categories. The available range of brightness contrast depends on the technology, modified by the ambient lighting, screen reflectance, and other characteristics. Important dynamic information that must be read requires a ratio of about eight-to-one between the dynamic foreground and its immediate background. The tendency to demonstrate technological advancement by ratios of sixteen-to-one or higher must be discouraged, as they are far from optimum for human use. Low brightness contrasts may be compensated for to a limited extent by other codings such as increases in size or changes in font design.

The full technically available contrast range should be specified first. The darkest usable condition may not be when nothing is generated since characters may then seem to float in a void, and whatever has to be generated to prevent this effect then constitutes the lowest practical brightness level. Other categories of background information can then be portrayed with the minimum permissible brightness contrast ratios above this base level. The principle of transparency can be helpful. Superimposed alphanumeric information requires the recommended brightness contrast ratios, using the brightest commonly occurring background level to calculate the applicable ratio. Important dynamic information must never be misread because of insufficient brightness contrast. Although a ratio of six-to-one may be permissible with well-designed, adequately sized characters, this principle should not normally be compromised, particularly because the brightest parts of the background, such as danger areas, restricted flying areas and regions of adverse weather, are likely to be of most operational importance.

Depending on the available technical range of brightness it may be possible to portray dynamic information using more than one brightness contrast ratio, and thereby employ brightness contrast itself as a coding dimension. In air traffic control this can be a means to adapt to different task and job requirements. Simple examples are to increase the brightness of data that are the subject of specific training exercises or of data with greater priority or urgency, or to decrease the brightness of data that the controller should know about for consultation or reference but which relate primarily to the responsibilities of others.

Brightness contrast can be an efficacious coding because quite small changes in it can successfully distinguish between adjacent columns or blocks of data, It is much less suitable for presenting single items or characters with different brightness contrasts, since they do not stand out unless the differences between the ratios are large. The maximum brightness contrast ratio for large amounts of data should not be more than about ten-to-one, where the latter refers to the brightest common background. It may not be desirable to use the full range of brightness contrast for dynamic information because other more important information has to be superimposed. The most obvious example is the small symbol that denotes the position of the cursor or marker, which must remain clearly visible wherever it is, preferably by being the brightest item on the whole display because then it can remain quite small.

All of the above assumes that the brightness contrast is not significantly degraded by reflections in the display. In the past quite drastic measures have sometimes been needed to circumvent this problem, such as compulsory dark clothing, painting reflected walls dark and banning bright items from them, and cowling light fitments. Sometimes statutory requirements such as clear exit signs have resulted in obtrusive display reflections. Modern display technology now offers solutions to this problem through appropriate coatings, filters and phosphors. Glare and reflections in the display reduce the perceived brightness contrast of any information viewed through them, to the point in extreme cases of rendering the information completely unreadable so that the controller must make a head movement or move the seat to read it. This is clearly unacceptable, and the prevention of glare and reflections acquires high priority, although it can be very difficult to avoid under all circumstances reflected glare from the sun in air traffic control towers.

Hitherto, the brightness contrasts within the display have been considered, but brightness contrast applies also to the display in relation to its surroundings. A well-designed air traffic control environment should contain no pools of light or of darkness, and the total luminous flux should be similar but not identical wherever the controller looks so that scanning induces no gross changes in pupil size. The furniture and decor can be specified with a freedom not applicable to the information on cathode ray tubes. The immediate display surround should not be in a saturated colour, nor differ grossly in brightness from the display itself when it is in use, but should be an integral feature of the decor of the room which includes its walls, ceiling and floor, and all the surface textures of the furniture. The whole room and its contents should have the appearance of having been designed as a visual entity. The brightness contrast between each display and its surround should not be large but should be visible so that the framework never looks like a continuation of the display itself.

Opting for a mid-grey display background with the objective of being able to present both lighter and darker information on it runs the risk of being the worst compromise if neither lighter nor darker information can attain sufficient brightness contrast with the background for optimum legibility. Normally the background then has to be either lightened or darkened so that either lighter information or darker information can reach the recommended brightness contrast levels, with the contrary kind of information being restricted to codings for which lower contrast ratios suffice. Even if the range of brightness is large enough for both light and dark characters to be read against an intermediate background, this may merely result in a contrast ratio between the light and dark characters that is excessive for human use. This need not rule out light or mid-grey backgrounds entirely for they can have other benefits, but their brightness contrast problems can counter other benefits if they are not resolved.

11.5.6 Texture

Texture is little used as a coding in air traffic control, although it seems to have possibilities to differentiate between background areas. A suitable texture difference might suffice to separate airways from other regions on a radar display without an additional line. Texture changes may be an effective and subtle way to show distinctions within area fills. Principles for texture discrimination are emerging. Julesz (1981) describes the principles of textons, the basic units of texture perception, which are illustrated by Gregory (1987). Much of the research on textons has utilized them as a tool to explore basic principles of human vision (Marr, 1982), but they could provide a key to the understanding required for the successful application of texture discrimination as an information coding.

11.5.7 Inversion

Inversion is used occasionally but not widely as a coding on air traffic control displays. The evolution of radar displays has resulted in light information against a dark background, although dark information against a light background is now technically feasible. This grossly increases the total luminous flux or light output from the display, and a change to it must never be contemplated without reference to all the other displays viewed regularly by the controller in the same workspace. The reason is that gross differences between displays in their luminous flux induce significant changes in pupil size as the controller scans them, which is experienced as tiring. Because the mechanism of adjustment is quite slow, it lags behind the eye movements so that the pupil size is seldom optimum and stable during frequent scanning.

The issue of employing inversion as a coding tends to be addressed for each display in isolation, whereas the main human factors problems relate to its integration with other displays. The question of inversion can be asked separately about the radar display and tabular information displays, but should not be answered for either of them separately without reference to the other. General recommendations about the inversion coding of all the information on a display should be mistrusted, for the pros and cons are often finely balanced, and sometimes the requirements of specific circumstances should prevail. Because luminous flux can affect the perception of other displays, it is important if major changes in traffic demands may change it significantly. This happens on tabular information displays of electronic flight progress strips: a dark display background results in a prevailing impression of a dark display and low luminous flux if there is little traffic and therefore only a few light electronic strips, and of a light display and high luminous flux if there is heavy traffic and therefore many strips. The range in total light output is smaller and easier to keep compatible with other displays if the electronic flight strips consist of light characters against a darker background rather than the converse, but subjective preferences may favour the retention of the familiar light coloured strips when they are converted into electronic form. It is important to bear such factors in mind when inversion is considered.

Inversion as a coding can also be applied to single items or to small information clusters, or to structure data. If a very small number of characters or of data blocks among many are inverted in that they are depicted as light characters against a dark background or vice versa when all the other characters on the display are depicted according to the opposite principle, this can be one of the most visually distinctive and successful ways to maintain attention on the inverted information for some time. However, it has some potential disadvantages: one is that inverted characters may not retain their legibility, and it may be necessary to thicken dark characters to read them (Figure 25: see colour section after page 238). With single characters or small groups of them inversion must be obtrusive to succeed, whereas with large blocks of characters, inversion can be less distinctive against the background and still be efficacious.

11.5.8 Other codings

A selection from the numerous other visual information codings is mentioned here because they have been or could be used for air traffic control purposes. Some air traffic control information is dynamic. Aircraft, however represented, appear to move across radar displays. In early primary radars, positional updating coincided with the sweep of the beam, behind which the information was most recent and had the best contrast. In synthetic radar displays, it is not always possible to tell how recently the positional information on each aircraft has been updated. Movement can be an effective coding to attract attention or to indicate speed or heading, provided that its magnitude is substantially above the visual threshold. A controller may glimpse movement without knowing what has moved, how far or in what direction. Continuously updated displays may move too slowly for the movement to be detected. Although mere movement is very noticeable, the size and direction of movement are less discernible, and changes in movement or in the rate of change of movement require prolonged viewing to detect. While genuine movement may be practical as a coding occasionally, spurious apparent movement never is. Displays which appear to present a void rather than a surface may be particularly liable to illusory apparent movements which should be designed out. If various labels on a radar display are updated randomly or in fixed sequence, this can generate apparent repetitive ripples or moiré patterns which are distracting and should be prevented.

Flashing or blinking, or any regular variation in the brightness of displayed items, is a usable coding to draw attention to information but it must be used sparingly and for short durations as it is distracting. Flashing as such conveys little information beyond the need to attend. The recommended rate is about two cycles per second, with approximately equal durations of the flash on and off. Several flashing rates are not recommended for coding, since the differences in rates have to be gross before they become reliably discernible. Progressively faster flashing rates can denote increasing urgency, but increasing intensity is usually better, and an additional auditory coding is better still. Flashing must be restricted to important events because it is so disruptive. No air traffic control display should ever have information flashing on it under normal conditions.

One coding that can effectively remind a controller about items on a display for a considerable time places some form of marker beside them or round them. Slashes, dashes, asterisks or other shapes, or visual boxes round items can all serve such a purpose. A visual box can be particularly useful for required actions with two or more stages. For example, the whole box round an aircraft label can show that it is straying from its scheduled route, and part of the box can be retained after the pilot has been appropriately instructed, to remind the controller that the pilot must still be requested at some future stage to return to normal navigation.

In some circumstances information can be coded by being offset. The commonest application of this principle is to flight strips when a paper or electronic strip is cocked to one side to stand out from the others visually as a reminder that an action is outstanding. Offsetting is immediately visible without being too distracting for others. Some displays can be zoomed or expanded selectively or as a whole. This can be used sparingly as a coding dimension on a short-term basis by expanding important items until they are noticed and reacted to. Codings should always be considered in conjunction with others, because of the desirable principle of redundancy in coding of air traffic control information. If a distinction is of major operational significance it is prudent to show it concurrently in more than one way. Some of these less used codings can be suitable to achieve coding redundancy.

Two broad kinds of coding are different in nature from the above, but nevertheless valuable. One is coding by an absence of information: taken to its extreme, nothing is shown if no action is required. More commonly, much less is shown if no action is required. For example, on a radar display an aircraft may be shown only by a symbol and an adjacent alphanumeric callsign. The absence of other information denotes that no action by the controller is required on that aircraft currently or in the near future. When all is well, the display is relatively empty, and the provision of extra information itself acts as the coding that requires action. An implication of this approach is that the computer and not the controller decides on the need for human intervention, but as a coding the principle can work well and seems likely to become more widely adopted. Already only aircraft within designated height bands appear on some radar displays.

In contrast to coding by not showing information is the coding of highly complex information. This applies to much graphical data which are expected to become more common in air traffic control. Among recent attempts to develop taxonomies for classifying graphical information are the five categories of graphs and tables, maps, diagrams, networks and icons, their main differences being characterized by the amounts of spatial information and of information processing effort (Lohse *et al.*, 1991). The visual coding dimensions of graphical information can be much more complicated than those considered above (Foley *et al.*, 1990). Graphical data vary greatly in their inherent meaningfulness. Some that are related to spreadsheets, flow charts, models, menus and listings may require detailed study to make sense of all the information they contain and all the relationships within them. Others such as maps of familiar countries and logos may be instantly recognizable if they are familiar. Still others such as printed music may be immediately meaningful to some and totally meaningless to others. Information coding must develop and evolve to take better account of these visually complex graphical structures.

11.6 Clarity and stability of information

At one time the clarity and stability of air traffic control information were major human factors issues. They no longer are because of advances in display technology and because greater processing and synthesizing of information have allowed more control over the factors that govern its clarity and stability.

Displays should not shimmer or flicker. Their refresh rates should be high enough to preclude any perceptible oscillation on them in all air traffic control environments, which in practice requires a rate above about 90 cycles per second (Bauer, 1987). The manner of display generation should not induce perceptible strobing of the display. There are now no adequate reasons for introducing into any air traffic control environment displays which lack clarity because they smear, employ badly drawn characters, are too dim or are mismatched with bright air traffic control environments such as towers, because many displays are now marketed which do not carry such penalties. Therefore, all air traffic control displays should be visually stable and clear. If they are not, the equipment may be old and due for replacement. Information that is not clear enough to read is inherently misleading and could be dangerous. It is a false economy to postpone the installation of modern displays that remove this problem.

11.7 Legibility and readability

Legibility and readability are sometimes treated as synonymous but they often are separate and successive processes. Legibility refers primarily to perceptual clarity, and is influenced by the method of display generation, the choice of appropriate codings, the application of the human factors guidelines for their correct depiction in relation to the task requirements, the ambient lighting, and eyesight standards. Failures of legibility are attributable to inability to perceive information correctly because of psychophysical deficiencies such as inadequate size or contrast or because of specific design deficiencies whereby items or characters are too similar and confusable perceptually with each other. Errors caused by poor legibility originate in insufficient perceptual discrimination, for whatever reason. Legibility has been studied commonly by systematically degrading stimuli to increase the quantity of resultant perceptual errors, which assumes with only tenuous justification that the errors under degraded conditions typify those under the less degraded conditions more usual in real life.

Enough is now known about the conditions and requirements for legibility to virtually guarantee it, provided that all the pertinent ergonomic recommendations for coding and presenting information have been followed correctly and scrupulously. Confirmation of legibility should need only limited empirical testing as verification, with perhaps a few minor adjustments. Findings on legibility do not readily transfer across display technologies without verification, particularly between cathode ray tubes and other displays such as electroluminescent panels, or between luminous and non-luminous displays. The cure for inadequate legibility is to change physical attributes of the information, the display or the environment.

Readability includes comprehension and understanding. Legibility can vary greatly between characters on the same display, but readability refers to information more generally rather than to its minutiae. Because little of the information on air traffic control displays is in the form of continuous text, many standard measures of readability cannot be applied to it. Ways of assessing readability include measures of eye movements and visual search patterns, errors, omissions and delays, level of understanding, speed of comprehension, memory and recall, and compatibility between what is read and what is already known. Of course poor legibility impairs readability, but so do poor layouts, formats or codings and inappropriate display contents. It is much easier to establish optimum legibility than optimum readability, and it is prudent to settle for acceptable readability rather than to attempt to prove that it is optimum.

11.8 Qualitative information

Most air traffic control displays contain far less qualitative information now than formerly. Digitized information about positions, heights, speeds, headings, identities, reporting points, times, distances and separations is essentially quantitative. Information derived from calculations, computations, predictions and extrapolations also purports to be quantitative. Qualitative aspects of information refer, for example, to its reliability, validity, trustworthiness, accuracy, precision, or frequency of updating. Raw radar displays contained significant qualitative information about clutter, signal-to-noise ratios, coverage, fading and decay rates which guided the user on how trustworthy they were likely to be. The typical modern air traffic control display contains no information at all about how trustworthy or accurate the various categories of information on it are, makes no provision for conveying such information visually to the controller even when it is available within the system, and cannot indicate when temporary conditions have made some kinds of information less reliable than usual.

This might not matter much but for the human propensity to base qualitative judgements on visual appearance. Data portrayed with crystal clarity tend to be treated as more accurate and reliable. Improvements in display technology can mislead the user into ascribing enhanced quality to the information when new displays have merely replaced old ones without any commensurate improvements in the information on them. It is time to take more seriously the issue of whether indications of information quality should again appear on air traffic control displays. If a technical failure has left information still usable but much less reliable than it normally is, the options are often to show nothing or to declare the display unserviceable, yet it could still be used safely if some loss of trustworthiness could be shown.

Sometimes users' ignorance about qualitative aspects of the data can lead to their failure to appreciate the significance of technological advances and adapt to them (Sen and Boe, 1991). When satellite-derived data are introduced, some of the benefit from the greatly enhanced positional information will be lost if the controller continues to treat its accuracy in hundreds of metres rather than single metres. Displays of oceanic traffic may look much the same when they contain accurate satellite-derived information as they did when they depicted information of poor quality requiring large separation standards. The means of gauging the quality of information nowadays tend to be indirect, reflected by such indicants as separation standards. Visual codings need to be devised and proved to convey different appropriate degrees of trustworthiness in the information portrayed.

11.9 Display innovations

This major topic can be mentioned only briefly here. All display innovations have human factors implications (Stokes *et al.*, 1990). Several issues arise with any new display technology. Do the ergonomic display principles developed for other technologies transfer to it? Is the visual environment required for the innovation compatible with the environmental requirements of existing equipment? Do new options offered by the innovation pose difficulties in the training or selection of controllers? Does the innovation actually deliver the expected benefits for air traffic control which justify its consideration? What are the new forms of human error associated with the innovation, which must be found and removed or circumvented before the innovative technology can be employed safely operationally?

The graphical displays mentioned above, which can take many technical forms, exemplify display innovations. Another is that aspect of human–computer interaction commonly referred to as WIMPs, which is not a universally agreed acronym. To Stewart (1991), WIMPs are windows, icons, mouse and pop-up or perhaps pull-down menus, whereas in some contexts including air traffic control WIMPs are most commonly windows, icons, menus and pointers. This more generic interpretation of WIMPs covers those human–computer interactions whereby aspects of the display are manipulated as windows, information is coded in the form of icons, dialogues are conducted in which the human selects an option from a computer-presented menu, and the human uses some form of electronic pointer or marker to identify or designate a chosen item. These more generic meanings are independent of the particular display technologies, icon designations, menu structures, methods of pointing, and input devices, but most of their human factors implications are far more specific. For example, the implications that arise from pointing with a mouse are not necessarily those with any alternative pointing device (Stewart, 1991).

WIMPs represent a developing trend in the human factors implications of workspace designs and specifications. Early studies sought to optimize displays or input devices. Then it was realized that the optimum might be task dependent, so that the best visual display or input device might not be the same one for all tasks. Some residual uneasiness was occasioned as interacting effects started to emerge, so that the optimum display might depend not only on the display characteristics and the tasks but on the choice of input device, or the optimum input device might depend not only on input device characteristics and the tasks but on the displays and the forms of feedback they provided. In short, there could not simply be an optimum display or an optimum input device, even for a single task, but only an optimum combination of display, input device, and interactions between them, which remained task dependent.

WIMPs take some aspects of this thinking a stage further. They imply that the choice, usage, location, superimposition and manipulation of windows holding data, that the choice of symbols, codings and icons presenting data, that the choice of the structure and content of menus guiding and directing the user, and that the choice of pointers and input devices for implementing human actions are all interactive, to be treated as a whole and optimized only as an interacting set for the particular tasks. The human factors approach is evolving away from the simplistic notion that there is one best way to perform every function and that it can be discovered by combining the optimum display, the optimum input device, and the optimum human–machine interface design, each of which can be ascertained separately beforehand. It is recognized that this approach may still be too simplistic since the usage of WIMPs depends on such user characteristics as being naive or expert, novice or experienced, concepts which require careful definition (Fisher, 1991).

There are too many display innovations to list all the human factors implications of each one, but their general human factors implications can be mentioned. As a general rule it is prudent to treat human factors recommendations that have been developed and proved for one display technology as hypotheses that require testing and verification for other technologies, unless the recommendations demonstrably rely on universal visual principles that are independent of specific technologies. Any display principles which are unique to a technology have to be identified and proven separately for it. Of particular concern are how the display would appear if it was faulty and whether the user would be able to recognize malfunctioning. Any new display technology removes some kinds of human error by making them impossible, at the cost of introducing some new kinds of human error with perceptual or cognitive origins. These must be identified and circumvented, and their effects minimized.

With any new technology a major effort is usually necessary in order to be fair in evaluating it. Its proponents identify closely with its claimed advantages, and insist that any evaluation demonstrates these fully. Comparable thought has seldom been given to what the disadvantages might be, but a fair evaluation must address them with equal thoroughness, to identify and preferably quantify them. The purpose is not to disparage the new technology but to minimize unwelcome surprises during subsequent development and commissioning, and to make the decision on whether to progress with the device as well informed as possible. Human factors efforts can seem biased against technology because of their insistence that advantages and disadvantages be explored with equal thoroughness, whereas the primary concern of others is with the advantages. It can be very difficult for those who have become wedded to a technological advance or breakthrough to accept that it does not necessarily convey human factors benefits in air traffic control. Its technical excellence may not serve air traffic control better than tried and tested but duller technology, and its reliability may be unproven. If technology is new, it is inherently unlikely to be optimum for human use immediately, and further iterative development cycles can be expected, which should precede comparisons with alternative technologies. Novel technologies should not replace

HURN 136.5
TIME SEQUENCE
EDD1 PAGE 6
1610 Q03
ORT IBY FAW MID
A B7207 VC8 070 EGNH R1 R1N EGSH R130
B GBARE C414 060 38 30 EGBB A1 R1 IBY EGWT
C NA002 B747 290 29 34 R37 EBBR FULL
D B7106 VC8 140 27 LFPB R1 A1 EGFF R150
E BA012 B747 120 26 31 KORD UR37 UR1 EGLL
ORT IBY FAW MID
PRESS ANY KEY FOR NEXT COLOUR CHANGE. (INCLUDES FULL AND HALF BRILLIANCE)
COLOUR SEQUENCE:- BLACK, RED, GREEN, YELLOW, BLUE, MAGENTA, CYAN, WHITE.

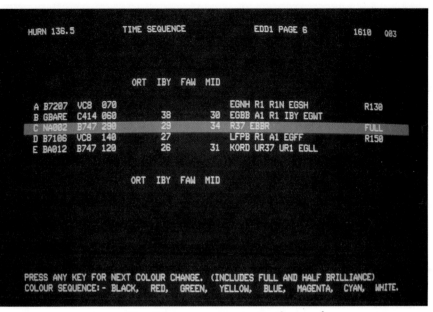

Figure 18. Brightness contrast using the same hue

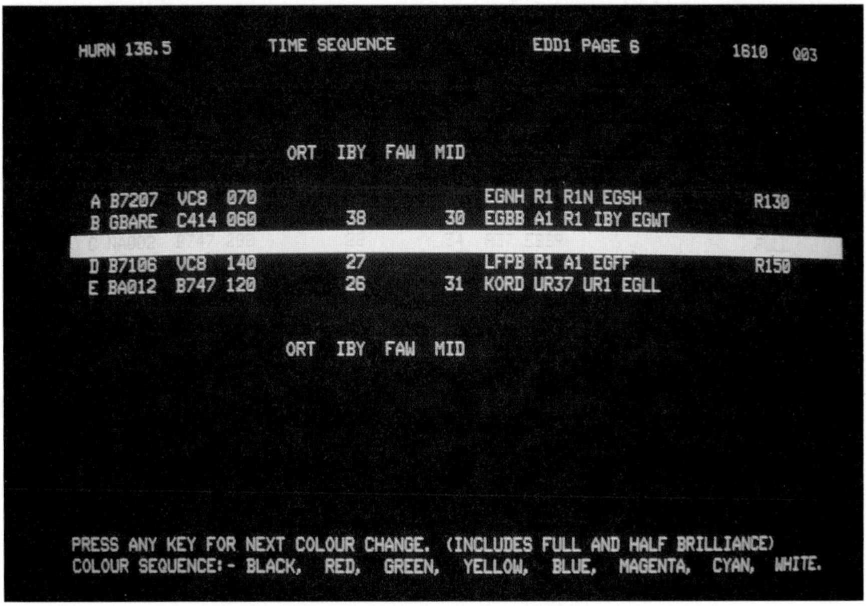

HURN 136.5
TIME SEQUENCE
EDD1 PAGE 6
1610 Q03
ORT IBY FAW MID
A B7207 VC8 070 EGNH R1 R1N EGSH R130
B GBARE C414 060 38 30 EGBB A1 R1 IBY EGWT
D B7106 VC8 140 27 LFPB R1 A1 EGFF R150
E BA012 B747 120 26 31 KORD UR37 UR1 EGLL
ORT IBY FAW MID
PRESS ANY KEY FOR NEXT COLOUR CHANGE. (INCLUDES FULL AND HALF BRILLIANCE)
COLOUR SEQUENCE:- BLACK, RED, GREEN, YELLOW, BLUE, MAGENTA, CYAN, WHITE.

Figure 19. Colour contrast without brightness contrast

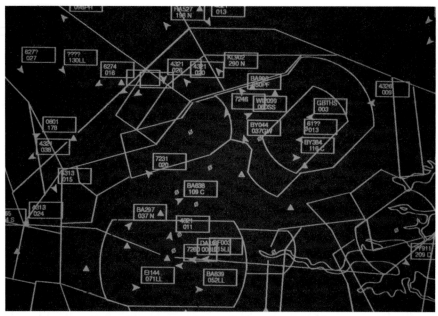

Figure 20.   A monochrome display of a terminal area

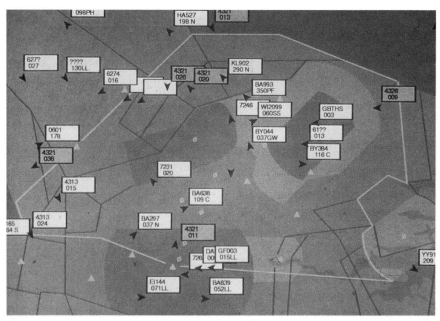

Figure 21.   The same information as in Figure 20, using colour and transparency for backgrounds

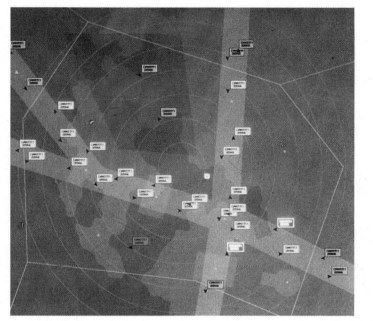

Figure 22. *Transparency for backgrounds, with colour and monochrome block labels*

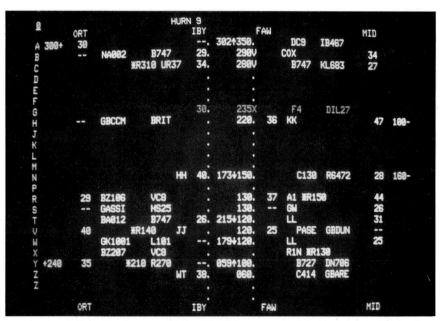

Figure 23. *Spacing of data prevents full visual organization by linear horizontal colour coding*

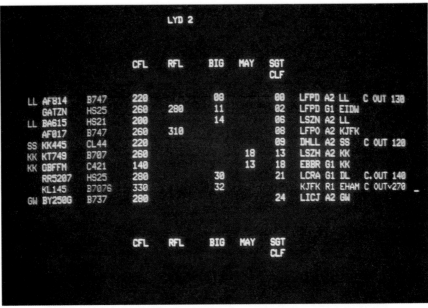

Figure 24. Small differences in brightness between columns of data

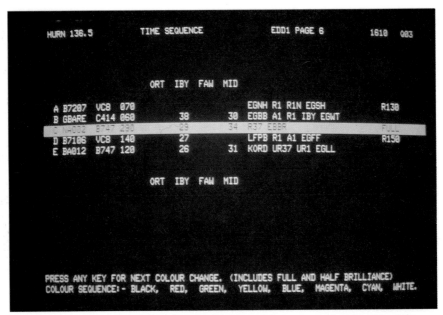

Figure 25. Inversion showing typical slimming of characters and reduced legibility

older ones until they really are better, unless they carry no major penalties but have other significant benefits such as lower costs or increased reliability. The ultimate criterion for their acceptance must be safety. Any potential hazards to safety are therefore the aspects of new display technologies that must be studied most thoroughly and cautiously.

12

Input devices

12.1 Location within the workspace

Input devices, which are often called controls in the earlier literature (Hopkin, 1970), convey information from the human to the machine. They are the main means for the controller to initiate actions, respond to system events, and interact with the system. To be fully successful they must therefore be capable of conveying at the right level of detail every kind of information that the controller may ever need to enter into the machine: otherwise the controller may know what should be done in given circumstances but lack the means to do it. As computer assistance and automation increase, the options available to the controller become progressively more constrained by the capabilities of the input devices. There can be no air traffic control without them, short of complete automation. If the input devices have been well chosen, designed correctly and optimized for the tasks, their users should gradually become unaware of them as they acquire skill in using them and gain familiarity with them as tools. To the extent that input devices continue to obtrude on their users' attention they are not optimum, for their mechanisms are overriding their functions.

Input devices are not interchangeable. Different devices are best suited to different purposes and tasks. Task analyses and comparable techniques should provide sufficient guidance on what the tasks and functions will be to allow initial recommendations about the most appropriate input devices to be made on the basis of existing evidence about the respective strengths and weaknesses of alternative devices, but there are further practical air traffic control constraints to be considered. All the required devices must be accommodated within the workstation and meet the applicable reach distance requirements. If an input device is to be shared, it must be possible to position it within the reach distances of all who will share it. The successful concurrent usage of two input devices is contingent on careful planning of the tasks, and on a judicious choice of devices and of their detailed specification and location.

Generally it is more probable with input devices than with displays that human factors findings about them will extrapolate validly from other contexts to air traffic control (Greenstein and Arnaut, 1987; NASA, 1989). Obviously if a task is unique to air traffic control, there may be no comparable findings elsewhere that could be extrapolated, but much of the ergonomic evidence about input devices in human factors handbooks is universally applicable in so far as it deals with fundamental anthropometric constraints, with reach distances, with the avoidance of physical fatigue, and with recommended locations, separations, spacings, gains, sensitivities and relationships to promote efficiency, to minimize human errors, and to prevent tiredness. This evidence can be applied, but should always be verified (Sherr, 1988; Shneiderman, 1992).

Usually there are so many factors to be considered in the location of input devices within air traffic control suites and workstations that it is more realistic to seek a good practical compromise rather than an optimum or ideal. After the tasks have been broadly defined and appropriate input devices chosen, their fuller specification can be developed by detailed design that matches their characteristics more closely with the task requirements. The normal sequences of tasks and of actions within tasks suggest the respective importance of various input devices, their order of usage, the significant relationships between input devices and displays, and common interactions among input devices. The location of input devices should reflect and support these identified links.

Any requirement to split and amalgamate jobs in order to adapt to gross changes in traffic demands constitutes a major complication in the layout of input devices. At the very least it implies a requirement for flexible communications, and the input devices must permit such flexibility. If the functions of one controller are divided, almost all the input devices that were needed for the combined job may still be needed for each split one, but needed by two controllers at the same time. This tends to encourage the splitting of jobs by opening new workstations rather than by additional staffing on suites, so·that each additional controller retains exclusive use of a full set of input devices. The problems that arise when two or more controllers share the same input devices begin with configuring the workspace, since what helps one controller may hinder others whose tasks are different, and any selection of displayed information is very unlikely to be optimum for everyone. However, adding extra workstations requires the split jobs to be integrated sufficiently for each controller to know what the others are doing. The physical splitting of the work of one workstation between two workstations resolves some problems, but the implications for the location of input devices and for communications in particular must be accommodated.

It is not unknown in air traffic control for an initial check to reveal that all the input devices that seem to be required will not fit into the workstation or suite. In this event, the responsibilities and tasks of each controller, the workspace design, the relationships between displays and input devices, the types of input device, and the level of detail of tasks should be re-examined, to consider the practicality of combining input devices, of choosing more flexible or more compact ones, of condensing or simplifying functions, and of employing the same device in different roles at different times, although this last option is never really recommended since it can introduce new kinds of human error that have to be prevented by suitable interlocks. It may also be worth considering whether more devices could be housed in a redesigned console, and the substitution of different types of device with less stringent requirements, such as greater permissible reach distances or satisfactory usage with the non-preferred hand. If the input devices will still not fit in after improvements from the above methods have been made, then some input devices may have to be positioned beyond the recommended reach distances, and it may become necessary to design the suite so that the seated controller can move easily along it.

A set of simple principles guides the location of input devices within a suite or workstation. The most important devices should occupy the prime shelf space in front of the controller. Devices that are used less may occupy more remote positions, but must remain within reach. Only devices used very rarely, such as those for setting up a workstation, may be placed beyond normal reach. Violation of these principles can result in postural and occupational health problems for some controllers. The relative location of several input devices on the shelf is influenced by the following factors: the direction and magnitude of the physical forces required to operate them, the kinds of feedback available from them, their physical dimensions, the kinds of function for which they are suited, their relative frequency of use, the common patterns and sequences of usage of various devices, concurrent usages of more than one device, and any obscurations of one device or restrictions on its use by the simultaneous activation of another input device.

Provision must be made for requirements other than input devices. On the console shelf, space may be needed for hard copy, to place elbows and arms, and to support the hand while using some input devices. Devices must be positioned correctly in relation to the displays to which they refer. Appropriate feedback that an input device has been activated correctly has to be provided by movement, touch, sound, or vision, on an associated display or within the input device, or by both means. If this is achieved by appropriate changes on the display that correspond to activation of the input device, their efficacy depends partly on the respective positioning of the input device and the display and on the degree of detailed correspondence between their layouts.

In almost every air traffic control environment each input device will be used for numerous tasks. Its location must therefore represent a compromise which reflects that some tasks are less important than others. Although safe and efficient task performance may depend more on human adaptability and proficient training than on the optimum layout of input devices, which could probably not be recognized even if it was achieved, nevertheless the provision of an efficient and satisfactory layout is both important and influential. Deficiencies of layout are a major determinant of some common kinds of human error.

Comparatively recently, moveable input devices, which controllers can position where they will, have been in vogue (Edmonds, 1992). The practicality of this is device dependent: for keyboards it is easy and for rolling balls difficult. Some devices, such as the mouse, work by moving the whole device. This mixture of principles has some attractions, but only if the number of input devices can be severely limited and their dimensions do not preclude other essential requirements, such as space for hard copy. Users usually benefit from some guidance on other factors affected by the positioning of input devices.

12.2 Physical characteristics

The physical differences between input devices influence the suitability of devices for the tasks within the workstation. Differences in their physical size require different amounts of shelf space. The forces needed to operate them differ in direction, duration and magnitude. Some devices can be used in conjunction with other input devices, and others cannot. The physical appearance of devices differs in terms of colour, shape, contrast, reflectivity, and other attributes, and devices vary in their visual harmonization with the decor and their immediate surroundings. The kinds of feedback that can be provided to users are not the same for all input devices. They differ in gain, sensitivity, and other operational attributes related to task requirements. Physical attributes, such as their weight, robustness, and cooling requirements, impose different constraints on their positioning. They differ in the physical and functional relationships that there can be between them, in the kinds of relationship that they can have with displays, in the complexity of the functions for which they can be used, and in the nature, range and mix of tasks that they can fulfil. They interact differently with physical environmental factors such as ambient lighting.

Sometimes new input devices encounter problems because they cannot provide feedback in forms familiar to their users. Keys with a positive stop seem thereby to confirm that they have been activated (though this may be spurious), whereas touch-sensitive input devices with identical functionality but no physical response to pressure may require additional feedback in such forms as a light or a click to denote activation. Not only the kinds of task should govern the choice of input device, but also the appropriateness and immediacy of the feedback. A continuous device like a rolling ball, which looks and feels the same however it is moved, implies forms of feedback that are remote physically, such as a cursor or marker on a display. The feedback has to be appropriate for the task, particularly when a controller must know that one input has been completed successfully before beginning the next one.

A listing of desirable physical attributes of air traffic control input devices includes the following: compactness; fluency in use; compatibility between different input devices; self-evident relationships to displays; modest forces required in their operation; positive unambiguous feedback; dexterity, speed and accuracy in use; appropriate location in the console; within recommended reach distances; unambiguous labelling of their functionality; conformity with all recommended separation standards applicable between devices and within each device; appropriate gains and sensitivities for the task requirements of accuracy, precision and speed; and the ability of users to develop high levels of skill in using the device.

12.3 Types of input device

12.3.1 Keys and push buttons

Keys and push buttons have always been common as data input devices in air traffic control, and seem set to remain so despite their technical obsolescence. A single push button often has a unique function, but when push buttons are assembled into keyboards their functions relate to their layout, and several keys are often required for each function. Air traffic control information expressed alphanumerically requires an alphanumeric keyboard of some kind, and the standard QWERTY keyboard layout is adopted extensively in air traffic control, not because it is ergonomically ideal or particularly suitable for air traffic control tasks but because of its familiarity and its widespread marketing throughout computer-based systems requiring alphanumeric data entry. Wherever it is adopted, the standard ergonomic recommendations on the size, shape, appearance, labelling, forces, travel, feedback, layout and separation of keys are valid for air traffic control applications.

Because space on the shelf is at a premium in most air traffic control workstations, and because other tasks such as the selection of communications channels may seem to require further keyboards, various stratagems to make keyboards as compact as possible have been studied. One involves the amalgamation of all keying functions into a single keyboard. Another utilizes chord keyboards whereby a function requires the concurrent activation of two or more keys. A third stratagem aims to extend functionality without extending the keyboard: this can be done either by allowing the function of a key to change with the task context such as phase of flight, or by listing options each of which consists of quite extensive alphanumeric data that can be selected and entered into the computer by a single key. Complex keying functions of this kind rely on careful labelling of the keys to prevent errors and ambiguities and add to training commitments, but they can be successful if well planned. Within the same keyboard and the same workspace, the same conventions for the layout and coding of keys should be adopted throughout.

The increasing conduct of air traffic control through human–machine interfaces has extended the roles of keyboards. They are now used more to enter data into the system and discard data from it, to update and manipulate data, to access and retrieve further data held in the system, and to conduct human–machine dialogues and select menu items. The functionality of keys becomes more complex and fluid, and their meaning more context dependent. The range and appropriateness of their usage rely more on a combination of training and the user-friendliness of the interface design. Keys may acquire extra functions in directing, guiding, reminding or assisting the controller, and even by preventing some actions by selective inoperability. The technical feasibility of such extended roles has not yet resulted in firm policies or recommended practices concerning them.

It is essential to know at once whether a key or push button has been successfully activated or not. Feedback must be subjectively instantaneous. If a button selects one of two states (e.g. on/off), it should be apparent from the appearance of the button itself which state it is in: it may remain depressed, be internally lit or acquire a different label, but it must look different. If the button is within a keyboard, the feedback to confirm its activation preferably combines a displayed event and the tactile feel of the key against a stop, particularly if tasks may require repeated activations of a key in rapid succession. The tops of keys should not shine or reflect ambient light. They should be slightly concave and not smooth enough to feel slippery. Keys should not be embossed, except to facilitate discrimination by touch between rarely used embossed keys of vital operational significance and standard unembossed ones.

12.3.2 Off-screen continuous input devices

Several familiar input devices are of this type, ranging from the early joystick through the rolling ball to the more recent mouse, puck and graphic tablet. Whereas a push button or keyboard can be functionally self-sufficient, although nowadays it often refers to displays, continuous off-screen input devices cannot function without an associated display which is physically separate from them. Judgements of their functionality therefore require evidence not only about them as input devices but about the associated display content and their relationship with it. In air traffic control contexts, continuous devices usually move a cursor, symbol or marker across a display screen to a required position, which may be of direct operational significance, for example as a waypoint on a route, or of indirect significance, for example because it is superimposed on the electronic label of an aircraft. It will be apparent that the optimum payoff between speed and accuracy of positioning is not the same for the two examples mentioned, but that to some extent speed and accuracy can each be bought at the cost of the other. A requirement for greater flexibility in this has prompted the study of gain, the relation between device movement and marker movement, and some examination of non-linear relationships whereby the movement of the marker depends not merely on the magnitude and direction of the device movement but on its speed as well. Results to date are best treated as inconclusive and should not be generalized to air traffic control tasks without verification.

A few simple practical precautions have to be taken with these off-screen devices. Only with the joystick used linearly is it possible to tell by looking at the device where the marker is likely to be. The rolling ball in particular gives no information at all about this, and the others may yield very little, depending on how they have been set up. It must not be possible to lose the marker from the screen. When the marker has been positioned, it is often necessary to enter the coordinates of the marked position into the computer by pressing a key or button, which may be incorporated in the device itself. If it is, pressing it must not also move the marker from its position. The screen should indicate how accurately the marker has to be placed, for example, by highlighting a label as it is captured so that the controller does not waste time in achieving unnecessary accuracy. Devices differ significantly in how well they can be used with the non-preferred hand, and this also depends on the accuracy required. For great accuracy, the limits become perceptual and the input device should be engineered so that the smallest possible increments in movement correspond approximately with the applicable just noticeable difference. Accuracies in positioning a marker cannot exceed what can be perceived on the screen, although, subject to other conditions, the operator may sometimes be able to correct a visible discrepancy on the screen by imperceptible movements of the input device.

Several further kinds of function relevant to air traffic control can be fulfilled by continuous remote input devices. Regions of airspace can be defined using their drawing mode. They can be employed to specify the dimensions and positioning of windows in electronic displays. They could be used to construct icons, though this is not recommended. Where they can drag items of information across the screen, they can collate, assemble, integrate, package or recode information, given appropriate software. Such a facility could conceivably be extended to set up communications links. Although some tasks in other contexts have encouraged operators to use more than one continuous input device concurrently, this has always incurred significant problems and major training commitments, and it is not recommended for air traffic control where there seems no obvious need for it.

12.3.3 On-screen input devices

Various kinds of input device allow data to be entered directly into the system via a display screen that is sensitive to some form of contact or stimulus. The input occurs either because the screen responds to an external stimulus brought to it, as with a light pen, or because the screen itself is sensitive, as with touch-sensitive displays. At one time the former principle was widely considered for air traffic control tasks, but it has now yielded almost entirely to the latter principle. Devices such as light pens which have to picked up and put down do not combine well with other input devices, and there are few applications in air traffic control for which they seem really suited. This section therefore deals mainly with touch-sensitive displays, which fit some air traffic control needs quite well (Stammers and Bird, 1980).

A touch-sensitive display responds to touches of its screen, which are sometimes by a hand-held object but usually directly by a finger. An immediate practical problem is the accumulation of finger marks, which often have to be removed regularly with a solvent because they can impair screen visibility, may be emphasized by the ambient lighting, look unattractive and unhygienic, and may eventually make the screen insensitive to touches.

A touch-sensitive surface has the dual functions of a display screen and an input device. Unfortunately, its recommended mounting angle would be 75 degrees or more from the horizontal if it were solely a display, and horizontal or nearly so if it were solely an input device. Attempts to find the best compromise have a long history in air traffic control. Sears (1991) described experiments in which an angle of 30 degrees from the horizontal seemed to represent the least fatiguing compromise, thereby confirming the much earlier finding by Orr and Hopkin (1968) with less developed technology, but the issue remains unresolved: no compromise is ideal under all circumstances. Raising the display towards the vertical makes it tiring and uncomfortable to use for data entry but the display is clearly visible: tipping it back nearly flat is less tiring for data entry, but parallax distorts displayed positional information and ambient lighting often curtails its visibility. The most practical compromise is to confine their usage to functions that require relatively few touches and never to employ them for prolonged continuous data entry tasks. They are therefore better for tasks like the selection of a labelled communications channel by a single touch than for tasks like extensive and frequent updating of alphanumeric information one character at a time.

It is not possible for the controller to tell from touch alone what data are being entered into the system. There are no moving parts and no equivalent of feedback from key movement. The controller therefore requires some form of feedback that the surface has been touched in the correct place. This can take many forms, including wires embedded in the display surface, protruberances or other surface irregularities that can be felt, engraved surface markings or symbols, grids of engraved surface lines or electronically generated ones, electronic markers on the touch-sensitive surface or on a related display, electronic labelling of surface areas if the controller is intended to look at the touch-sensitive surface while using it, and auditory signals to signify the input of an item. Feedback must be immediate and show what is entered as it is entered. It can be linked with associated changes in the display such as the presentation of a new set of labelled options, although they may signify only that an entry has been made and not that it is correct. The information that is entered into the system by a touch is usually the information displayed at the touched position, but the touch action itself often constitutes a related system event. Examples are the connection of a communications channel, or the initiation of a dialogue about the aircraft whose labelled callsign has been touched.

The region of the display that must be touched to make a desired data input has to be obvious. It must never be possible for one touch to activate more than one function, for example because the finger straddles two separate regions and activates both. If it is possible to touch the surface and activate no function, there must be sufficient feedback to show when and why this has occurred (Sears *et al.*, 1993). Consistency of operation is essential, in terms of required positioning, pressure and duration, so that once these have been learned they always succeed. The performance of tasks with touch-sensitive surfaces is very vulnerable to any apparent arbitrariness in their functioning, and only a few instances of failure to make an expected input or of apparent malfunctioning result in gross reductions in the speed and efficiency of their use.

12.3.4 Voice activation

Even when controllers have extensive computer assistance, they still speak to pilots to convey to them air traffic control instructions, which they also enter into the system through data input devices. This looks like duplication of effort, and in some respects it is. The application of machine recognition of speech, whereby the controller's action of speaking to pilots also enters the message contents into the system, has therefore been hailed for a long time as a promising development of obvious benefit to air traffic control. The circumstances seem propitious because air traffic control has a limited vocabulary, rigid message formats and syntax, and language structures and terminology which have already been refined to be highly discriminable, at least by other human beings. It was anticipated that initial applications in training and evaluations would be the forerunners of operational acceptance. Initial progress was rapid so that by 1979 direct voice input was becoming competitive with keyboards for entering certain limited categories of air traffic control data into the system (Connolly, 1979). The lack of subsequent progress calls for some explanation, particularly because of research and progress in other applications (Simpson *et al.*, 1987).

Among the most recalcitrant initial problems were technical ones, such as difficulties in recognizing the middle digit in the three-digit strings so typical of air traffic control messages, and reliance on context to distinguish between "to" and "2" or between "for" and "4". The common cold drastically impaired machine recognition of what was said, and so did the changes of pace and pitch in speech associated with strain or high workload, the very circumstances under which failures of recognition could have the most adverse consequences. Controllers talked not only to pilots but to colleagues and others in the work environment, and sometimes to themselves. Much of what they said was irrelevant for machine recognition, consisting of preliminary discussions and negotiations, courtesies, requests for information, irrelevant chatter, and bilious remarks which could pose an interesting dilemma of whether it was ethically acceptable for a machine to recognize such words. The serious point was that there would have to be provision to switch voice recognition off. When the machine failed to recognize what the controller said, it proved difficult to indicate successfully how speech should be modified to gain machine acceptance. The resultant irritation and frustration were clearly counterproductive.

The developments of new automated sources of data, of increased automation of the transmission and updating of information between air and ground, and of air traffic control procedures that rely on human–machine rather than human–human dialogues, combine to make speech less important for air traffic control in the future than it has been in the past, with the corollary that applications of automation to it, including speech used as data input directly, can no longer have the impact on air traffic control that they might have had years ago. Speech as direct data input in air traffic control is perhaps being overtaken by other events, leaving training and evaluation as the main beneficiaries of its ultimate development. Lurking in the background is the legal problem of whom to blame if the machine misrecognizes what is said to it.

12.3.5 Other input devices

Only a few of the numerous other input devices available are used regularly in air traffic control. When they are, they should normally conform with standard ergonomic recommendations for their usage, in terms of their location, applications, tasks, appearance, sensitivity, labelling, conventions, forms of feedback, and relations with each other and with displays. Switches with two or three positions and rotary switches are common, particularly for setting up workstations, for selecting and manipulating the categories of information to appear on displays, and for adjusting physical attributes such as display brightness. The combination of visible switch positions plus appropriate labelling must always make the available range of switch settings apparent and the chosen setting or state unambiguous. Some combination of spring loading, auditory and visual feedback, and the feel of the switch when operated, must together guarantee that no switch can be left inadvertently between positions. There must be no possible ambiguity about whether a switch has been activated or not. The functions of each switch should be clarified as much as possible by its labelling and positioning in relation to the displays and other input devices that it affects. If tasks require the successive or joint activation of several switches, this must be clear from their labelling, context, and relative positioning, and from shared operational or visual codings and conventions that reinforce their inclusion within a functional group or set of switches.

Knobs usually look different from each other. Discriminable differences in their shape and feel are also vital if knobs in close proximity may be operated without looking at them. Knobs have fewer direct applications to air traffic control tasks than to other tasks in air traffic control environments, such as maintenance and equipment calibration. In a sense, plugs and sockets can be construed as input devices. The sockets for headsets should grip the plug firmly enough to provide unmistakable feedback that the plug is in the socket and functioning. There should be a means to release the plug without tugging. A channel, groove or clips should house the cables for the headset and telephone, so that they do not straggle across the shelf and inadvertently catch on or trigger other input devices.

Footswitches are common in air traffic control, particularly to switch communications channels on and off independently of other input devices. Only one footswitch per workstation is recommended. It should provide touch and kinaesthetic feedback of successful operation, supplemented by a visual indication or perhaps an auditory one in the case of communications channels. It can be an on/off switch, dabbed with the foot to switch it on and to switch it off, or a depress-to-operate switch which remains on as long as the foot pressure is applied. In the latter instance, the requirements for adequate heel support, for a short foot travel of perhaps 3 cm where the foot is in contact with the switch, and for a small operating force of between 5 and 10 pounds become particularly important. The data on foot switches in human factors handbooks should be applied. Adjustable footrests can be a complication in air traffic control. The best stratagem is usually to have the foot switch on the footrest and to raise and lower it with the footrest. The footrest must incorporate a groove for the heel, correctly located in relation to its mode of operation, to provide adequate support and to prevent the foot from slipping across the footrest whenever the footswitch is operated.

12.4 Sensitivity

In the older literature, the main devices for the controller to interact with the system through the human–machine interface are called controls rather than input devices, and display–control relationships describe associations between the information displayed and the means to manipulate it. At the time, systems did not necessarily include computers, and controls could be mechanical rather than electrical or electronic. In computer systems, workspaces can become adjuncts to the system in some respects and systems may be far more distributed than they once were, but the input devices may remain similar to the former control devices in appearance and usage, though not always in system functionality for they may be much less closely dedicated to particular tasks or roles. Most questions about the choice of input devices or controls refer to specific input devices or to a categorization of devices. They often seem to imply that the most important influence on safety and efficiency is an optimum match between the tasks and the type of input device. However, although this match is essential, it is not sufficient because a suitable input device must meet further human factors requirements, among which are appropriate sensitivity and feedback.

Sensitivity has different meanings with different devices. An over-sensitive key or touch display can be triggered inadvertently by too little pressure, whereas an insensitive one may require excessive pressure to function at all. One connotation of sensitivity therefore refers to the force or pressure required to activate the device. For continuous input devices, sensitivity refers to the gearing or to the constancy of gearing, that is to the characteristics of the relationship between movements of the input device and perceived corresponding displayed changes. An insufficiently sensitive device requires excessive movements for minor display changes whereas an over-sensitive device produces excessive display changes in response to very small movements. Gearing that is not constant can take several forms. Examples are a switch to select alternative sensitivities, or non-linear associations between the movement characteristics and the resultant display changes, usually achieved by displayed accelerations and decelerations driven by the pressure, the duration or the magnitude of the operation of the input device. Occasionally such expedients may become essential, but they signify that different tasks at a workstation make different demands on speed versus accuracy.

The optimum sensitivity of continuous input devices depends to some extent on task requirements, and particularly on the payoff between speed and accuracy of performance. The achievable payoff depends partly on the type of input device: for instance, the negative relationship between speed and accuracy, whereby either can be achieved but only at the cost of the other, is much closer for a rolling ball than for a joystick. Quite large movements of the marker from one aircraft label on the radar display to another are made whenever a dialogue with one aircraft ends and with another begins. For this task, the operational requirement is for speed of movement but not great accuracy if placing the marker anywhere on the label suffices to begin the dialogue. It is time-wasting and infuriating for the controller if a much smaller area must be marked before the dialogue can begin, particularly if small input device movements move the marker a lot and thus promote rapid movement of the marker across the display while making its accurate placing difficult. The tasks and the available gearing should not require both rapid movement and precise placing of the marker with the same input device.

Optimum gearings, if they exist, have not been established for most tasks, and there is some uncertainty over how much they would improve task performance. With highly inappropriate gearings, certain tasks requiring speed or accuracy become physically impossible. In air traffic control, any requirement for great accuracy in positioning a marker warrants close examination of the reasons for it. If the marker does not have to designate a particular position exactly but merely the position of an aircraft or of a label, relatively small movements of the input device can produce relatively large marker movements. If great accuracy is required, large device movements and small marker movements can provide it, but slowly. If both requirements are essential, the options are task redesigns, or resort to gearing changes, multiple input devices or display zooming. With a joystick, its total arc of travel should move the marker across the whole display. Rapid spinning of a rolling ball should never be necessary. The area on the shelf over which a mouse moves should remain quite small, and clear of other input devices. As a rough guide, it should take no longer than about a second for an experienced controller to reposition a marker on a display with the required level of accuracy. A basic principle of sensitivity is that operation of the input device should seem natural, helpful and unobtrusive. It must never become distracting, for it is only a tool. The controller's attention should be on the tasks and not on the input device, which is not optimum for its purpose if it consistently distracts the controller.

12.5 *Mutual compatibility*

It should not be possible to mistake one input device for another. Devices should look and feel different unless they function as a group and rely on similarities, for example of alignment, to indicate agreement and compatibility between them. Such conventions of compatibility are common in some contexts, but not in air traffic control. Every input device in every state should carry sufficient designation for its functionality and state to be obvious and not speculative. This aids the controller's memory but also guides and warns others in the workspace. Labelling can help to discourage random exploratory tinkering when there is nothing to do. Important input devices, particularly those close to each other, should not feel the same when they are operated. They can be distinguished from each other by differences in the force required to operate them, in their distance or direction of travel, or in their forms of feedback. If necessary, more drastic steps should be taken, because similar input devices should never be adjacent if the accidental operation of one of them instead of the other could be potentially hazardous.

Input devices are never chosen in isolation, partly for ergonomic reasons. All have to be positioned efficiently within the same workspace. They will often be used in close succession, and sometimes simultaneously if the tasks require it, and the types and sensitivities of the chosen input devices must collectively satisfy such needs. As far as possible, unwanted incompatibilities between input devices have to be designed out, as they can be very difficult to remove or circumvent without high costs after the equipment has been installed. Many devices fulfil different functions within several air traffic control tasks, and they must not be optimized for one task without regard to the others. If it proves impossible for an input device to fulfil all its intended tasks satisfactorily, this warrants re-examination of the choice of input device and its sensitivity, and of the allocation and grouping of tasks, because although attempts to improve the compatibility of the input devices can be made they are likely to be palliatives that do not achieve satisfactory performance of all the tasks.

The following practical checks can be made on the mutual compatibility of input devices. Any devices intended for concurrent use must be intended for use by the same hand or by different hands, and positioned accordingly. Devices frequently used in succession should be positioned to facilitate smooth transition between them, without drawing the controller's attention away from the task to the means of performing it. Input devices should be ready for immediate use, and not have to be picked up, moved, switched on or adjusted beforehand. The forces required to operate input devices that are used in conjunction in the same workstation should not differ grossly in magnitude or direction, but quite small differences in force or direction that are not obtrusive or inconvenient should be encouraged as indicants that the correct device has been chosen. Different tasks should not impose different and incompatible sensitivities because they require both very gross and very fine movements of the input device. The functioning of input devices should not add significantly to any delays in task performance. It is desirable not to have many different types of input device within a single workstation. All the input devices within a workspace should share the same expected directions of movement and conventions of operation. Where more than one conventional layout may be applicable in principle, as with the layout of a ten-numeral keyboard, a single layout should be adopted universally throughout air traffic control workspaces.

Within the workspace, every visible but unobtrusive feature of the layout, such as its interior detail, background block colourings, common colour codings, and engraved and electronic lines, can reinforce the planned compatibilities and links among input devices. Every input device must indicate its operational state unambiguously and provide adequate feedback about when it is switched on: this can be particularly vital with switches for communication channels so that the controller does not start messages before the channel is open, nor clip the end of messages by switching off the channel too soon. Input devices should be chosen and positioned to be suitable for the hand which will use them. All devices are not equally suitable for the non-preferred hand, and those on the left of the workstation should have been put there partly because they can be used efficiently by the non-preferred hand for the required tasks. Task-induced changes from one input device to another should be minimized, but not at the cost of complicated instructions and procedures or of high memory load because a given input action can have many meanings and functions depending on the task, the stage reached or the form of feedback. Touch screens with different labelling at different times that the controller must initially read in order to use them are best suited to simple data entry sequences that can be learned easily, so that the controller can then look mostly at the radar rather than at the data input device. Basically the cure for incompatibilities between input devices is prevention. If the deliberate or inadvertent operation of input devices together or in succession could be operationally hazardous or could damage the equipment, then it is sensible not to rely on human memory or on labelling or positioning within the workspace to prevent this, but also to employ mechanical or electronic interlocks so that potentially incompatible devices cannot be operated together.

12.6 Relationships between input devices and displays

The acknowledged influence of the relationships between input devices and displays on task performance and efficiency has increased in recent years, but interpretations of the nature of the relationship have also changed. For air traffic control applications, it may be necessary to evaluate the input device, the display and the relationship between them as a single entity, so that the independent variable in experiments should be the combined display and input device instead of either of them alone. Recommendations then refer to combinations of display and input device, since the efficacy of either depends so much on the other and on how they are combined. Questions about what kind of input device to use become unanswerable without some reference to displays and relationships. Expressed more pessimistically, any benefits expected from an improved input device or display can be nullified by unoptimized relationships between them. The findings from the many experiments comparing and measuring input devices without reference to associated changes in displays or in relationships require circumspect treatment and scrupulous verification for air traffic control applications. The most appropriate combination also remains task dependent.

When paper flight progress strips were the displays and the record of controller actions, the current problems regarding the relationships between displays and input devices did not arise. Radar displays required controls to set them up and adjust them, and the input devices to select and amend the information on secondary radar displays introduced quite complex relationships between the displays themselves and the input devices that could control their appearance and content and could relate actions to displayed events. Not only were the relationships crucial for tasks, but they were often safety critical. Most forms of computer assistance, such as aids for prediction, problem solving or decision making, rely on displays for essential information and on input devices for taking action, and the workspace design combines and integrates them by showing the relationships between them. The more sophisticated and automated the workspace becomes, the more numerous and complex the relationships are, and the more vital it is to represent them correctly. Although current air traffic control workspaces vary considerably in the need to show the relationships between displays and input devices in detail, this need is growing as systems become more sophisticated, and it may introduce novel forms of relationship.

Workspace designs based on task analysis and other techniques indicate broadly the functions to be fulfilled by input devices, and the information necessary for those functions and for appropriate feedback. The relationships between displays and input devices reflect task requirements through the practical links between them in both directions, when displayed information prompts an input of data and when displayed information shows the consequences of an input of data. The relationships can serve as memory aids. They can remind the user about facilities and options available, can propose which should be used, and can provide feedback on the outcome. The relationships between displays and input devices can also constitute a form of job aid. This becomes apparent when the reason that an input device is never used or that displayed information never influences decisions lies in failure to show the applicable relationships between displays and input devices because of inadequacy in task design, in workspace layout, in labelling, in specifications, in information portrayal, or in training. Perhaps the information has been noted but the controller does not know what to do because the workspace design fails to indicate what the relevant input action would be, or perhaps an input device provides no appropriate recognizable form of feedback that would encourage the correct action. Comparable evidence can be obtained when an input device is used inappropriately. Failure to use an input device may not be attributable solely to the device itself but also to the relevant displayed information and its relationship to the device. The workspace design should make the functionality of every input device self-evident, and render all the interactions between input devices and displays obvious.

There are several practical ways in which good relationships between displays and input devices can be promoted and demonstrated. An obvious one deals with physical positioning. Input devices should be alongside or adjacent to the displays to which they refer. The workspace design can reinforce a relationship by applying the same unobtrusive but slightly different background hue to both the display surround and the supporting surface of the input device, or by employing equivalent shared codings. Exceptions may be input devices which are seldom required, so that the input devices to adjust a vertical radar display can be above or beside it provided that they do not thereby become associated with another display, thus reserving the optimum input device positions below the display for more important and frequently used devices for performing tasks with the display and for amending information content.

Basic principles should be applied; for example, no part of the display should be obscured by the operation of its associated input devices. A further principle is to adopt the same layouts for displays and input devices. Touch-sensitive surfaces embody this principle, but the same spacings and layouts should be applied to electronically generated labels and to their remote keyboards. Their absolute spacing need not be copied slavishly though it should be approximately the same, but their relative positioning must not be varied. The optimum spacing between adjacent electronic keys should be that recommended for keyboards, whereas the optimum spacing between adjacent labels of the keys on a remote display must be influenced by the content of the labels. The same conventions should be followed for movements of displays and of input devices so that the display changes made by input devices are always predictable. Movements that are clockwise, to the right, forward or up, should denote increases, in accordance with standard ergonomic practice. Wherever possible, displays and input devices should be aligned. It may be best to avoid switches if there is any possibility of international usage, because their conventions are contrary in different countries: down means on in the United Kingdom and off in the United States, for example. Common labelling can link displays and input devices. Aspects of the labels on the display that can be applied to input devices include label content, colours, and other physical attributes such as typefaces. The more similar the input devices and the corresponding displayed information look, the more effortlessly they can be associated and the fewer the errors of misassociation will be.

Physical links between input devices and displays can be denoted by engraved lines, mimic diagrams or other principles. These links are associated more with maintenance and fault finding than with air traffic control itself, but could be applied to it. It is sound practice to employ as few conventions as possible to show the relationships between displays and input devices within the same workspace, to employ them consistently, to make sure that the chosen conventions can never be incompatible with each other, and to group the usage of different conventions according to a simple and memorable logic.

The most direct links tie display and input device movements together, and are so obvious that they scarcely need mentioning. Examples are the movement of a joystick, rolling ball, mouse or other device to drive a cursor or marker across a display. The relationship should be so self-evident that the distinction between the input device and the display vanishes and the relationship becomes so close that it results in a compelling impression that the input device drives the marker directly. Such relationships exemplify such well-designed unobtrusive directness that the user perceives only a unitary process and becomes unaware of any contrived linkage or relationship.

13

Communications

13.1 Information transmission within the system

The safe and efficient conduct of air traffic control depends on the communication of information. Few human factors studies in air traffic control have taken communications as their main theme, though most have measured communications in order to achieve other objectives. Recent communications developments such as data links have prompted renewed interest in their human factors implications for the pilot and the air traffic controller.

Studies of communications can be classified in several ways. Since prerequisites for communication are a sender of information, a receiver of information, and a channel between them, human factors studies can examine the human sender, the human recipient, effects on the human of characteristics of the transmission channel, or aspects of the information itself. Possible themes representing the above four kinds of study could be the sender's intentions, the recipient's interpretations, the consequences of typical distortions during transmission, and the match between information content and task requirements. Another four-way classification of communications applicable to air traffic control is information transmitted from human to machine, usually through an input device; from machine to human, usually through visual or auditory displays; from human to human, usually through speech, gestures or signals; and from machine to machine directly. In the last instance, any human factors implications are indirect, and result from consequent changes either in the information presented or in the information to which the human can gain access. A further classification of air traffic control communications, which also happens to be fourfold, deals with information from air to ground and from ground to air, which is discussed in Section 13.2; information from ground to ground, which is discussed in Sections 13.3 and 13.4; and information from air to air, which is the direct concern of air traffic control only in so far as pilots are able to glean air traffic control information from it.

The information transmission within an air traffic control system can become complex and elaborate. Through modern sensors, a great deal of information is transmitted, stored, updated and processed by numerous automated computations. There is far too much information available for any human to absorb all of it, and it is far too detailed for most air traffic control tasks. A perennial human factors communications problem in air traffic control is to choose the best selection of categories of information, in terms of their content, formats, level of detail and updating, so that collectively they represent the aircraft in a region in the most appropriate way for the controller to handle them as air traffic. Some of the most difficult decisions concern information in the system which the controller has no need to know. Communications issues include the anticipated utilization of communications channels so that there is neither over-provision nor overloading. A familiar issue affecting the validity of real-time simulations is the discrepancy between the smooth and predictable functioning of simulated communications channels and the unreliable, highly utilized and noisy real channels which may sometimes force controllers to seek alternatives to them. It is not always apparent to the user when communications are faulty, and a failure may be attributed wrongly to an input device or display.

To the controller, the most familiar aspect of communications is speech. Its human factors aspects are discussed in Section 13.5. Speech is waning as a medium for air traffic control communications as more information is gathered and sent automatically. Human factors problems arise because this progressive replacement of speech with automatically transponded data can be an advance in quantitative terms but a retrograde step in qualitative terms. Communications have developed through dialogues, menus and other means of information exchange through human–machine interfaces. It is not always clear how much of the extensive human factors literature about similar information exchanges in other contexts can be extrapolated to air traffic control. The current comparative paucity of detailed human factors knowledge about air traffic control communications limits recommendations on the design of dialogues and menus in order to retain much of the traditional functionality of spoken air traffic control communications. Meanwhile, it is prudent to treat recommendations based on data from elsewhere as hypotheses which require verification for air traffic control.

13.2 *Information transmission between air and ground*

As the technical means for transmitting information between air and ground have increased in variety and complexity, so have the human factors issues associated with them, to the extent that air–ground communications have now been identified as a distinct human factors research topic (FAA, 1990; Benoit, 1991), in addition to communications within cockpits or within air traffic control, both of which have received some previous human factors study. The forms of communication have to match associated forms of information communication within the system. For example, if a controller with a radar display issues a change of heading to a pilot, the radar provides feedback that the instruction has been obeyed; or if the automatically transponded flight level of an aircraft appears on the aircraft label on the radar display the controller can check that the aircraft is at its prescribed flight level, whereas if a flight level instruction is issued to an aircraft in mid-ocean beyond radar coverage, speech may be the only feasible means of communication to verify that the instruction has been followed.

Initially most of the information about aircraft routeings was spoken by the pilot to the controller, who used it to provide the pilot with a safe route and flight level, clear of other air traffic and any known hazards. Because of the reliance on speech, the main human factors problems concerned procedures to minimize misunderstandings and auditory confusions, to link spoken messages optimally with associated activities such as the annotation of flight progress strips, and to make the spoken messages and the flight strip layouts as standardized and compatible as possible in order to prevent error and to provide an unambiguous record of the instructions given and agreed. The annotated flight strip and the recorded spoken messages became constituent parts of the official record of the flight and its progress. Although various practices and procedures and the available navigation aids influenced the updating of information about the progress of flights and offered opportunities for amendments, essentially the initiative of when and what to communicate rested with the pilot and the controller, working within the standardized rules.

New means of transmitting information raised new human factors issues. The advent of radar and the automated transponding of information between air and ground both resulted in information that was visual rather than auditory. The main objective became the unambiguous coding and presentation of the communicated visual information rather than the prevention of auditory confusions. This also generated a crop of fairly orthodox human factors problems, for which satisfactory air traffic control solutions had to be found. The alphanumeric characters within the labels on radar displays must never look so similar as to be mistaken for each other. Aircraft that are close to each other on a radar display must be clearly differentiated by other means, including positive identification of each separately. Lacking other means, a controller might request an aircraft to change heading and note from the radar display which aircraft did change and thereby identify it positively, but this is not feasible without radar. Changes in the information available thus introduce commensurate changes in the means to authenticate and verify it. New forms of information may not only require new means of verification, but may invalidate existing verification methods.

Future air traffic control systems will be able to transpond from air to ground far more information than the human can absorb or understand. Appropriate information has to match the task requirements in level of detail, quality, coding, formats, accuracy, precision, frequency of updating, and trustworthiness, so that the controller is presented with enough but not too much information for each task, portrayed optimally in ways that minimize the propensity for human error. In principle this seems straightforward, although a lot of work. In practice it is certainly a lot of work, but not straightforward

(Kerns, 1992). The controller employs almost every specifiable category of information for different air traffic control purposes at different times. These several purposes seldom require the same level of detail, and any compromise must therefore tend towards too much detail for any given purpose. Attempts to be more flexible either require the system to sense and adjust to the controller's tasks and needs, or let the controller choose the level of detail by the extra task of selecting it.

Task analysis can determine the quantitative information that is required, and ergonomic principles can suggest optimum forms of its portrayal in relation to the tasks, but the user also needs to know how far to trust it and cannot usually judge this by looking at the information. This issue of the quality and trustworthiness of information is emerging as important. What does the user need to know, and how could qualitative information be coded? It is possible to tell at once if speech is not clear and to ask for the message to be repeated or set about verifying it in other ways. There is no equivalent procedure or confirmation for information transponded automatically. A primary radar display contains qualitative information about clutter, fading, and the signal-to-noise ratio that secondary radar lacks. The circumstances under which information should be verified may not be apparent. Modern displays contain no information about how much processing they have undergone. The quality of the communicated information may have been vastly improved but with negligible effects on its appearance and on its usage by controllers. Changes in information transmission may be much less significant for the user than they should be if the nature and extent of the changes remain unknown and are not apparent from the information itself.

As the number of aircraft and the data about each increase, and until forms of computer assistance replace many human communications, channels and frequencies tend to become overloaded at high traffic levels. Pilots may also have to change frequencies more often wherever the region controlled by each controller has been partitioned because of the traffic loading. The propensity of the channel to become overloaded results in pilots in effect queueing to talk to the controller, and in concern with how long it takes to transmit messages (Cardosi, 1993) and to respond to them (Cardosi and Boole, 1991). In some current air traffic control systems the controller is talking or is being talked to practically all the time. Messages may be condensed, although the controller may try not to condense them too much lest the speed of the messages outpaces the ability to keep the annotation of the flight strips up-to-date or to make the corresponding data entries. This may be the only means by which some residual control over workload can still be exercised by a busy controller whom several pilots are trying to contact. A criterion of successful input devices is that the controller must be able to enter messages with them at least as fast as the messages are spoken, because in busy traffic there will never be time to recover from an accumulated backlog of messages.

In future, some pre-sequencing of the spoken messages on the same communications channel may become necessary, whereby pilots are heard in the order in which they try to contact the controller, although there would have to be some allowance for operational priorities so that routine communications could not block emergency ones. When channels are very busy, the question arises of how much of the communicated information is actually essential. Some time is gained by curtailing chattiness and courtesies and by choosing succinct message formats. It could be hazardous to gain time by dispensing with acknowledgements or confirmations, or by switching channels too rapidly and cutting off vital information. Although reducing the redundancy in air traffic control communications seems a sensible ploy, some apparent duplication of information is a vital safeguard, and a quest for less redundancy without full appreciation of all the consequences could hazard safety. Redundancy as a topic is discussed in Section 13.10.

13.3 Coordination

The coordination requirements can become the limiting factor on some types of air traffic control in high traffic densities. Where this can occur, reductions in coordination can seem an effective means to increase system capacity, but the evidence suggests that the implications of reductions can extend far beyond the reasons for making them. Coordination refers to controller activities that impinge on the activities of other controllers and therefore must be liaised and coordinated with them. An action by one controller to resolve a problem must not generate problems for other controllers. The removal of one hazard must not lead to others. A decision by one controller must not retrospectively invalidate decisions already taken by others.

Decisions about traffic in airspaces under the jurisdiction of several controllers require the agreement of all of them before they are implemented. There are many examples of coordination. A military controller must coordinate with civil colleagues before routeing a military aircraft across a civil airway without permanently allocated military flight levels. Departing and arriving aircraft using the same runway must be coordinated with each other. An aerodrome controller must coordinate with others before releasing departing aircraft. A terminal area controller must coordinate with others before clearing a climbing aircraft into an en route sector. Aircraft en route must be coordinated with oceanic controllers before embarking on trans-oceanic flights. A long flight across many sectors must be coordinated with the controllers in charge of those sectors before the full flight plan is agreed.

The extent of coordination depends on the airspace structure. In tacit or overt form it is required wherever the air traffic control responsibility passes from one workstation to another. The apparent cure for excessive task demands of reducing the region of airspace and hence the amount of traffic for which each controller is responsible could be applied indefinitely as long as there were workstations available, were it not for the associated massive increase in coordination requirements which rapidly cancels all other benefits and sets the capacity limits. In addition to coordination, negotiation can also limit capacity. A common example affects oceanic traffic where, from the pilot's point of view, there is scope for negotiation. The pilot, who is often implementing airline policy, wants the optimum flight profile on the best route to avoid severe weather and adverse winds, to conserve fuel, and to have a quick flight. This is helped by beginning the oceanic crossing at the optimum flight level with a high fuel load, typically at about 33 000 feet, and then gaining height progressively as fuel is burned. Since this is nearly optimum for most aircraft, many pilots will request it. The busy oceanic controller cannot satisfy every request and maintain separation standards. A pilot who is offered a non-preferred initial flight level or route may try to negotiate a better one. This takes time and complicates the coordination processes.

A common military air traffic control practice is for the control responsibility for an aircraft to remain with the same controller wherever it flies. To achieve this, the military controller has much more coordination work than the typical civil controller, with the result that each military controller can handle fewer aircraft than each civil controller.

Despite many advances in other aspects of air traffic control, coordination procedures often remain quite manual, and continue to rely extensively on the telephone in systems which otherwise include extensive automated assistance. Coordination tends to be wholly manual or wholly automatic, and attempts to integrate the two have not been particularly successful. Sometimes they have incurred additional manual tasks, such as the updating of computer databases with information on changes negotiated during coordination activities. The feasible coordination procedures are set by the system and workspace designs, although the main design decisions are not reached for coordination reasons. Changes are made for other reasons, and the consequences for coordination are then accepted. Most system designs represent a policy to establish how much coordination there has to be, rather than a policy to ascertain how much coordination there ought to be.

Coordination is much influenced by the division of control responsibility, by changes in staffing in response to traffic demands, by whether individuals or teams perform tasks, by team roles and structures, and by whether team members sit back-to-back or alongside each other. Decisions that displays will be shared affect coordination because the feasibility of temporary display changes for coordination reasons depends on how many are using the display. Changes to accommodate one controller doing coordination are unacceptable if the tasks of others who are using the same display for other purposes would be interrupted.

There are two distinct kinds of coordination. One essentially concerns planning, either pre-flight planning or the planning of the later stages of a flight already airborne. It involves obtaining clearances and agreeing the position, height, and speed to be achieved by an aircraft at the time when the control responsibility for it is handed over. The other kind of coordination is more tactical and short term. An example would be making the appropriate arrangements to agree to a request from a general aviation aircraft for special clearance to overfly an airport. Coordination can be considerably affected by policies on the air traffic control service that is offered. In this example, the policy could be to grant such requests for clearance whenever possible or to insist that the aircraft remains outside the airport control zone. Both options require coordination activities, but different ones.

Coordination requires clear standardized procedures that result in agreement between controllers about all aspects of the flight, the instructions and the decisions taken. The coordination procedures must be designed to be completed quickly, safely, and without ambiguity or misunderstanding. What has been agreed must be honoured. Coordination does not consist of a series of unrelated decisions about one or two aircraft, but of additional decisions within the context of previous coordination decisions which must not be invalidated or undermined by the new decisions. This implies that there must always be a clear and available record of previous coordination activities and of their outcome.

13.4 Handovers

A variety of air traffic control practices have evolved for the handover of control responsibility as an aircraft leaves the airspace for which one controller is responsible and enters the airspace for which another controller is responsible. Most civil flights involve handovers. Exceptions may be local training flights confined to the vicinity of an airport, and flights for which no air traffic control service is provided. In some countries, silent handovers are now quite common, whereby the transfer of control responsibility occurs tacitly at a fixed and agreed boundary between sectors unless it is prevented by some formal overriding action. Silent handovers are feasible only if no air traffic control intervention is required, but often in en route level flight no new instructions are needed at a sector boundary. Where it is essential to know that control responsibility has been handed over, it may also be necessary to inform the pilot of a change of frequency even though no control action is required. Some initial forms of computer assistance imposed a formal procedure for the handover of control responsibility because the computer had to be instructed to redirect information about the aircraft to the new control position. If these forms of assistance precluded silent handovers they were perceived as unhelpful because they added to the controller's work instead of reducing it.

The transfer of control responsibility for an aircraft is called a handover in some countries and a handoff in others. A further terminological confusion is that in some places a handover implies the reallocation of responsibility for an aircraft on which the information is complete, whereas a handoff denotes a more exceptional need to transfer responsibility to another controller because the current controller is unable to deal with it for reasons such as excessive workload or equipment failure. A handover in air traffic control can also refer to the transfer of the responsibilities of a workstation from an outgoing controller to an incoming one, for example at the end of a shift. The incoming controller normally has the initiative to assume responsibility and may need to build a mental picture of the traffic before doing so, but the outgoing controller has to decide what the incoming controller needs to be told beforehand that is not otherwise apparent. The handover of responsibility between controllers can be almost instantaneous when the traffic is light and there are no problems, but can take some time in heavy traffic while the incomer builds a picture of the traffic from the radar, the flight strips, the other information displays and spoken messages, prior to assuming control responsibility for it.

Handovers can occasionally become limiting factors on the traffic-handling capacity of the system. Extensive anecdotal evidence and some more limited objective evidence suggests that sometimes a controller may be able to handle extremely high workloads safely but not be able to hand them over. In practice this is seldom a real constraint because controllers know long beforehand when a change of shift or a rest break is due and can plan in advance to hand over their responsibility then. Nevertheless it may not always be feasible to hand over responsibility for heavy traffic instantly at other times. A practical measurement of workload or of the traffic-handling capacity of a workstation in simulation can be to demand an immediate handover without warning, and measure its feasibility, the consequences and penalties, any extra tasks, and the actions of the outgoing and incoming controllers in effecting the handover.

13.5 Speech

In principle, speech has great flexibility as a medium of communication. In practice much of this flexibility has to be curbed in air traffic control in the interests of standardization, intelligibility, completeness, and the prevention of misunderstanding and error. Speech is immediate, but compared with most other forms of communication it is also slow. It is free from the constraints imposed by even the most adaptable displays, input devices, and human–machine interfaces, through which the only functions that can ultimately be fulfilled are those for which some provision has been made in the design. All speech has to be intelligible to other human beings, and many of the products of speech have to be converted into data entries to the system.

Although speech in air traffic control seems different from other forms of communication, its actual application is similar enough to other forms of data entry to be compared directly with some of them in terms of efficacy for conveying air traffic control information (Prinzo and Britton, 1993). Speech is not necessarily more efficient, quick, direct or safe than other alternatives. Recordings of speech are less compatible with the rest of the system, and are more difficult to interpret in relation to other system events, than recordings of functions using the human–machine interface are. For some tasks and functions, speech is useless or very limited; it is difficult to describe verbally such functions as steering a cursor or matching complex visual patterns. However, much essential air traffic control information is in the form of words, and speech seems an obvious way to convey it. This applies not merely to single messages and their acknowledgement but also to discussions, dialogues, coordination and liaison, and to some planning activities.

Speech is subject to some characteristic sources of confusion and misunderstanding that are relatively independent of its content. Some of these are associated with individual human attributes such as accents and speech in a non-native language, both of which apply to air traffic control and have to be overcome mainly by training. Others originate in speech against high background noise, again a practical problem in some air traffic control environments. Still others refer to measurable speech distortions that characterize particular speech communication channels or particular items of equipment such as telephones, headsets or loudspeakers. The above sources of speech distortions in communications processes may induce misunderstandings, errors or false expectancies, or render spoken words unintelligible. Efforts are made in air traffic control to cure all the above kinds of distortion at source, but without full reliance on their success. Parallel efforts are made to compensate for any distortions of speech that cannot be removed at source by devising air traffic control words, language, syntax, and formats that allow for the characteristic distortions of speech during its transmission and for the resulting phonetic confusions by being robust and resistant to them, and thus attempt to minimize their consequences for safety and efficiency.

The inability of the air traffic control system to accept speech directly as an input has prompted extensive study of the possible applicability of automated speech synthesis and recognition. Forms of automated assistance with poor acceptability have not replaced speech but have required the products of speech to be entered into the system to keep the flight information in the system up-to-date. Such apparent additional tasks involving message duplication in spoken and data-entered forms are generally unpopular with controllers, but may not be wholly disadvantageous if they further entrench memory and understanding or help to prevent errors because identical errors in both media are improbable. Nevertheless, the extra work normally more than outweighs any such potential benefits, particularly since a claimed purpose of the assistance is often the reduction of controller workload.

Phonetic assessments of speech intelligibility often have to be supplemented by measures of the understanding of speech content. The purpose of speech in air traffic control is to convey information correctly and efficiently to the listener, a process that includes the listener's correct hearing and understanding. The standard speech discrimination tests have limited applicability to air traffic control since they can only show that a speech channel is inadequate, whereas comprehensive understanding of the controller's frame of reference and existing knowledge may be required to gauge the consequent misunderstandings and misinterpretations associated with the combination of the particular distortion with air traffic control speech. Intelligibility may often be improved by changing the frequency modulation or by reducing bandwidths since background noise tends to interfere with higher frequencies most. The listener's greater confidence in a clear speech channel affects all speech heard through it, and applies equally to messages that are correct and to those that are not.

There is extensive anecdotal evidence, though relatively little experimental evidence, that controllers and pilots base judgements about each other and about their colleagues on speech. These judgements cover general impressions of the quality of the air traffic control service or of the flying, as well as judgements of individual pilots and controllers. Listeners believe that speech conveys usable information about individual confidence, competence, professionalism, uncertainty and unease. Controllers and pilots use this information to request that a message be confirmed or repeated, or to modify their actions. For example, a controller who judges that a pilot seems unfamiliar with local air traffic control procedures and practices will give that aircraft extra attention since there seems a higher probability than normal that it will behave unexpectedly.

Although dichotic listening tasks in which one source of speech is directed to one ear and another source to the other are sometimes feasible, the ability to perform such tasks declines with age and some potential for confusion is always present. In air traffic control there can be direct speech between controllers in the same workspace and indirect speech through telephones and radio telephones, but it is preferable for the controller to be listening to only one speech source at any given instant. Ambiguities from more than one concurrent source of speech are not only of hearing but also of understanding. Multiple speech sources can involve rapid attention switching between them, distractions from new messages, the inadvertent combination of parts of different messages, the need to hear the entire message to ascertain its relevance, and the misallocation of information to the wrong source. If the task and function designs do not preclude simultaneous spoken messages from different sources, a facility to suppress sources temporarily and selectively may have to be provided if the interference between the sources impairs intelligibility or understanding. Some partial monitoring of further speech sources may be feasible and advantageous, so that the controller using a headset or telephone nevertheless remains aware of other relevant spoken messages within the control room. There must also be some provision to override speech channels in an emergency although only a dire emergency justifies interrupting the controller at once, since tasks interrupted when half done may remain uncompleted and be forgotten thereafter.

Air traffic control speech is a product of much training and learning. To a naive listener, air traffic control messages are almost unintelligible. Someone unfamiliar with the characteristics of a communications channel may be able to understand very little of what is said using it. Even if the words can be made out, they do not make much sense without a knowledge of air traffic control. The correct interpretation of speech relies heavily on the experience, knowledge and training of the controller. Predictable and standardized patterns of communication benefit performance and team functioning in aircrew (Kanki *et al.*, 1991), but any findings about speech obtained in such other contexts require verification before they can be applied to air traffic control.

The amount of qualitative as distinct from quantitative information in speech has been seriously underestimated in air traffic control and elsewhere. Most proposed developments in automated assistance for air traffic control communications consider quantitative aspects of air traffic control messages exclusively, and contain no information that corresponds with the qualitative information in speech. It would seem prudent to check that it is safe to dispense with such information before doing so, but the most telling evidence is often anecdotal rather than objective or experimental. Many individual pilots confronted with an emergency have thanked air traffic controllers afterwards for the professionalism and confidence of the air traffic control service that they received, as well as for its actual substance, the implication being that the pilots were able to concentrate on the airborne emergency without having to worry about its air traffic control aspects which they were confident would be handled safely. These confidence judgements were not based only on the air traffic control actions but on the calmness, authoritativeness, professional confidence and competence which the controller conveyed to the pilot through speech.

Many aspects of speech can be the basis of such judgements. The soundness of the judgements themselves has not been tested. Some pilots may attribute to controllers a self-confidence that does not exist. The point is that these qualitative judgements by the controller are the basis for many actions and decisions: to repeat a message; to seek confirmation sooner than normal; to give extra attention to a pilot who seems uncertain; to speak messages more slowly and more distinctly if there is a hint of uncertainty; to adhere with particular rigour to standard message formats; to omit inessentials and pleasantries if a listener sounds exceptionally busy; and to keep reconfirming very unusual information. All qualitative aspects of information are not necessarily beneficial. A curt or angry response may discourage further spoken enquiries. An action that should have been challenged may not be queried. A controller may hesitate to remind a pilot if previous reminders have been rebuffed. But aviation safety is more important than irritability, and the irritability itself is a form of qualitative information and feedback to which any further speech content can be adapted. Without speech, such information is unavailable.

The progressive replacement of speech by the automated transmission of information has significant consequences for human errors. Many characteristic errors associated with spoken messages will become impossible, but new forms of human error will emerge that were impossible before. Most of them will be unrelated to speech. A potential benefit of automated assistance is often increased willingness to query a message that is not understood. Whereas a questioner may be reluctant to address an apparently simple question to a human being for fear of seeming stupid and losing face, the same questioner knows that it is the computer that is stupid because it has not made its meaning clear and so is more willing to go on requesting clarification or further information since this incurs no loss of face. For efficient and collaborative dialogues it may be necessary for the human to ascribe all failures and misunderstandings to computer inadequacies rather than personal ineptitude. The meaning of direct speech between controllers is supplemented and interpreted in relation to information about the speaker, including gestures and status. These kinds of information are lost when data are transmitted automatically, and it is necessary to confirm beforehand that this loss can be sustained and has no adverse effects on safety or efficiency.

13.6 Air traffic control language and terminology

The specialized terminology of air traffic control is not large compared with those of some other disciplines, and considerable progress has been made towards the international standardization of its language and vocabulary. Most maturing professions evolve their own technical terms, and air traffic control is no exception. The technical terms of some disciplines such as medicine and chemistry are instantly recognizable as such because they are neologisms with carefully defined meanings. Unfortunately both human factors and air traffic control share the contrary propensity to borrow as technical terms words in common currency, and to assign a narrower and more specific meaning to them whenever they are employed in human factors or air traffic control contexts. This leads to the confusion that it invites. People unfamiliar with air traffic control who overhear air traffic control dialogues may fail to appreciate that many words do not carry their everyday meanings but specific technical meanings familiar to controllers but not to others. Inexperienced pilots may be misled if they ascribe everyday meanings to air traffic control terms, in ignorance of their technical meanings. The experienced controller must be on the alert for such misunderstandings which can require additional confirmation of messages.

Most air traffic control language is not in the form of sentences, and therefore many of the recent developments in linguistics are not applicable to it. Although air traffic control language has a pattern, and many vocabulary items have symbolic significance, it has not proved fruitful to apply semiotic concepts to air traffic control language either. There have also been problems in ensuring that human–computer dialogues in air traffic control employ its technical terminology correctly, and that concepts do not vacillate between their technical and everyday meanings. Air traffic control language and terminology, though standardized, retain some flexibility, particularly in the pace and details of spoken messages. If there is time for courtesies, they are exchanged; if there is not, succinct standardized messages shorn of verbosity are transmitted more quickly and acknowledged peremptorily. A consequence of the fluidity and adaptability of detailed air traffic control message content in accordance with immediate needs is that some system measures of the utilization and occupancy of transmission channels may be uninformative as indices of capacity or workload.

Air traffic control employs numerals and the alphabet. The widely used International Civil Aviation Organization (ICAO) alphabet evolved through an extensive programme of research (Moser, 1959), and is about as robust as an alphabet can be in remaining intelligible to the listener despite noisy backgrounds, distorting transmission channels, mispronunciations, misplaced accentuations, foreign or regional accents, and other sources of phonetic confusion, though even the ICAO alphabet can be defeated by background noise, garbling, and gross distortions. The criteria of good human intelligibility do not necessarily transfer to good machine recognition, and the ICAO alphabet is probably not optimum for automated speech synthesis or recognition. The alphabet is: alfa, bravo, charlie, delta, echo, foxtrot, golf, hotel, india, juliet, kilo, lima, mike, november, oscar, papa, quebec, romeo, sierra, tango, uniform, victor, whiskey, x-ray, yankee, zulu.

Some human factors evidence on air traffic control terminology that could apparently remove potential sources of confusion is not applied because the reasons against it have seemed more compelling. An example concerns the allocation to flights of the callsigns which feature in almost every air traffic control message because they are the means by which aircraft flights are identified. A callsign consists of a mixture of letters and/or numerals, and the ICAO alphabet is used for the letters. Controllers worry about the possibility of confusing the callsigns of different aircraft, although actual hazards attributable to callsign confusion are rare. The extensive human factors evidence defines the kinds of similarity between callsigns that can lead to confusion in this context. The cure must lie in forward planning to prevent the prior allocation of similar callsigns to aircraft that will be in the same airspace at about the same time. These plans may conflict with established practices, such as the addition of a callsign suffix to denote a back up flight along the same route, and the allocation by an airline of similar callsigns to its aircraft on the same route so that the callsign also serves as a route designation. Because the aircraft of each airline share a common callsign prefix, phonetic discrimination between their callsigns is more limited than it would otherwise be. For these and other reasons, the probability of aircraft with similar callsigns being in the same airspace at the same time is greater than chance would suggest, or than it need be. On this topic, the requirement is not for further research, but for the political will to apply the human factors evidence that is already known.

13.7 Message formats

Most air traffic control message formats have become entrenched and familiar, and compelling reasons would be needed to consider changing them. It is the normal practice to send all the items of information to be conveyed to each aircraft within a single message that begins with the aircraft callsign and ends with a positive indication that denotes completion of the message and requires acknowledgement of its receipt. Many of the potential sources of ambiguity in air traffic control message formats were identified and resolved long ago, especially where they could hazard safety in exceptional circumstances. While such potential hazards can never be wholly eliminated, they have been much reduced by experience, subject to strict compliance with message formats in which the content, phraseology, sequences, and phonetics are all standardized. A major reason for caution is that well-intentioned changes in message formats could reintroduce obscure sources of error in rare circumstances, some of which might take years to become apparent.

The message formats from ground to air and from air to ground are not the same. Most messages from ground to air consist of information and instructions, whereas confirmations and requests are more common from air to ground. When traffic is dense and controllers are busy, pilots are discouraged from quibbling over air traffic control instructions, although they retain responsibility for the safety of the aircraft. In such air traffic control contexts as trans-oceanic flight levels or routes, there may be quite protracted negotiations, conducted using standard message formats, between the controller who knows the routes and flight levels available and is obligated to provide an even-handed service to all aircraft, and the pilot who is concerned with fuel efficiency, company policy, and a smooth and expeditious flight.

A few guiding principles can help to ensure that message formats are satisfactory and safe. All messages should be spoken at a regular but not completely uniform pace without gross changes in pitch or loudness, particularly near the beginning or end of messages. Standard phrases and formats must always be employed, even when they seem unnecessary or superfluous. Acknowledgements, requests for confirmation or repetition, and sought clarifications, or any other actions to resolve doubt or ambiguity, must never be omitted. It is always better to ask, no matter how obvious the answer seems, than to make a false assumption. Attempts to pass too much information too quickly within a single message can induce errors, particularly if more than one item contains similar sets of numerals. No circumstances whatever can justify dispensing with standard procedures.

13.8 Automated speech recognition and automated speech synthesis

Several years ago, both automated speech recognition (also known as direct voice input) and automated speech synthesis, where the machine speaks, seemed to have obvious applications in air traffic control. The messages were relatively standardized, the vocabulary was limited and specific, the formats and contents were closely prescribed, and some speech was becoming redundant because the same information had to be spoken to the pilot and entered into the system. If a machine could recognize the speech, the manual data entry could become superfluous. Even a superficial examination of air traffic controllers revealed that they spent a great deal of time talking and that much of the conveyed information was quantitative. The automation of speech also seemed to offer a solution to political difficulties over the reconciliation of the adoption of the national language for internal air traffic control with international air safety requirements. Many foreign pilots would fail to understand overheard messages not spoken in standard air traffic control English, but perhaps what the controller said in his or her native language could be converted automatically and transmitted to the pilot in standard ICAO English.

There has not been much further progress in recent years. The technology still looks promising but its operational adoption seems to recede rather than advance. The technical complexity entailed and the human factors problems to be overcome are now better appreciated. They include dealing with non-standard messages, the implications for safety of the associated reduction in redundant information, and the reduced roles for speech because of fewer tactical instructions and more automated transponding of data. The work to optimize the human intelligibility of air traffic control messages does not guarantee comparable machine intelligibility. Even with a limited vocabulary, machines seem to have particular difficulty in recognizing the speech of some individuals, or of the same individual with a head cold or when using different equipment. It has proved difficult to achieve consistently levels of reliability in speech recognition comparable to those with keyboards, and human exasperation with failure is counterproductive.

The emphasis has now turned towards practical applications that are not safety critical, as in air traffic control training where the technology could issue a limited number of instructions in response to a controller's actions or could recognize a limited number of spoken messages and respond appropriately. Tangible savings could accrue if the equipment replaced humans simulating pilots while controllers practised basic procedures. Any practical applications of automated speech might require some modifications of wording or special forms of speaking to minimize potential confusions. Voice recognition might aid the security of the system and of the information within it: the controller speaking his or her own name could set up the workstation in the state preferred by the controller, but could also prevent unauthorized access to it.

The human factors implications of automated speech synthesis and recognition are not confined to their technical aspects. Human–machine and human–human relationships are not the same, particularly in terms of loss of face. Any significant national differences in this respect are within the province of cultural ergonomics, which deals with cultural differences in acceptable human–machine relationships among other matters. The effects of spending much of the working day conversing with machines on the ability to converse with humans warrants examination, because automated speech will not replace all human speech in air traffic control. Those who do relate best to machines may not be best in teams, and the implications for the selection of controllers would have to be pursued. The prevention of task duplication by direct voice input may not be a boon under all circumstances, especially where automated cross-checking of entries in different modes is feasible.

13.9 Other auditory information

Speech is not the only form of auditory information in air traffic control, though it is by far the commonest. Speech must neither interfere with nor be impaired by other auditory information. Some auditory warnings such as fire alarms are a standard feature of workspaces, but air traffic control environments contain other sounds that interact with speech. There is noise from aircraft, from operating equipment such as keyboards and switches, from communications channels, from environmental equipment such as fans, from people moving about, and from ancillary functions such as cleaning and maintenance. Some sounds are beneficial and are introduced deliberately for that reason: an example is clicks that provide feedback of successful contacts or inputs in devices that have no moving parts and operate silently. The ideal background noise level is no higher than about 55 dB, so that the levels of all speech and of all auditory feedback can remain correspondingly low but may always be heard. Auditory warning and alerting signals pertinent to air traffic control must be discriminable from external signals such as fire alarms against the background noise levels, from the room itself and from headsets, telephones, and loudspeakers.

If auditory warnings are employed sparingly as they always should be, they can be very efficacious in alerting the controller, but at the cost of disrupting the current tasks. Care should be taken over presenting even auditory warnings of emergencies during any message or conversation rather than on its completion, because an interrupted message is a potential source of error. The standard ergonomic guidelines for the discriminability of warning signals apply to air traffic control and should be followed (Hellier *et al.*, 1993). Any warning signals which function only as alerting devices and carry no intrinsic meaning should be used very sparingly, and be supplemented or combined with concurrent visual indications of their meaning. Sometimes urgent auditory information demanding immediate attention arrives in the form of speech.

Several principles apply to all auditory warnings. The meaning of every warning must be clear, preferably as an intrinsic aspect of the warning itself rather than as the learned significance of a sound such as a buzzer which requires supplementary information on the nature of the warning. Warnings should be given only if the urgency of the situation warrants them. They should be linked closely to information about appropriate actions, particularly if they are rare. They should be mutually discriminable within the air traffic control environment. They should never last for a long time. Handbook recommendations about auditory warnings may have to be modified to compensate for background and extraneous noises and for the characteristics of headsets and other equipment. If auditory warnings must be heard above speech, it may be desirable to adjust the amplitude of the warning, depending on whether the speech channels are in operation or not. Loudspeakers may be used to draw attention to an emergency that affects everyone in the control room, but they are not suitable for other warnings. The telephone may be the best means to attract the attention of a controller temporarily absent from his or her workstation but still within the workspace. The practicality of auditory warnings through the headset depends on whether the headset must always be worn while on duty, and on whether it normally covers both ears or only one ear because the controller listens with the other to the sounds and speech within the control room. Warning signals are always more successful in a good acoustic environment.

13.10 Redundancy of information

When air traffic control communications are described in terms of information theory or expressed in alternative quantitative forms, considerable redundancy of message content is usually found. Apparently the same information could be conveyed more quickly and efficiently by curtailing this redundancy, but the consequences for safety are not known. The principle of presenting to the controller only the information required for the tasks in forms that can be absorbed in the time available has the incidental effect of reducing redundancy, which would be far greater if everything available was presented. Information categories that seem fully redundant but are not, can remain unrecognized until they have been removed deliberately or inadvertently, whereupon it becomes apparent that some of the information that controllers formerly used has gone. This has happened before with many kinds of qualitative information and could happen again with the replacement of speech by data transponded automatically and with the resultant loss of information about the confidence and competence of speakers.

The idea that less redundancy in air traffic control communications is on balance beneficial is an assumption that has not been proved. Many of its implications are predictable. If tasks are never repeated, functions are never duplicated, repetitions and acknowledgements are curtailed, updates are automated or requested less frequently, and message formats are shortened by pruning redundant information, then time should be saved and system capacity should increase. On the other hand, there might be more errors and misunderstandings and less certainty, and part of the time saved might have to be spent in gathering and cross-checking other information. Nevertheless, the potential savings might seem substantial enough to encourage those trying to increase traffic-handling capacity by reducing the time that each controller needs to devote to each aircraft. What has not yet been quantified is the extent to which air traffic control efficiency and safety are currently enhanced because errors are detected and remedied through redundant information, because more than one sense modality is involved, because there is some duplication within tasks, because functions are cross-checked, or because data are routinely confirmed. The redundancy within such procedures does not render them infallible, but on balance their widespread adoption stems from their perceived helpfulness in preventing human error. Incident investigations commonly identify, as contributary causes, laxity in following standard procedures to confirm, duplicate, and otherwise act redundantly.

Almost all generalizations about redundancy seem ill-advised. Probably some air traffic control tasks and functions would indeed benefit from reduced redundancy of information if it could be achieved by automated cross-checking without loss of safety or efficiency. Equally probably other tasks or functions would suffer because reduced redundancy would bring more human errors, a lower probability of detecting them, and a more superficial level of controller understanding. Unfortunately our current knowledge and measurement tools are inadequate to discriminate between the tasks that would or would not benefit from less redundancy.

13.11 The quantification of information

Attempts to quantify air traffic control information flows and communications have many applications, including the specification of task requirements during system design, model building to describe system functioning and to predict the effects of changes, and information flows in relation to distributed or centralized system architectures. The broader issues are not primarily within the province of human factors except for the human factors implications of their solutions in terms of the communications facilities, the level of detail of the information, and the data available for access and manipulation at each workstation. A particular problem is information derived from independent sources and not fully integrated for presentation to the controller, so that for example information on an electronic display and on a radar display that purports to be the same contains discrepancies of detail because the displays are not updated at the same rate or at the same time or with the same incremental steps.

The main human factors problems associated with the quantification of air traffic control information relate to its details rather than its broad categories, to the quantity and quality of information that the controller needs to gain access to, and to its presentation and coding. Attempts to quantify information treated hitherto as qualitative should be pursued, particularly if it might otherwise be removed in the guise of technological progress. An example may illustrate some human factors complications associated with quantification. It is easy to express quantitative measures of the occupancy of communications channels in absolute or percentage terms, and apparently possible to predict incipient overloading from the percentages and patterns of occupancy, for high percentages will lead to occasional queues and consequent delays. However, verbal communications have a propensity to expand or contract to occupy the time available. High occupancy times alone cannot indicate if this is happening or if occupancy times could be reduced with no loss of essential information. Detailed examination of the information transmitted, and of its level of redundancy, pace, formats, and succinctness, is required to furnish such evidence. All the useful information has to be identified in detail to discover how much channel occupancy could be reduced or how much more information could be sent through it. Even then some suspicion would linger that not every useful kind of quantitative and qualitative information had been identified and that its compression might affect its usage in unexpected ways.

There are some alternative approaches. One is to quantify information in terms of air traffic control categories and headings, using air traffic control terminology. This can show how the controller's time is shared among various air traffic control activities, which can be subdivided until the desired level of detail is reached. Although this approach can yield quite a full description of air traffic control, it becomes so complex to include all the interrelationships that few objectives can justify the effort entailed in such detailed quantification. Another approach is to employ psychological headings such as statements, requests, questions, confirmations, acknowledgements, instructions, dialogues, negotiations, and the like. That different tasks require different proportions of these communications activities can be of interest factually but not for most planning or research purposes. Such categorizations can inadvertently encourage misguided policies, such as the reallocation of time from routine data entry tasks to decision making, which sound sensible but are actually naive and impractical, for they misunderstand the integrality of air traffic control tasks and imply wrongly that time saved on one task can be reallocated to any chosen alternative. A further problem can be an absence of suitable codings, for example, of the various dimensions which define the different characteristics of traffic flows.

It is essential to define the purpose of any quantification of air traffic control information beforehand and to select the categorizations and the level of detail accordingly. An inappropriate form of quantification can be very wasteful; quantification intended for a basic taxonomy of errors could be useless for other purposes such as identifying the main limitations on capacity or deciding if additional communications channels are needed, and it should never be presumed that the same kinds of quantification can achieve several different objectives. Simplistic attempts to quantify air traffic control information for such purposes as the estimation of system capacity do not merely misuse scarce resources but are actively misleading if they omit vital factors or ignore important qualitative influences.

14

Forms of computer assistance

14.1 Data manipulation

14.1.1 Data gathering

In the early years of air traffic control, data gathering was a human activity. All the information about a flight was written on a strip of paper. If different controllers had the responsibility for different phases of a flight, either the strip was handed on or separate strips were prepared. The controller at a work-station put the strips into the sequence in which the aircraft were flying along the route, or ordered them according to an alternative principle such as air-craft heights or destinations. Changes agreed between the pilot and controller during the progress of a flight were annotated on the strip by the controller. Navigation aids that provided fixes and gave bearing or distance information could also supply data about the air traffic, although they led to some prob-lems in synthesizing the different kinds of information. The data gathered by radar could be presented to the controller as a plan view of the traffic, but a plan view alone can be confusing unless each aircraft can be identified so that its position on the radar, its flight progress strip and information spoken by its pilot can be correlated. Further kinds of data were therefore gathered, such as positive evidence of aircraft identity squawked automatically in response to selective interrogation from the ground, which is much better than requesting an aircraft manoeuvre in order to identify it. Such data, once gathered, can serve other purposes.

The applications of automatically gathered data depend on their quality, which may not have to be as high for human factors applications as for other applications such as automated computations because human attributes limit the extent to which the quality of information can influence human per-formance. For example, the minimum changes that can be made on a second-ary radar display depend on the number of pixels and the screen size. However, their human consequences depend on the minimum size of changes that are perceptible, the minimum changes that have any effects on tasks, the rates of the updating of information compared with the rates of human information processing, and the effects of the data processing on the human ability to detect the onset of changes. There is no point in gathering for human purposes information that would be below the applicable visual or auditory thresholds and could not be seen or heard if it was presented.

Technological advances provide new data sources such as data links and satellites, and practical recommendations are made about data transmission, content, detail, updating, verification, and incremental changes, based primarily on technical feasibility, storage capacities, computational requirements, and regulatory and legal constraints. All human factors needs can normally be fully met by a very small proportion of all the data gathered, and the system contains far too much data for the controller to process, even if all the data could be presented. Selection of the most apposite data for the controller is correspondingly crucial.

Data gathering as a human activity in air traffic control has progressively decreased and is expected to decrease further, although data gathering by traditional manual methods persists where traffic is light or no advanced equipment has been installed. Nevertheless controllers generally learn less now than formerly from data gathering, and the building of the traffic picture and the maintenance of understanding have to rely on other means whenever the data are gathered automatically. An issue is the extent to which controllers need to remain aware of data that have been gathered and are accessible, but are not presented. In more automated systems it becomes impractical for the controller to initiate the gathering of data, except by the traditional methods of speaking to pilots or other controllers. The controller may be keen to discover the serviceability of automated means of gathering data if the integrity or correctness of the data arouse suspicions.

14.1.2 Data storage

Since the computer storage of data became comparatively cheap, it is seldom a significant limiting factor on the system. It is practicable to determine what data should be gathered on the assumption that any consequent data storage problems can be overcome. The human factors problems associated with data storage are of a different kind. The human controller is presented with, or has access to, the highly selective compiled synthesis of the data in the system that is believed to constitute the information required for all human tasks. The large data storage capacity of the system, which may induce the controller to ascribe any remaining problems to difficulties of data accessibility, cannot guarantee that the requisite information for every task has been gathered and stored.

Almost unlimited capacity can encourage the storage of all available information rather than of all information with known uses. The more data there are, the more complex the means to identify and access particular data items tend to become. Structuring the storage to facilitate accessibility can pose problems that must be resolved. Air traffic control cannot afford the common dilemma for the system user when long and inconclusive searches for data are futile either because the sought information was not stored or because the means to access it could not be discovered. From the controller's point of view, the storage of large quantities of data may have little relevance unless their information content is known and they can be accessed. The data storage requirements for human factors objectives and for automated functions differ greatly.

14.1.3 Compilation and synthesis

The main human factors effects of improved data gathering and storage occur through consequent advances in the treatment of data (Lane, 1991). These advances change human tasks and procedures and promote air traffic control objectives by fostering efficiency, smoothing traffic flows, enhancing safety, increasing capacity, improving controller performance and productivity, or benefiting communications. The intentions of data manipulation are to provide up-to-date information in appropriate forms and levels of detail to satisfy air traffic control needs. The compilation and synthesis of data appropriate for human tasks and needs can be very different from the compilation and synthesis of data appropriate for automated functions and for computer assistance based on calculations, and this disparity between human and machine functions is growing. Systems that can store and handle more data are designed to fulfil automated functions that require more data, but controllers cannot process much more information than they already do process. Therefore the amount of information provided for human tasks should not normally expand with the amount available or stored. Although greater amounts of data require more condensation and perhaps more simplification as they are compiled and synthesized for human use, their more condensed and simplified forms must nevertheless continue to represent adequately the more complex stored data from which they have been derived.

Initial processes of data compilation and synthesis that were intended to meet human task requirements reduced the quantity of data, smoothed and integrated data, clarified relationships, and specified the criteria for changing or updating data. An obvious requirement was to synthesize the plan view on the radar display with the tabular information about the same aircraft, and with all other information about them in such forms as labels, electronic flight strips and windows, in order to integrate the information about the location of each aircraft with its identity, flight level, heading, route, state and intentions. For the controller to assemble this information manually, much cross-referral and data compilation was involved, and the automated manipulation of data to effect this synthesis could be of great benefit, though it must be reliable. Unfortunately, the best efforts to date have achieved only a partial integration.

The removal of human errors in compilation and synthesis must not be negated by new automated errors or new human errors which are artefacts of the kinds of data manipulation employed. From time to time technological advances in air traffic control provide an opportunity to re-examine the compilation and synthesis of air traffic control data in relation to its human usage: a current example is the re-examination of the compilation and synthesis procedures for flight progress strips, prompted by the advent of electronic strips.

Nowadays some of the main human factors issues from the compilation and synthesis of data concern calculations based on the data. The available database can contain details of each planned and current flight, of air traffic control rules, standards, practices, procedures, instructions and tasks, of many information processing capabilities, and of other relevant information such as temporary restrictions and weather. How should these data be manipulated and what calculations should be done with them to promote air traffic control objectives? An implication of the stated intention to increase system capacity partly by reducing controller workload is to compile and synthesize data to support or reduce time-consuming human roles like solving problems, reaching decisions, making predictions, issuing reminders, and scheduling work activities. The brunt of current efforts is therefore directed towards supporting the controller's higher mental functions, since compiled and synthesized data can be applied to compute solutions to current problems and to plan future traffic flows efficiently and safely. In principle the compilation and synthesis of the data could become so advanced as to remove many currently significant practical limits on the automated and the human functions that could be fulfilled. The key issue then becomes what should be done, rather than what can be done. However the issue is resolved, some human factors problems will arise in integrating human and system roles.

14.1.4 Distribution and retrieval

From a human factors point of view, distribution concerns the availability of the required information at the locations and workstations in the system where it is needed for the tasks. Retrieval refers more specifically to the accessibility of data through appropriate devices within the human–machine interface, and to the helpfulness of the designs of tasks, of the labelling of input devices, and of other tools in facilitating the retrieval of required data.

Trends towards the greater standardization of air traffic control workspaces in the interests of cost savings, flexibility and interchangeability have encouraged the provision of similar hardware and facilities at each workstation. Differentiation between jobs then relies primarily on software changes, including selective data manipulation. The furniture and hardware of every suite may look the same so that every workstation appears identical until it is switched on, whereupon it is distinguished from others by display content, and by the labelling of functions and facilities such as communications channels.

For successful data manipulation, the controller must know all the available data sources or be able to discover them easily, must understand how to retrieve data quickly from them, and must recognize the forms in which the data will appear after they have been processed. The controller's knowledge covers all the tasks for which the data may be required, and how and when to use the data for the tasks. The data must not just be available when needed, but packaged in optimum forms and levels of detail to match the tasks. This involves prepackaging, preplanning and preclassification of the data in relation to the task requirements, so that the data retrieved are always right for the tasks. The level of detail must not be so excessive that the controller cannot find what is required, nor so restricted that the controller must seek additional information. Nor should significant amounts of the retrieved data be so irrelevant to the tasks as to constitute clutter. The data retrieval processes must not themselves be cumbersome or obscure.

Similar considerations apply to the distribution of information. In communications and in coordination and liaison tasks, controllers may need to know what information colleagues are using. This implies some knowledge of how information is distributed and made available for different tasks within the system. It may be vital for controllers to know who else is using the same information, particularly if they can change it. In packaging information for retrieval for various tasks, coordination requirements must be included and met. This aspect of data manipulation is often neglected in air traffic control systems. Particular difficulties can arise either when different tasks within the same workspace have different data requirements, or when similar tasks in different workspaces have different data requirements because they deal with different airspace regions. Without appropriate data input devices and menu selection, the correct balance can be elusive between selectivity of information and the provision of all the information needed for all the tasks at a workstation. The controller should never have to resort to protracted data entries or to extensive searching through menus to find what is wanted. Not only has the information to be correctly distributed and labelled for retrieval, but the controller may have to be reminded about what is available and how to retrieve it, which is an aspect of data presentation.

14.1.5 Data presentation

The presentational aspects of data manipulation have received most human factors attention because the interactions between the human and the forms of computer assistance use the presented data. Even if all the preceding stages have been accomplished satisfactorily, deficient data presentation will result in deficient forms of computer assistance. Some forms of computer assistance portray data in novel ways, such as windows of tabular data, icons, colour codings, electronic tabular displays, and the selective calldown or suppression of information categories. In menus and dialogues, the data presentation includes the options and the categories that the controller may manipulate.

Task analysis procedures, if followed correctly, include the specification of the data to be gathered for human tasks, from which the data storage requirements can be deduced. The stored data are compiled and synthesized into forms that match the task requirements and that are compatible with practical procedures for data retrieval and distribution. Issues of data presentation can then be addressed in terms of available technologies, environmental considerations, and appropriate sense modalities and codings. The outcome ought to be computer-manipulated information optimized for all the identified task requirements, with effective data presentation as the culmination of the other activities. In practice, these exemplary principles and procedures are seldom followed, and human requirements are often insufficiently heeded, with excessive reliance on technological advances and on putative rather than substantiated benefits. For example, if a decision to opt for colour displays is taken first, the applicability of colour coding to the defined tasks must then be considered whether colour is a suitable information coding for them or not, whereas the task requirements and the applications of computer assistance should be defined first, followed by the optimum forms of visual presentation of information chosen to match the defined tasks and technologies. Colour is then used optimally as the most efficacious coding, rather than non-optimally because it is available and must be applied somehow. It is fair to claim that this problem has tended to become less serious since stronger evidence has been demanded to prove the claimed benefits of changes.

Most of the ways in which computer assistance can extend human functions depend on its capacity to handle far more data than the human can comprehend or than can be presented. Among vital distinctions that have to be made clear by data presentation are that predicted future situations are never confusable with real current ones, and that the presented options are the only ones or further unpresented options are available. Some indication of the criteria for choosing options and for showing their different consequences is needed. If options have been excluded because of circumstances or the activities of other controllers, suitable presentations to inform the controller of their existence are essential. The data presentation should also show the status of information in relation to tasks, and particularly whether procedures are optimal or mandatory. An obvious presentational deficiency of many forms of computer assistance is the absence of qualitative information, often because suitable codings to depict qualitative characteristics of information such as its reliability and trustworthiness have never been developed. Solutions of such human factors problems have to be devised and proven before the controller can apply the presented information optimally.

These remaining problems of presentation stand out because so much successful human factors work on the presentational opportunities offered by computer assistance has been accomplished. Many principles for the successful presentation and coding of symbolic, graphical, alphanumeric and pictorial information on displays are proven and validated, to the extent that direct recommendations can be based upon them. If the information is right for the tasks, optimum or at least satisfactory forms of presenting it that meet all known human factors requirements can often be specified. The main practical problem is to educate those who plan, design and procure air traffic control systems on the availability of the extensive human factors database.

Feedback is an essential feature of information presentation, to let the controller know if instructions are obeyed and if expected consequences materialize. Through feedback, the controller learns from experience. Manual systems incorporate standard acknowledgements and confirmations as feedback, but computer assistance rarely provides comparable independent forms of feedback or means of verification. Some computer assistance in other contexts automatically compensates for certain system deficiencies, but the system user remains unaware of this because the cues are too tenuous to be noticeable or there is no provision to inform the user about them. In such circumstances no feedback can take place. These compensatory activities may be a useful form of computer assistance, but a user who is unable to recognize them may realize that a situation hitherto hidden by the computer assistance has become serious only when the computer assistance can compensate no longer and presents the already serious situation to the user without any history of its development. Particular care is needed to prevent this kind of problem in air traffic control.

14.2 Assistance of human tasks

14.2.1 Problem-solving aids

Many of the above forms of computer assistance for data manipulation are not specific to air traffic control but have counterparts in other large human–machine systems. However, the cognitive functions of the controller are often quite specific to air traffic control so that the forms of computer assistance devised and proved for them have to be correspondingly specific, for example for en route traffic (Carlson and Schultheis, 1993). Even where principles can be borrowed from elsewhere, details of their application must be adapted for air traffic control.

The simplest problem-solving aids merely indicate the existence of a problem. There are many examples of this kind of aid in air traffic control. Among the commonest is conflict detection, which shows on a display the aircraft between which the minimum permitted separation standards will be infringed at some future time according to computer calculations based on their current positions and planned flight paths, unless the controller intervenes. Action or close monitoring by the controller may be required, but a conflict detection aid makes no recommendations. Another simple problem-solving aid applies when an aircraft is not where it is supposed to be, for example because it is straying outside an airway or from its planned route to an extent beyond predefined criteria of tolerance, and the aid displays this to the controller. A further simple example indicates unavailable options, such as flight levels temporarily fully occupied on trans-oceanic routes. These and comparable aids draw the controller's attention to a problem by codings that are compatible with other task requirements and possess an appropriate degree of urgency in relation to other functions and timescales. For example, the controller's attention may be drawn to a problem on a radar display by increasing the size or intensity of labels, by changing their colour or backgrounds or brightness contrasts, by underlining or drawing a box round the labels or by flashing them, by changing their typeface or case, by changes in symbology, or by any other codings that suit the tasks and the physical environment. Few problems are so urgent that the controller must be interrupted immediately under all circumstances. A controller who apparently ignores the initial indications of a problem may be presented with a succession of more intense and obtrusive stimuli culminating in an auditory alarm, but only as a last resort and never routinely.

The next level of computer assistance for problem solving is not merely a prompt but supports an aspect of the problem-solving process. It may automatically provide further information, or offer menu options, or illustrate applicable constraints, or indicate possible benefits or penalties associated with alternative solutions. Conflict resolution is an example of this kind of aid. So is the correction of a deviation from an aircraft's planned route by a computed revised heading to a suitable waypoint on its original route. Other examples of this kind of aid are indications of when an option will become available or of how much leeway there is in modifying the earlier stages of a flight so that the aircraft can achieve a future designated en route time slot. Such aids do not actually solve problems so much as make calculations, sometimes complex ones, which genuinely help the controller to solve problems.

Some forms of computer assistance may allow the controller to solve problems in a different way. An example is ghosting (Mundra, 1989): where aircraft are on two or more routes in a terminal area that converge into a single final approach route, ghosting can provide computer assistance for the controller in determining their most efficient order of convergence and in maintaining safe separations between them during their convergence. In ghosting, each aircraft is depicted normally on the radar on the route it is actually on, but is also depicted, in a much less intense and more ghostly fashion, on the routes it is not on, showing where it would be in relation to the convergence point if it were on the other routes. The controller can then see before convergence the separations that will apply after convergence if no corrective action is taken, and can adjust separations accordingly before the routes converge. Because the computer calculations are much superior to the controller's unaided estimates, ghosting can aid problem-solving if the calculations are based on data of high quality, although as an aid it does not actually solve problems but illustrates clearly the problems to be solved.

In a different category are more complex forms of computer assistance that do offer solutions to problems (Volckers, 1991). The controller may be free to accept or reject the solution offered, and more than one alternative computed solution may be presented with or without an indicated computer preference, but the actual process of problem solving is automated to the extent that the controller does not actually have to gather the relevant information and formulate a solution. The controller usually has to judge if solutions should be checked and how to do so, a dilemma that does not arise unless ready-made solutions are offered. Examples are conflict resolution with a presented preferred solution, aids that allocate aircraft to slots, ghosting that recommends an optimum sequence, and aids that recommend a flight level. Most forms of computer assistance of this kind are envisaged rather than current. In human factors terms, there is a big difference between forms of computer assistance that aid problem solving by calculations, extrapolations, predictions, or presented analogies and models, and forms of computer assistance that actually provide solutions, even if the latter require human acceptance or rejection. The associated human information processing is different, and so are other aspects such as the legal implications. Common human factors problems include over-reliance on the aid and poor cooperative problem solving (Layton *et al.*, 1994), and the effects of differences between controllers in their knowledge and experience (Amaldi, 1993).

14.2.2 Decision-making aids

In human factors terms, aids to decision making may seem to parallel aids for problem solving since the outcome of solving a problem is often to make and implement a decision. However, the detailed processes differ sufficiently for the two forms of assistance to be treated separately. Decision-making aids also have to take full account of air traffic control rules, regulations, instructions and procedures, and any aids to decision making must therefore allow for the associated constraints. Because the legal responsibility for the correctness of decisions remains with the controller, most aids therefore have to assist decision making rather than make decisions. The idea that machines should implement some air traffic control decisions without the knowledge of either the controller or the pilot is a feature of some air traffic control planning in the longer term, but is not imminent, and indeed the prospects of it appear to be receding rather than advancing at the present time. Meanwhile automated aids to decision making take more modest forms, such as the selection and presentation of appropriate menu or dialogue options in response to the traffic situation or to previous actions or decisions by the controller. Even with such simple forms of decision aid, the controller may remain unaware of the machine decisions that are a direct function of the controller's previous activities.

The simplest decision-making aids alert the controller about the need to take a decision, but do not contribute further towards it. They may represent the outcome of computer-assisted or manual problem-solving activities. The most basic decision for the controller is whether to intervene in a situation or not. A closely associated decision in air traffic control concerns the optimum timing of any intervention. There are many examples of this combination of simple decisions: whether and when to request minor speed changes to maintain consistent separations between consecutive aircraft within a traffic flow, and when to notify a climbing aircraft of an extension of its clearance towards a previously requested flight level. The simplest decision aids generally refer to the tactical control of single aircraft.

At the next level, decision aids support the decision-making process directly. They may introduce additional information or highlight existing information automatically, and are thus often associated with presentational changes. In some sense they affect either the information that the controller uses or what the controller does. Requests from pilots may not be answerable immediately but may require further information or confirmatory checks or the usage of presented information in new ways before the controller can decide whether the request can be granted. The main reason is that the request will typically have air traffic control implications beyond those that the pilot may be aware of, which rightly influence the decision. The outcome of the decision making by the controller, which may take the form of granting or refusing a request, is regularly underpinned by a series of subdecisions based on the additional sources of information consulted, including subdecisions about which further sources of information should be consulted.

Where decisions are formulated for the controller to accept or reject, a human factors problem that always arises concerns the permissible grounds for overriding them. The most obvious basis for the controller to substitute a human decision for a machine one would be the controller's possession of knowledge that invalidates the computer decision and which the computer could not have used in its formulation. This apparently simple human–machine relationship, which sounds so sensible and rational, is actually very difficult to achieve, because it implies that the controller knows or can readily discover the entire basis of the computer decision, not only the factors which were taken into account and had some influence on the decision but the relative weightings of those factors, and not only the factors that were ruled out straightaway as inapplicable or irrelevant but the further factors that initially were taken into account only to be assigned a nil weighting subsequently.

Because of the respective human and machine capabilities for information processing there is always some probability of overruling a computer decision with an inferior human one based on far less evidence, more biased evidence and more incorrectly weighted evidence. It is relatively easy to conceive but difficult to supply in practical form the kind of computer assistance that would be of most benefit to the human controller in these circumstances. The controller requires forms of computer assistance for decision making that would permit interrogation of the machine about its decisions. In particular it must be possible for the controller to ask whether designated items or categories of information have or have not influenced the machine-formulated decision, and to require the machine to recheck the decision by recomputation that includes this additional designated information if it has formerly been excluded. The consequent human factors challenges, particularly for human–machine interface design, are formidable, but the outcome would be some progress towards the optimization of collaborative human–machine roles and relationships in air traffic control.

Another major human factors problem associated with decision aids has proved to be the reduced human knowledge of the current air traffic control situation that often accompanies the provision of decision aids. The reasons for this are now understood sufficiently to implement a policy to ameliorate the main consequences if it were requested. In system terms, a decision aid that leaves the final decision with the controller but automatically does all the preliminary processing for it seems similar to a corresponding decision formulated by the human personally on the basis of information chosen because it seems relevant to the decision. In human terms, these two processes with similar or even identical outcomes are grossly different, because they impose very different demands on the controller in terms of the amount and level of human information processing required, very little for the automated decision, and a great deal for the human decision. Because much less information has been processed, the controller knows much less about the situation and is placed at a disadvantage in adjudicating on the merits of the computed decisions. To restore the controller's knowledge requires the reintroduction of active human participation in the decision-making processes, which tends to negate the purpose of the aids and of broader policies to reduce human workload. This dilemma has no easy resolution, though the evolution towards strategic rather than tactical human roles in air traffic control may be helpful. Endsley (1994a) has recently reviewed the literature on dynamic human decision making, Amaldi (1993) has considered some of the implications for skills, and Klein *et al.* (1993) examine rapid critical decision making.

Most computer assistance in the form of aids to decision making by controllers performing strategic air traffic control tasks is still in quite early stages of development. Despite frequent policy statements about future strategic roles for controllers, most of the forms of computer assistance actually provided remain remarkably tactical. They apply to aircraft individually rather than to traffic flows, even where strategic forms of computer assistance would be expected to find their initial air traffic control applications, as in oceanic control or traffic flow management. Initial attempts to devise strategic forms of computer assistance can be exemplified by the segmentation of traffic flows into time periods with maximum flow capacities, with which the actual future traffic as predicted from currently airborne aircraft and filed flight plans can be compared. On the basis of these comparisons, flows or flights may be reallocated to other time segments with spare capacity. These encouraging developments are still at the stage of devising tasks, deducing information requirements, formulating and proving appropriate forms of information portrayal, and integrating such roles and functions into the wider air traffic control system. The ultimate balance between tactical and strategic functions of such systems should have a major influence on the kinds of computer assistance that will be most effective, but this balance also is still under discussion.

14.2.3 Prediction aids

Limitations in the ability of controllers to make accurate predictions of future traffic scenarios on the basis of current ones have suggested for a long time that an appropriate form of computer assistance for controllers must be prediction aids, provided that the evidence on which the predictions are based is of satisfactory quantity and quality. Improvements in the sensed data and in their processing have progressively reduced the technical constraints on predictions, by making them more accurate and reliable so that they contain fewer errors and false alarms, and by extending the time horizons within which air traffic control predictions can be of practical value. With experience and familiarity, unaided controllers can improve their own extrapolations from the current traffic scenario considerably, but not to the extent of matching in speed or accuracy the capabilities of machines to extrapolate on the basis of the high-quality data now stored and frequently updated in many air traffic control systems.

A consequence of evolving from tactical to strategic control is a change in emphasis from solving tactically problems that have arisen towards preventing strategically problems from arising. This change implies more concern with future traffic scenarios and less with current ones, and this in turn implies more displays of the future traffic and fewer displays of the present traffic. Therefore the commonest form of future prediction aid for air traffic control is not expected to be indications of future states on displays of current traffic, or even displays of current traffic which can be rolled forward in time to show how future scenarios are expected to develop from current ones, but displays which consist entirely of future traffic and do not contain any dynamic information about current traffic at all. There may be displays of the current traffic in the same suite and even in the same workstation. If there are, it is obviously vital that the separate displays of current and of future traffic cannot be mistaken for each other.

Meanwhile, the commonest forms of prediction aid superimpose future information upon displays of current traffic, usually by indicating for each aircraft the relationships between its current and future states, and perhaps by indicating intervening states also by means of lines or other visual connections between the two states. The implication of this is that prediction aids often function as future snapshots, in which the entire traffic within a region such as that covered by a radar display is depicted at some chosen future time, so that all the relationships among the aircraft traffic can be perceived as they will be at that time, instead of having to be visualized.

Prediction aids which are applied to single aircraft have limited value in air traffic control. An exception is the use of a prediction aid to attempt to detect the future path of a rare blundering aircraft that turns sharply towards a parallel approach path to an airport during final approach. Even when predictions apply to single aircraft, as when the future track of an aircraft turning onto final approach to a runway is predicted, this is primarily of interest in relation to other traffic in the vicinity, and is only of interest in its own right if it allows the controller to issue a revised instruction to the aircraft which expedites its arrival. Otherwise the future tracks of aircraft converging onto their common final approach path are predicted by extrapolation so that the safety and legality of the separations between them can be verified. The provision of the prediction aid will normally permit finer adjustments to be made than would have been possible without it, provided that the computations are complex enough and are based on high-quality data. This example can incidentally reveal that alternative kinds of computer assistance can serve similar objectives, because a prediction aid in the form of computer-assisted approach sequencing and an aid in the form of ghosting may have similar objectives and be about equally successful, though they take very different forms and are applied in different ways.

14.2.4 Menus and dialogues

In air traffic control, the functions of menus and dialogues are exercised through the human–machine interface (Edwards, 1991). Human factors guidelines for the design of menus (Paap and Roske-Hofstrand, 1988) are generally applicable, as menus and dialogues in air traffic control provide cognitive control (Norman, 1991) and can be approached through task analysis (Phillips *et al.*, 1991). Combined with data input devices that may include touch-sensitive surfaces associated directly with them, menus and dialogues can be a flexible and adaptable alternative to other forms of data entry, and offer advantages of compactness. Their value and efficacy depend considerably on the thoroughness of their specification, on the success with which they meet task requirements, on the ways in which they assist learning, and on the extent to which they can be modified to meet changing needs and circumstances, either by being self-adaptive in that they adjust to the usage that is made of them or by being comparatively easy to reprogramme. Well-designed menus and dialogues instruct controllers about available options and direct them towards the most appropriate ones. They can function as teaching aids, support human memory and remind the controller of relevant information or uncompleted actions. They can prevent certain kinds of human error or query the correctness of certain human inputs or instructions.

The use of menus and dialogues permits the presentation of many options that would be cumbersome or even impossible to present in any other way. Their structure guides and directs controllers, and can be adapted to indicate computer preferences and priorities. They can offer support and encouragement for human choices because the inclusion of those choices among the presented options, and the anticipation of the need for those options which their inclusion in the menu or dialogue entails, confirms their viability and validity as choices. They can permit choices to be explored in advance of commitment to them, together with prior identification of many of their consequences. Their structure can help to show interactions between options, and mutually exclusive options can be depicted as such or may be inferred from omissions from the choices presented.

Menus and dialogues can thus assist human tasks in air traffic control in many ways, but they bring a mix of advantages and disadvantages, in common with most forms of computer assistance. They can reduce human choices, thwart human wishes, curtail skills, and seem obstructive. They can impose greater rigidity of task performance by making it more difficult to perform more than one task concurrently, to stop and resume tasks, or to share tasks. Their use can curtail team functions since they are seldom adapted to be used by teams rather than individuals, and for the same reason they can make it more difficult for supervisors to intervene or participate actively in dialogues. Under the guise of increased flexibility they can actually reduce it, through concealing that some options have disappeared by not presenting them or through concealing a true paucity of key choices by providing many branches that frequently reconverge.

Fundamentally, when menus and dialogues become widely used features of the human–machine interface, the controller's functions tend to be restricted to the facilities and options that they provide. It may be impossible for the controller to fulfil many functions except through the menus and dialogues and using the options which they offer. If they are rich in options, present the options effectively and in timely and task-related sequences, and do not seem to curtail flexibility or defeat human intentions or ingenuity, this may be generally satisfactory except in a few unusual circumstances.

A major human factors problem arises whenever the options offered by the menus and dialogues fail to meet the requirements of a controller who wishes to follow a course of action that apparently is not permitted or has not been foreseen in the specification of the menus and dialogues, and that cannot be implemented by innovative adaptation of the menus and dialogues or by circumventing them. The most critical circumstances are emergencies when actions that are normally forbidden provide the optimum or the only viable solution. It is peculiarly unhelpful if a dialogue merely indicates that a proposed action is invalid or illegal since the controller knows this already but needs to discover how to override the dialogue's prohibition or circumvent the limitations imposed by the menu and dialogue designs and options. The controller wants to be able to use the dialogues to communicate the human objectives to the machine and enlist its collaboration and support in achieving them. An emerging criterion of good menu and dialogue design is that it can accommodate human intentionality.

14.2.5 Prompts and reminders

Computer assistance in these forms aids human memory. In principle it is prudent and productive never to rely exclusively on unsupported human memory unless there is no choice. To employ computer assistance in support of human memory therefore seems an attractive application of it, to counter a known source of human fallibility and support a known human limitation. The most potent reminders are potentially distracting, and therefore all forms of computer assistance for this purpose have to strike the correct balance between reminding and interrupting. Sometimes they may have to become progressively more strident if they are ignored. To be helpful rather than irritating, prompts and reminders need to be appropriate. There should not be so many of them that controllers develop immunity to their signals or try to switch them off or disconnect them.

Their efficacy is crucially dependent on good preplanning of the need for them and on the circumstances and timing of their introduction. If they relate to current tasks and functions they should be apposite for them, but if the substance of the reminder is unrelated to current activities then its coding should make this clear. Wherever possible, the codings should not merely draw attention but be informative about the nature of the reminder. The stimuli employed to remind and prompt have dual functions for attention and memory. They should be self-sufficient as memory aids, without any requirement to supplement them with additional details of the action required as a result of the prompt or reminder.

One of the most beneficial kinds of reminder concerns tasks that are incomplete, either because the controller was interrupted and failed to resume them or because there is a natural pause within them to be followed by an outstanding action required to progress or complete them. An example of a task with a pause occurs when an aircraft has strayed from its track and been given a new heading to a fix on its original route; there may be an outstanding requirement to instruct the pilot to resume normal navigation on reaching the fix. The codings for these stages in the task need not be obtrusive: one suggestion is to draw a box round the radar label of the aircraft that has strayed, and remove most of the box while the aircraft is returning to its normal route, leaving only a label underlining as a reminder of the instruction to resume normal navigation.

Some features of menus and dialogues also function as prompts or reminders. The commonest is when a dialogue awaits formal completion, and its continued presentation is itself the reminder of the requirement to complete it. For important actions, auditory signals can give a prompt, and a ringing telephone or a buzzer can be construed in this way. So can the flashing of information on a display, but in all these instances the stimulus is so distracting that it must be reserved for important reminders, and the initial action in response to the reminder must cancel it. Prompts and reminders often aid communications. They can remind the controller that a return call is awaited, or signal that a call is being held until it can be accepted.

14.2.6 Commands, instructions and directives

The main forms of computer assistance hitherto have been for machine commands, instructions and directives to the human, and not for human commands, instructions and directives to the machine. There has also been more limited computer assistance for the human controller when the recipient is a pilot or another controller. The ultimate reasons for retaining rather than replacing human commands are often legal ones of responsibility, although technical feasibility and some human factors implications have also been influential. As long as the legal responsibility for decisions about the conduct of air traffic control remains vested in the human controller, any forms of computer assistance that would remove that responsibility or would curtail the manner or extent to which it could be exercised will encounter resistance. The dilution of responsibility is generally not favoured by the controllers themselves, who wish to retain the means to exercise the responsibilities they have, but those who specify and design the systems also tend to shy away from accepting more responsibility for they believe that their own responsibility can only cover the system design and cannot be extended to include the ways in which the system is actually used, which they have no control over. Thus the issuance of commands, instructions and directives by the machine would be acceptable only if the human controller retained full control over every aspect of them, which would seem to defeat their main benefit. Compared with all the preparatory work in formulating and verifying commands, instructions and directives, the work required to issue them is quite small, so that the benefits of computer assistance for these functions would be correspondingly small.

However, computer assistance of tasks of these kinds is expected to spread in the future (Hockaday, 1993). A natural progression from the automated transmission of data between air and ground is to convey instructions automatically about what to do with the transmitted data. Initially these would mostly be technical instructions not intended for a human recipient, but current and pending information that includes recommendations or predictions about future traffic states implies instructions to the pilot, although they may not be expressed overtly as instructions. Similar forms of computer assistance to the controller that present recommendations for resolving problems or making decisions do nominally leave the final choice with the controller, but the controller is really expected to treat these recommendations as commands, directives and instructions, for they generally are the best options available and it would be pointless to ignore or reject them all the time. In fact, the controller is trained to choose the computer-preferred solutions rather than compelled to choose them, but the distinction is a fine one in that it seldom has much effect on what actually occurs but allows the claim that the responsibility remains with the controller. More forms of computer assistance of these kinds are planned, in which the controller actually obeys computer instructions most of the time but nominally is not mandated to.

Computer assistance that included the issuing of commands was originally planned to be almost full automation, in that the outcome of ground-based computations would be transmitted automatically to the aircraft and implemented automatically, bypassing both the controller and the pilot. Although this concept has not been wholly abandoned in the long term, its prospects have receded because of the many kinds of practical problem that it leads to. Among the main human factors ones are the location of responsibility, the impossibility of effective human intervention in emergencies, issues of public acceptability, and vagueness in defining what the associated human roles could be. The introduction of some direct routeing for aircraft seems a more practical possibility that will rely on extensive computer assistance to check its safety and feasibility but will begin and end with human rather than machine functions, whereby the human pilot requests a direct route from the controller who makes appropriate checks and then grants or refuses the request or offers an alternative partial direct route instead of the requested one. Requests for direct routes are not initially intended to be granted automatically without human intervention. This typifies the mix of computer assistance and human roles intended to satisfy responsibility requirements.

Obvious forms of computer assistance for tasks involving commands, instructions or directives would employ machine recognition of the human voice or human recognition of machine-generated speech, but the problems in evolving successfully from a technology which works well in principle for most of the time to a technology which is reliable enough to be applied with confidence in real air traffic control have proved to be far more formidable than was originally expected, mainly because of inadequate levels of recognition and unacceptable error rates. Most errors have been artefacts of the particular technology employed, and this has restricted the accumulation of general evidence. Even now, the initial applications of such technology are likely to be in training controllers, where tangible savings can be made from employing machines instead of people as surrogate pilots. Sooner or later, there will be computer assistance for commands, instructions and directives in these forms, but not for some time.

14.2.7 Task scheduling

Task sharing is intrinsic to air traffic control. Even when the controller is working exclusively on a problem or completing a task, something more urgent may occur at any time and take precedence. Task analyses and descriptions reveal the multiplicity of air traffic control functions, and the best descriptions address the integration of tasks into jobs and the sharing of tasks that are not mutually exclusive. The busy controller normally seems to perform several tasks at once, mostly by combining frequent attention switching with some automaticity of task performance, rather than by parallel processing of different information for different tasks performed concurrently. It follows that the controller has some scope for scheduling tasks. One of the measures of individual controller style refers to the characteristic task scheduling patterns of different individuals. Computer assistance can have two kinds of effect on task scheduling. It can propose schedules and remind the controller about tasks to be included, or it can limit the controller's freedom to schedule tasks by presenting them in the order preferred by the machine. In both instances, successful human–machine collaboration will depend on the matching of task scheduling with human tasks and needs and with the air traffic control objectives.

Among the main influences on task scheduling are the job and task designs; the designs of equipment and tools such as input devices and dialogues; the controller's training; the rules, practices and procedures of air traffic control; international and national requirements; local practices; and the expectations of pilots and of controllers. For example, in an air traffic control message between a pilot and controller, some of its parts such as the initial identification and final acknowledgement are fixed, but other parts such as the ordering of different items within it may retain some flexibility, or their flexibility may be limited by other influences such as data entry device layouts or dialogue structures which favour items in certain orders. The controller has partial but not complete control over the scheduling of communications tasks, for the timing of those initiated by the controller can be at the controller's discretion to some extent, but the timing of those originating elsewhere cannot be controlled at all unless the communications channel is already occupied.

The main forms of computer assistance with task scheduling therefore take the form of machine-initiated events. Some of the most obvious are machine prompts or reminders, but many concern forms of computer assistance such as conflict detection that are intended to induce some kind of controller intervention. All such aids are bound to influence task scheduling; that is not their primary intention, though it does suggest that their implications for task scheduling should be considered before they are introduced. A further influence of computer assistance on task scheduling can occur when the machine imposes an order of priorities that requires different information to be presented. The order in which the controller performs tasks must accord with these machine priorities, although this may not become very apparent. A feature of computer assistance in task scheduling is that much of it is not recognizable as such by the controller. Mandatory or peremptory machine requirements, even those which actually prevent other tasks from being done, can be cloaked under the guise of helpful suggestions and directives, which makes them more acceptable and helps to prevent impressions that the machine is scheduling the tasks too much.

14.2.8 Work organization

Computer assistance is beginning to be applied to assist work organization in air traffic control. This could offer considerable benefits if it can be done successfully. It seems feasible at three levels. Most globally, the prediction of traffic demands hours ahead would permit some adjustment of staffing levels and some broad reallocation of work. At the next level, computer assistance can be applied to the organization of the work within a centre or tower or other workspace, by anticipating the requirements and organizing the predicted work through efficient partitioning of it. At this level it could assist in dealing with unusual or specific requirements of work organization, associated with particular events such as national holidays, inoperative runways or locally severe weather, or with current emergencies and their aftermath. At the third level, the computer assistance in work scheduling can be helpful when the forms of computer assistance provided are themselves flexible and to some extent optional, so that they can assist the scheduling of work by varying the extent to which they assist the work.

Future forms of computer assistance seem likely to be more effective and influential than most current ones are for these purposes. Some may also delegate more air traffic control work to the cockpit (Ryberg, 1993), although that brings its own human factors problems (Begault, 1993). Developments such as direct routeing, adjustable or flexible regional or sector boundaries, and aids for strategic air traffic control functions, will benefit work organization. However, the most effective development would be forms of computer assistance that restore greater control over workload by enabling the controller to choose whether or not to apply computer assistance to alleviate workload demands. This kind of computer assistance, whereby functions are fulfilled by flexible combinations of human and machine, including complete autonomy of either, brings most benefit for work organization, provided that controllers have received proper training in how to use it, for overloading can only arise through the controller making insufficient or incorrect use of the computer assistance available to alleviate it. Although this rather idealized state is some way off, the introduction of strategic forms of air traffic control reduces the most severe time pressures, and hence offers better prospects for smoothing and controlling work demands through more effective work organization.

14.2.9 Optional and mandatory forms of assistance

Although this distinction has important human factors consequences, it does not seem very significant because it is not an obtrusive feature of most available or planned forms of computer assistance. In a sense, that is what makes it important. Many of the mandatory forms of computer assistance do not seem mandatory. They can convey the opposite impression because they actually present choices and alternatives, but they are nevertheless mandatory in that they demand that a choice is made and will veto attempts to make a choice which is not among the alternatives presented. In that sense much computer assistance is mandatory by nature. It is accompanied by a loss of flexibility and innovation, and that is why it is so important to include as many options as possible among those that it presents, for often there can be no others.

A contrary trend is to provide forms of computer assistance that are optional, in that the controller does not have to use them and can switch them off. Although the human factors principles that underlie this option are often worthy ones, in practice optional forms of assistance will not be of optimum benefit unless the exercise of the options is appropriate. This will be achieved only if the controller knows how to use the computer assistance to best advantage and if it has been optimally matched with human and air traffic control needs. An optional aid that can be quite helpful when traffic is not heavy, but is switched off in heavy traffic because it then becomes too much trouble, is not worth having as it violates the basic human factors principle that effective forms of computer assistance are those that are of most help when they are most needed.

Many of the forms of computer assistance for functions with no major human factors implications are mandatory, and rightly so. The performance of the whole system would be degraded if they were not used. This applies to most of the computer assistance for data collection and handling within the system. However, the benefits of making computer assistance optional for many human tasks seem compelling enough to treat it as the desirable norm, except for a few forms such as conflict detection where the safety implications are too severe and human fallibility is too high to dispense with it intermittently for any reason. A more significant criterion for choosing between mandatory and optional forms of computer assistance where a choice can be made is the need to keep the controller fully informed about the current traffic scenario by remaining thoroughly involved in the control loop. This implies active participation in tasks whenever possible, and positive steps to maintain knowledge, skills and procedures through familiarity, experience and practice. When there is little work, the controller should do it all. When there is too much, computer assistance should be enlisted. The implementation of this simplistic but valid policy implies that at least some forms of computer assistance must be optional, and that those that are optional must be effective in heavy traffic.

14.3 Air traffic control functions

14.3.1 Conflict detection

Conflict detection is an example of a form of computer assistance in air traffic control that is essentially simple, but is effective because the function can be performed better by the machine than by the controller. Data about the traffic under control are held in the computer and updated frequently. From the smoothed data from successive samplings of an aircraft's position, a record of the track of the aircraft is compiled, and extrapolations of its future track and of the expected time at any designated position along that track can be computed by combining these predictions from the stored track data with other relevant data such as its future flight plan and the flight clearances that have already been issued. Data about future aircraft tracks can then be compared automatically and systematically, to check whether any two or more aircraft will infringe at some future time the minimum separation standards applicable to aircraft in that phase of flight. Infringement implies that neither the expected horizontal or vertical separations between them is sufficient to guarantee their safety. When an infringement is detected automatically, it is signalled on the controller's display, normally by a coding reserved for the purpose. This coding is applied to both or all the aircraft in potential conflict, and remains as long as the potential conflict remains. This form of computer assistance is confined to detection and the signalling of what has been detected. It usually shows the location on the radar screen of the detected conflict and thereby implies its approximate time of occurrence, though it may show its predicted time of occurrence also since this is implicit in the computations.

The value of a conflict detection aid depends on its reliability, which takes two forms: no conflicts must be missed, but there must not be too many false alarms. Reliability depends primarily on the quality of the data on which the computations are based, assuming that the computational procedures themselves are correct. Quality is quite a complex notion, being a function of consistency, accuracy, precision, reliability, validity, trustworthiness, smoothing, and weightings of the data derived from several sources. The limits of extrapolation have to be set primarily by reliability, for omissions or false alarms will result from data degraded by random or uncorrectable error sources or from data extrapolated too far ahead. Although there will be policies on the applicability of conflict detection, it may be prudent to determine empirically the optimum predicted extrapolations that produce the best balance between the prevention of undetected conflicts and the generation of too many false alarms. To some extent, the way in which it is actually applied becomes self-correcting on the basis of experience wherever it is possible to delay responses until false alarms become less probable.

The main reason that conflict detection can be so helpful is that it essentially aids searching, and humans do not perform most search tasks well. It seems too obvious to mention that forms of computer assistance have the potential to be more helpful when they are applied to functions that the unaided human does poorly than when they are applied to functions that the unaided human already performs nearly as well as they can be performed, but this truism has sometimes been forgotten. For the controller, searching tends to depend on other activities and checks rather than be an independent function, because the timing of many other tasks is largely imposed on the controller by the job requirements, the traffic, initiatives by pilots or other controllers, or system-initiated events, and searching tends to fill some of the gaps. Except in old displays with a sweeping beam, searching patterns are not structured by the display characteristics and can become less systematic because of this. Searching can be tedious when for most of the time there is nothing significant to discover.

The principal benefit of conflict detection is not to detect conflicts that otherwise would be missed. The unassisted controller does not miss conflicts and cause danger. After all, there have been no conflict detection aids in air traffic control for most of its history. However, the unaided controller is less consistent than the computer in the timing of their discovery, and is less able to detect them early. The machine relies on complex and frequently repeated systematic calculations and indicates any detected potential conflict at once, whereas the unaided controller must rely on much poorer data based on radar traffic patterns plus knowledge of routes and flight plans and communications, and is less able than the machine to make accurate extrapolations or to work systematically. Thus conflict detection is a good example of computer assistance in air traffic control because it matches the human controller well by compensating for human limitations, by doing a routine task better and far more often than the controller can, by enhancing the consistency and predictability of system functioning, and by showing each problem clearly and immediately in forms that the controller can apply directly. Its acceptability as an aid for controllers is also linked to this last point, because it leaves the executive actions with the controller and does not impinge on controller skills and responsibilities.

14.3.2 Conflict resolution

Conflict resolution, the sequel of conflict detection, is an example of a more complex form of computer assistance in air traffic control. It is a future aid rather than a current one. It does not permit the same simple kinds of verification that apply to conflict detection: a detected conflict can be recognized at once, but the merits of any proposed solution are not so immediately self-evident. Conflict resolution encounters the human factors problems associated with the computer assistance of cognitive functions, whereby issues of responsibility, acceptability and skill arise (Ball and Ord, 1983). Although conflict detection and conflict resolution are linked functionally, the human factors problems associated with them differ.

Conflict resolution relies initially on the processes of conflict detection, but relates the detected conflict to current and predicted data on other traffic already in the region or soon to join it. Most forms of conflict resolution in fact formulate more than one possible solution for each detected conflict, and then apply a set of criteria to recommend a best solution or to list several recommended solutions in a preferred order. The controller can accept or reject the single recommendation, adopt the first or another preference from several presented, or formulate an alternative solution to the computed ones. There may be facilities to test the controller's own solution in terms of its safety and efficiency by applying to it the criteria used for the computed recommendations, but the controller may be unable to specify the human preference in terms that can be entered into the machine and checked by it. This depends on the specifications of the conflict resolution aids, and on the facilities for checking provided through dialogues and menus within the human–machine interface.

Although some misgivings are normally expressed about the efficiency, safety and trustworthiness of automated conflict resolution aids, in fact such forms of assistance are not introduced without very strong evidence of their safety; it would be foolhardy to do otherwise. Therefore the controllers' initial wariness of the trustworthiness of such aids can usually be overcome, partly by appropriate initial training to demonstrate that the solutions offered are sound and can be relied on, and then by operational experience during which trust is built as the solutions prove to be consistently practical and efficacious. At some point, the human factors problem tends to evolve from too little to too much trust in conflict resolution. The training for this and similar forms of computer assistance has not yet succeeded fully in instilling the optimum degree of trust in them. Complacency as a human factors problem is inherent in most advanced forms of computer assistance. It can induce losses of human insight and human competence, but these can be countered and concealed by a reliance on computer assistance that becomes total in that the controller becomes progressively less able to perform the corresponding functions unaided.

Such advanced forms of computer assistance as conflict resolution are intended to be used and not to be switched off or extensively overridden, although the controller may be reluctant to look away from the radar for long enough to use them (Cardosi *et al.*, 1992). They can be genuinely helpful, and they can and do take account of far more evidence of more varied kinds far more quickly than the controller can. These characteristics are the direct causes of three significant human factors problems associated with them: how to tell if they are not functioning properly, how to verify the recommendations, and how to discover which factors and evidence have or have not been included in the formulated recommendations. In so far as the correct functioning of conflict resolution is not fully understood, any malfunctioning of it will not be fully understood either. Although there may be provision to indicate some possible malfunctions to the user, it is impossible to foresee and indicate every conceivable malfunction, so that there is always some possibility, albeit remote, of the user employing defective equipment unknowingly. Because a main reason for the computer assistance is its superiority over the comparable human functions, the controller cannot verify the computer recommendations by repeating the functions manually for there will be neither the time, the data, nor the requisite human capabilities. Any feasible forms of human verification will therefore be confined to gross simplifications of the computer processes. The controller may possess information that seems to invalidate a recommended conflict resolution and may wish to ascertain whether that information was in the computer and applied to its conflict resolution, but the interfaces and dialogues required for appropriate interrogations are difficult to devise and are unlikely to be sufficiently flexible.

A dilemma is how far the controller can be or should be responsible for conflict resolution recommendations which are impossible to verify completely. These issues of responsibility are linked with the acceptability of these advanced forms of computer assistance which fulfil their intended objectives well and can be a real help to a busy controller, but seem to whittle away responsibilities and induce some loss of corresponding human skills that are no longer practised regularly.

14.3.3 Direct routeing and free flight

Direct routeing, a variant of which is called free flight, is a development that is favoured by pilots and airlines, as it could offer a more efficient and cost-effective air traffic control service to them. It applies best in regions of airspace with a high proportion of traffic in long-distance level flight within radar coverage, and is less advantageous where a large proportion of the traffic is climbing or descending, as in parts of Western Europe. From the perspective of the airline user, it is direct routeing because an air traffic control clearance is given to fly the shortest distance between two positions along a route which is in effect a segment of a great circle. From the air traffic control perspective, it is sometimes described as random routeing, which is not strictly correct but denotes that the route of an aircraft may be independent of any fixed air traffic control route structure or navigation aid. An aircraft pilot usually requests direct routeing between position fixes associated with the departure and arrival airfields. These fixes refer to transitions between en route traffic flows and the more constrained routeings within terminal manoeuvring areas around airports, where departing and arriving traffic must be kept apart. Ideally, direct routeings for en route flight lead smoothly to clearances through the terminal area to the airport. Free flight usually refers to the entire flight between the departure and arrival airfields.

Requests for direct routeing or free flight are checked using computers to see if granting them at the envisaged time would induce potential conflicts with other aircraft traffic on direct or fixed routes, or would mean that the aircraft would meet other difficulties such as adverse weather. All the calculations depend on data extrapolation, and most requests need a large volume of conflict-free airspace even when the data are of high quality, the computations are correct, and the subsequent flight will adhere closely to the direct route. The controller has to rely on computer assistance in granting a direct routeing because much of the information on the later stages of the flight is not yet available or cannot be extrapolated with sufficient accuracy by the human controller. Direct routeings lose their appeal quickly if they have to be revised frequently, especially when amendments to them result in penalties of delays, diversions or fuel usage that outweigh the benefits from following a direct route.

When a direct routeing is flown, computer monitoring of continued safety of the flight is required. Among essential checks are that there have been no intervening changes in circumstances and that the actual flight conforms closely in space and in timing with the flight intentions that justified direct routeing initially. The current strong pleading of the advantages of direct routeing seems at variance with strivings towards more strategic forms of air traffic control, because direct routeing is a highly tactical form of control applicable to aircraft individually, as an alternative to controlling them as traffic flows. Of particular human factors concern is that aircraft on direct routeings require individual treatment in the event of major system failures, whereas aircraft already in flows on regular routes can continue to be treated in some respects as items in separated flows while the normal air traffic control service is being restored.

The computations required for the widespread and safe application of direct routeing or free flight are complex. Its economic and other benefits are significant, so that the demand for it will probably increase. The reversionary modes of any system with extensive direct routeing will differ from those with most aircraft on regular routes, and the controller will need knowledge and training to cope with the associated problems, with their blend of tactical and strategic control tasks requiring forms of computer assistance that can be adapted for both. The combination of direct routeing with accurate on-board sensing of intruding aircraft nearby could increase the incidence of late tactical manoeuvring to avoid unpredicted aircraft in the vicinity. However, the value of direct routeing could be undermined quickly if extensive tactical manoeuvring became prevalent, because the computations on which the direct routeing is granted presume that tactical manoeuvres will not destroy the accord between the intended and actual flight. The human factors challenges in air traffic control from developments using advanced forms of computer assistance to offer new opportunities that are too complex for the human to verify are typified by the human–machine problems of matching roles and of interface design posed by direct routeing and free flight.

14.3.4 Automated updating of information

This is an essential form of computer assistance. The human factors issues associated with it are not about whether information should be updated but about the most appropriate forms of automated information updating. Under what circumstances and how often should information be updated? Should updating be with or without the controller's knowledge, and should it occur whether or not the information is actually being used? How useful is it for the controller to know when an update is pending? Decisions on these issues depend partly on the nature and operational significance of the updates, partly on the incremental thresholds that accord with human perception, and partly on the opportunities to boost the controller's memory and understanding that can be presented by updating. The controller's knowledge may be entrenched through manual updating of information.

Some updating simply prevents repetition. If a flight is a few minutes later than planned, entry of this information into the system once, automatically or manually, can automatically update the corresponding flight timings for every subsequent phase of the flight. Generally this is very helpful: it removes chores, and repetitive data entries are not an additional source of human errors, but an update may not be noticed unless the controller's attention is drawn to it. Whether this is necessary depends most on its operational significance, which can be difficult to predefine selectively except for very small or very large updates. When an update is indicated, this leads to the further human factors issues of whether acknowledgement of it should be required and how long the indication should last. An acknowledgement is an extra small task which may not suffice to denote that the significance of the update has been recognized. When an update is not signalled, or when it is indicated temporarily for a predetermined time whether it is acknowledged or not, the controller may or may not be aware of it. Because the treatment of updates must bear some relationship to their operational significance, some categorization of them in terms of their operational urgency and significance is clearly an essential prerequisite for matching them optimally with human tasks and responsibilities.

At some point, it becomes preferable to treat major information updates as new information. The delivery of a new electronic or paper flight progress strip can be construed as the provision of new information in terms of tactical air traffic control, but as an update of the flow of information in terms of strategic air traffic control. The concept of updating has different implications for tactical and strategic control, because the operational significance of various information categories and of updates to them does not remain the same. This points towards a guiding principle when updating information. If the updating has significant operational implications that the controller should be aware of, the achievement of this awareness through the direct involvement of the controller in the updating process itself should always be considered. The alternatives are to rely on fallible human attention and memory to notice the change and appreciate its significance, or to require the controller to process the updated information through separate and subsequent functions or tasks. If the update should become part of the controller's picture of the traffic, this is usually achieved more efficaciously if the controller processes the occurrence of the update itself through direct participation in it, than if the controller integrates the update into the picture more passively some time after it has occurred, when it is no longer a dynamic event. However, unless the controller can influence the timing of updates to some extent, enforced participation in them whenever they occur would be too disruptive. Active involvement by the controller in significant updates, coupled with some control over their timing, seems to offer the best compromise, to maintain the updating of the controller's picture with operationally significant information.

The automated updating of many kinds of air traffic control information can look less like a series of discrete events to the controller and more like a continuous process so that the information is always up-to-date. In effect this occurs wherever the updates are so small and frequent as to be unnoticeable, and radar displays continue to evolve towards this state. Updated alphanumerics are not so easy to present smoothly, but with electronic counters changes can look continuous when they are actually a sequence of small increments. When users become accustomed to updates that seem continuous, updating is less recognizable as a distinct function. Special provision may then have to be made for updates with particular operational significance, to restore their discreteness as events so that they are not treated merely as extensions of updating. Imminent infringements of separation standards already employ this principle, for example when the process of updating plan or flight level information triggers the separate process of conflict detection because certain predefined conditions have been reached during updating.

14.3.5 Full automation of functions and tasks

The full automation of functions that never were fulfilled wholly or partly by the controller is not of direct human factors concern, unless controllers have to know about them in order to do their jobs. The most probable reasons for knowing about them are associated with breakdowns and malfunctions, when the controllers may have to fulfil unfamiliar supportive roles. Examples are radar failures or unserviceable communications channels. In such instances, which are generally both rare and transient, the controller's objective is to maintain the safety of the air traffic control service. Other normal objectives such as efficiency are temporarily abandoned, and this allows controllers to cope better with emergencies. It argues for the retention of some flexibility even in highly automated systems because, for example, a controller will become fairly helpless without speech channels with pilots or basic flight strip information in an emergency, even if they are not used much in normal circumstances. Effective controller intervention in emergencies must rely on preplanning and not on improvisation, which becomes progressively less practical with more automation.

When a function or task in which the controller was formerly involved is fully automated, the controller no longer knows anything about it unless deliberately informed. The main reason for keeping the controller informed is the continued relevance of an automated function to functions that are still wholly or partly human ones. The familiar guiding principle is applied of providing controllers with all the information that is essential for their tasks, including information about automated processes if it is task-relevant. Information that was formerly available to the controller but is no longer available since the functions were automated will normally be missed. Controllers may report some loss of their picture or some reduced understanding, and try to compensate by seeking the lost information in the system and by accessing it when they find it, not to use it for any immediate control activities but to restore memory and understanding. Among the most significant human factors consequences of the full automation of functions which formerly had a human component are the effects on human information processing and understanding. To controllers such effects seem important although the measurable consequences for their remaining tasks may not seem significant. However, subjective impressions of reduced understanding are worrying and unacceptable, and lead to complaints and resistance.

Most attempts to automate manual functions fully in air traffic control have so consistently failed to capture the full range, complexity and integrality of the former manual functions, that the conclusion has emerged that such attempts are misplaced. This does not imply that full automation should never be attempted, but that the most effective forms of automation are unlikely to be direct replicates of the corresponding manual functions in their entirety, and should not seek to be. A human factors implication is that the most effective forms of automation are likely to require some rematching with the remaining human functions in the system that still interface with them and are most affected by them. Functionality for humans is not confined to performance, manipulation and physical attributes. It includes emblematic and symbolic attributes, the tacit communication and understanding of information, matters of history and record, observability and judgements, skills and responsibilities, forms of cognitive support and influence, attributes of information transmission, and the characteristics of tools (Hopkin, 1991c; Hlibowicki and Bowen, 1993; Smolensky, 1993). It is in such respects that rematching of the human and the machine is entailed by the full automation of any human functions or tasks. Most of the human factors problems arising from the full automation of functions are not from their automated aspects but from the failure to recognize their unautomated ancillary aspects. The greater significance of these latter factors can be traced through a series of papers on human factors aspects of automation in air traffic control, which reflect that the change has been in emphasis more than substance (Hopkin, 1985; 1989b; 1991d; 1993a; 1994a).

15

Common human factors implications of computer assistance

15.1 Monitoring

Monitoring has always been and will remain an essential human function in air traffic control, whatever the level of computer assistance may be. It is an active rather than a passive process, not a reversionary standby task to be done only when there is nothing more important to do but a vital continuous task that must never be neglected even if the controller is very busy. An important aspect of controller training is to counter the human propensity to become exclusively absorbed in a single urgent task until it has been completed. However urgent a task may be, an even more urgent one could arise at any time in air traffic control, and must not be ignored if it does. Monitoring must therefore be relatively independent of the general level of activity. While it may be more natural to monitor only when there is a pause in the other work, monitoring must continue at all other times as it is much more than taking stock. Although computer assistance can provide support for monitoring, it is imprudent to rely wholly on it for it may not be capable of detecting everything significant that could occur. The controller must keep checking that all is proceeding as planned, that nothing unexpected has occurred, and that no impending problem or emergency is developing anywhere within or near the region of airspace for which the controller is responsible.

In the workspace designs, every effort is made to inform the controller about what is happening and to indicate what is pending. An objective of monitoring that becomes particularly important when traffic is not heavy is to keep the controller's picture of the traffic up-to-date, as it provides the frame of reference for all the controller's plans, actions and initiatives. On the basis of monitoring the controller may judge his or her current understanding of the traffic to be insufficient and attempt to build a more thorough and detailed mental picture of it. Monitoring provides feedback that the controller's plans and intentions are being implemented because aircraft are making the instructed manoeuvres. Monitoring can include reviewing and entrenching the information about the traffic, checking that all is well and as predicted, anticipating forthcoming events and problems, assessing the quality of information and verifying it where possible, and discovering and correcting any errors, omissions or failures. Much of this monitoring does not necessarily involve significant overt activity or observable task performance, and the prevalence and importance of monitoring within air traffic control tend to be underestimated because the commonest measures of performance may fail to cover monitoring activities.

Although it seems too trite to mention, the efficacy of monitoring and the functions it can fulfil depend on what there is to monitor. To sustain attention indefinitely on nothing is impossible for the human, and very difficult when continuous verification confirms that all is as it should be and that there are no unexpected changes or events, as this seems to imply that the verification is unnecessary. Machines are superior to humans in monitoring roles wherever the criteria for significant monitored events or signals can be specified in machine-recognizable terms. The performance of machines as monitors is not subject to degradation because of inattention, mind-wandering, boredom, or a dislike of enforced idleness. Monitoring is not therefore a role that would be assigned to humans for reasons of efficiency or safety, and it is unwise to predicate the safety of an air traffic control or any other system on human monitoring, which is intrinsically fallible. The tendency for human monitoring roles to increase rather than decrease in air traffic control therefore warrants critical examination.

What the human controller is expected to monitor are aspects of the human–machine interface, other sources of information within the workspace such as communications facilities, and any visible or audible activities of colleagues, supervisors or assistants. Monitoring is confined to what is presented, can be accessed, or is incidentally available, for example by being overheard. An important implication of computer assistance is that it can change the kinds of monitoring that are feasible, through changes made for other reasons in the displayed information, in the input devices, or in other features of the human–machine interface. Changes in the layout of workspaces or workstations that affect the activities or events that the controller can see or hear may affect profoundly what the controller can monitor and be influenced by.

Forms of computer assistance are not normally introduced because of their consequences for monitoring, although a claimed benefit of them is often that monitoring, interpreted as an unproductive kind of drudgery, will be lessened and the time saved can be diverted to more profitable tasks such as decision making. These reasons are not wholly without substance in some instances, but many forms of computer assistance introduce new monitoring requirements and the time actually gained from the changes proves disappointing. Furthermore, roles formerly fulfilled by monitoring may have to be reintroduced in other forms, for example to maintain the level of detail of the controller's picture of the traffic. In systems with little computer assistance, the controller decides what to monitor at what level for what purpose. Forms of computer assistance which indicate automatically the occurrence of a pre-defined set of circumstances, such as conflict detection or an aircraft deviating from its track, may require less human monitoring of these events but have the incidental consequence that the controller knows less about what is happening. The significance of the link between monitoring and understanding is consistently underestimated, until monitoring aids are found to impair understanding because they change the content of what is available for monitoring, the amount of information processing required for monitoring, and the extent to which human monitoring is an independent process.

What the controller is expected to do when there are no active control tasks may remain vague. Training may not cover what the best monitoring strategies are or how monitoring should be done. The forms of assistance are driven by task requirements rather than those of monitoring. For example, they may draw the controller's attention to a change that has occurred, but not remind the controller to monitor information again. The efficacy of monitoring is task dependent in simulated air traffic control (Schroeder *et al.*, 1994), and probably in real air traffic control also, although the applicability to real life of laboratory findings about monitoring by controllers, such as those of Thackray and Touchstone (1989), has always been uncertain. The traditional monitoring functions and roles are among the most difficult to match effectively with current and envisaged forms of computer assistance in air traffic control.

15.2 Passive roles

A common outcome of the provision of computer assistance is the conversion of active human roles and tasks into more passive ones, of which monitoring is an example. The reason is that many technological innovations and forms of computer assistance seek to remove routine human activities and reduce the level of human involvement in those that remain, with the result that the conduct of air traffic control entails less overt controller activity that is observable by others. This last point applies to many kinds of activity, including speech with pilots and with colleagues, the handling and updating of data, the initiation of or intervention in control actions, and a number of actions that each human function requires. The central purpose of most forms of computer assistance that affect the controller's tasks has not been to render human roles more passive but to curtail the increased task demands on the controller associated with more air traffic, and hence to contain within acceptable bounds the consequent controller workload. If the controller must spend less time in dealing with each aircraft, one of the most obvious ways in which this might be achieved is to have fewer essential human functions to perform in order to control each aircraft.

Some of the most practical forms of computer assistance are therefore the tasks that computers perform automatically on the controller's behalf. They transpond data between air and ground instead of through human dialogues. They enter data into the system directly instead of requiring the controller to use data input devices. They update information automatically instead of requiring the controller to do so. They introduce new information and discard old information appropriately so that the controller does not have to remember to do these tasks or find the time for them. They position new information in the workstation according to a predetermined principle so that the controller does not have to look at existing information to decide where the new item fits best. They collate and cross-refer information from different sources and present it in more integrated and processed forms so that the controller does not have to gather and amalgamate it. They signal when a change occurs so that the controller does not have to keep checking on the changes that have taken place. They indicate when information has strayed beyond predefined levels of tolerance or accuracy so that the controller does not have to discover this by searching. They identify future potential problems, so that the controller can concentrate on solving problems rather than on finding them. They propose solutions to problems so that the controller does not have to choose and gather all the relevant information and use it to formulate a practical solution. They confirm the satisfactory completion of events so that the controller does not have to ascertain this or seek acknowledgement of their completion. They initiate a transfer of control responsibility and indicate when it has occurred, so that the controller does not have to implement it. In these and other ways, the controller's tasks and activities and routine workload are reduced by computer assistance. In practice, all such changes lead to more passive human roles, although this is neither a universal nor a necessary consequence of computer assistance.

Some of the earliest forms of computer assistance which seemed to offer reductions in human activity actually increased essential activity, often because they required duplication of functions whereby, for example, the same instruction had to be spoken to the pilot and also keyed into the system. The resulting scepticism over the claimed benefits of computer assistance and wariness regarding its acceptability among controllers have still not entirely disappeared everywhere. The origins of some additional tasks from computer assistance could be traced to the need to retrieve data from the computer that had formerly accrued from the performance of the corresponding manual task without any special accessing. Despite such exceptions, computer assistance has generally made controllers' roles more passive, as an artefact of decisions taken for other reasons rather than as a deliberate policy.

An important aspect of this tendency towards less human activity and more passive human roles is that there is therefore less for others to observe in the course of air traffic control. This applies to supervisors, colleagues, assistants, instructors or anyone else with a legitimate reason to observe the conduct of air traffic control. With extensive automated assistance for the controller there is much less total human activity, and much less of the activity that remains can be observed and monitored directly by others. If the notion of controller workload is misinterpreted in terms of some form of activity analysis, an over-optimistic and false impression can be gained of the success with which the workload has been reduced by the computer assistance, if that is its objective. Because there is less overt activity to observe and what remains is more difficult to observe, and because what is still observable is less interpretable as it consists mainly of human–machine dialogues that make sense only to the person conducting them, it follows that controllers cannot see or understand as much about the activities of their immediate colleagues as they formerly could. The more passive they are, the less there is for colleagues to observe, and the more tenuous the basis becomes for controllers to continue to make the extensive range of judgements that have hitherto relied for their evidence on the observability and openness of air traffic control. These judgements can range from career developments, promotions and retraining needs, through the trustworthiness, confidence, and competence of individual colleagues, to the provision of practical help and support by colleagues for a controller experiencing particular control problems for whatever reason. The increase in passivity can undermine the basis for all these judgements.

Not only can computer assistance convert active tasks into passive ones, but some of the active tasks that remain may become ancillary or supporting functions rather than true air traffic control tasks. An effect of increased passivity can be to encourage the controller to access information only to look at it and not to use it for any control action. The controller's intention is to refresh memory or check the information, an additional task resulting from passivity because it was superfluous when the manual tasks that have become computer assisted were more active.

15.3 Reported loss of picture

Controllers build a comprehensive mental picture of the current traffic scenario (Whitfield and Jackson, 1982). Typically the process of building the picture completely takes en route controllers about fifteen to twenty minutes after coming on duty because by then they have accumulated the full history of all the traffic while it has been in their sector, have contacted it, and know its current status and intentions. Their picture of the traffic is often very detailed: controllers removed from the air traffic control environment without forewarning and asked to recall as much as possible about the traffic scenario can surprise others by the depth and detail of their recollections, which reveal efficient methods of data gathering and structuring and effective use of working memory and recall. In practice, controllers build a more rudimentary traffic picture much more quickly whenever the control responsibility for a workstation is handed over. This can be almost instantaneous in light traffic or take up to a minute or two with heavy or complex traffic scenarios, and this picture is entrenched and becomes more detailed as the full picture is built.

The two human factors concepts that seem most closely related to the controller's picture are mental models and situational awareness, both of which are fashionable (Johnson-Laird, 1983; Wilson and Rutherford, 1990; Vidulich *et al.*, 1994). Mogford (1991) has considered the controller's picture in relation to mental models, and discussed both concepts in relation to air traffic control (Mogford, 1994). Salas *et al.* (1994) distinguish between the concepts. The neglect of the relevance of situational awareness in air traffic control claimed by Garland *et al.* (1993) did not last long (Endsley, 1993; Rantanen, 1994; Garland and Hopkin, 1994; Hopkin, 1994b), although many of the issues are still debated.

Research on how controllers build their picture has uncovered expected substantial individual differences in the kinds of imagery and memory structuring that they employ. The controllers' picture of their traffic depends partly on the dominant sensory modalities of their mental imagery, partly on the content and methods of their training, partly on the particular air traffic control jobs that they have done, partly on how they have been taught to think about air traffic control, and partly on the equipment and forms of information presentation with which they are familiar. Since the control of air traffic takes place in three-dimensional space it might be assumed that the controller's picture would be primarily spatial, but this is not necessarily the case and some controllers may think mostly in terms of time. Older controllers originally trained in procedural methods structure their picture so that the data on flight progress strips can be mapped more readily onto it, whereas those trained from the outset on radar may base their mental picture on a plan view and may have to recode any information structured differently before they can integrate it fully into their picture.

The maintenance of the controller's picture of the traffic is an active and continuing process. It provides the basic understanding of the traffic scenario as a whole on which planning, scheduling, predicting, solving problems and making decisions depend, and also provides the basis for checking that instructions are obeyed, that decisions are correct, and that plans reach fruition. The picture of the traffic is made meaningful by the controller's professional knowledge of air traffic control and of its procedures, rules and practices, and these aspects lie within the controller's picture.

When air traffic control has few advanced forms of computer assistance but relies on manual functions associated with speech, the annotation of paper flight progress strips and a radar display, many of the controller's actions not only depend on the picture but update and reinforce it. Because the requisite information has to be assembled actively whenever decisions are implemented or problems solved, even routine manual tasks such as the introduction of a new flight progress strip onto a board or the discarding of the strip of an aircraft that has left the sector require positive actions, the reasons for which are known. The recollection of these activities is strengthened both by the mental processes required to devise them and by the physical actions required to implement them. The manual events themselves, and the mental processing in formulating them, contribute positively to the building of the controller's picture and to its level of detail.

Any forms of computer assistance, even quite simple ones, that reduce routine controller actions and chores or that replace manual with automated data insertion, compilation, collation and updating, may be accompanied by reports from controllers that some of the benefits of the assistance were partly nullified by an experienced loss of the picture or by an impression that the picture had become more vulnerable to being lost. Controllers believe that their understanding of the traffic scenario has been impaired and that they know less detail about it. The upshot can be that some of the time saved by the computer assistance is spent by controllers in rebuilding their traffic picture to its former state by calling down information items onto their displays sequentially and thereby restoring some of the mental processing of the information about the traffic that the computer assistance has taken away.

Reports of incipient loss of picture had several independent origins, suggesting that the phenomenon was genuine, and they became prevalent enough to warrant investigation of them. A theoretical explanation for loss of picture in terms of levels of information-processing paradigms furnished a rationale for predicting the circumstances that could induce it (Narborough-Hall, 1987). The controller had to process far less information to accept computer assistance with a task than to perform the same task unaided, and therefore knew less information with assistance because less information was needed. For example, if the controller no longer had to key in information or write it on flight strips in order to update it because the system updated the information automatically and signalled that it had done so, the controller merely needed to know that the change had occurred successfully and need have no inkling or understanding of the underlying processes and computations.

This unwanted side effect of many forms of computer assistance highlights one of the major choices about how computer assistance should be employed. One approach views the maintenance of the controller's picture as essential, and therefore argues that the controller must keep enough active tasks to maintain a detailed traffic picture at all times, with sufficient understanding of the current scenario to respond to any emergency at once. The other approach does not view the maintenance of the controller's picture as essential, but argues that such a detailed picture of the traffic by the controller is unnecessary and that the associated penalties of reduced readiness are acceptable and can be paid. It is contended that the other benefits of computer assistance far outweigh the consequences of any loss of picture, the effects of which can be ameliorated by designing other system functions on the assumption that the controller's understanding has indeed been depleted and therefore the controller would need more time or help to recognize and deal with an emergency or perhaps should not retain any reversionary tactical role in an emergency. This latter approach seems a better match with future controller roles concerned with the strategic planning of traffic flows rather than with the resolution of specific tactical problems. At present, which approach is preferable remains unresolved. Both seem technically feasible but they have different legal implications.

Some of the longer-term air traffic control plans that foresaw much less tactical intervention by a controller who knew less about the current situation have been recast towards the retention of some tactical roles, with the implication that reasonable current knowledge of the traffic situation must also be maintained by the controller through continued active involvement in the control loop. In the very long term, to obtain full benefit from advanced forms of computer assistance some loss of controller picture may have to be accepted, although this may occur in any case because of the trend towards strategic and away from tactical human roles. Professional controllers perceive an incipient loss of picture as threatening, and try to compensate for it. It is a crucial issue in deciding the future human roles in air traffic control systems. The main influences on its occurrence are not the nature or sophistication of the computer assistance offered but the extent to which the controller remains within the control loop and has to process gathered information personally.

When the controller loses the picture of the traffic, this may occur suddenly or it may become increasingly apparent beforehand that it is going to occur but the controller is powerless to prevent it. What usually happens is that the portrayed information loses much of its meaning and operational significance, although it can still be perceived. Normally the only practical cure is to rebuild the traffic picture one aircraft at a time, which takes considerable time. Ironically, when this has at last been achieved the same circumstances have been restored that led to the loss of picture originally, so that expectations are increased that it could happen again.

15.4 Disguising human inadequacy

Compared with the loss of picture, this potential human factors consequence of computer assistance was acknowledged belatedly. It is most closely associated with computer assistance for cognitive processes such as problem solving, decision making, prediction, and task scheduling, and with the presentation of computed options for the controller to accept or reject in lieu of options that the controller formerly had to devise personally. For example, the computer may propose a solution to a detected conflict and the controller must accept it or formulate an alternative without necessarily knowing all the information on which the computer solution is based. When such aids are introduced, controllers start by being somewhat wary of them, but eventually learn to trust them. Such forms of computer assistance must be well proven beforehand, and would not be introduced if they could be inherently dangerous. In practice, controllers are trained to accept these computed solutions, decisions and predictions because they normally are the best options, and controllers gradually learn from experience that they are safe.

It is a small step from routine acceptance of such computer assistance to reliance on it, and another small step to over-reliance on it. Its basic forms encouraged reliance, for the controller cannot check the correctness of what is proposed as thoroughly as the computer has done. The controller cannot process as much information in the time even with ready access to it, and normally the controller does not have ready access to all of it. An easy habit to acquire is to accept the computer solutions, particularly since this is encouraged as the best use of the computer assistance. Reliance on the computer assistance and routine acceptance of it are resisted as long as there are doubts about its efficacy, but when trust has resolved these doubts not only are its presented options accepted routinely but the need to confirm and remember them declines to a vestige of its original thoroughness, for there is no compelling reason to remember an option that is never subsequently questioned or referred to.

Acceptance of a computer proposal can be by a very simple action such as pressing one key. What others observe is the controller behaving routinely and accepting computer assistance. When the corresponding decisions, problems or predictions are dealt with manually by the controller without computer assistance, there is much more for others to observe, and it becomes impossible to conceal indefinitely from colleagues and supervisors any decisions, solutions, or predictions that are frequently wrong or inadequate. The evidence is far too observable to be hidden, and indeed it is applied in many other ways to judge the controller's competence, trustworthiness, promotability and professionalism. Computer assistance for cognitive functions can remove much of the observable evidence on which such judgements are based. It is not possible to gauge the controllers' professional knowledge and competence from observable actions that consist of single key-pressings to accept computer options. To express the same point differently, a controller can become inadequate or incompetent without this becoming apparent to anyone else.

Many advanced forms of computer assistance possess this fundamental property that they can disguise human inadequacy and shelter human incompetence. This is of greatest importance when they are introduced into contexts such as air traffic control where hitherto it has been impossible for an individual controller to disguise for long from colleagues and supervisors any fundamental inability to do the job. This can no longer be assumed in air traffic control environments with advanced forms of computer assistance. The problem need not be insuperable. One alternative is simply to accept it, together with its implication that every controller cannot be relied on to have the ability to cope in an emergency. Another alternative is to confirm the abilities of each controller in the same way as for pilots, by regular independent off-line checks of professional knowledge and competence and ability to respond to emergencies, conducted in an air traffic control simulator. Periodic retraining could also serve a similar function. What is unacceptable would be to introduce advanced forms of computer assistance with no provision to check if professional competence has been retained, because of a failure to realize that professional competence could be lost. Hitherto the full implications of this issue have been avoided, but they must be faced because policies to resolve the issue are required.

15.5 The curtailment of team roles

Most forms of computer assistance in air traffic control are to support the tasks of individual controllers, yet many functions have been performed by teams or by individuals working within teams. The purposes of their introduction are to help each single controller and to improve the efficiency and safety of air traffic control, and the associated changes in team roles are among their incidental consequences rather than their stated objectives. These changes in team roles therefore have to be deduced and stated, for they are not part of the original plan or among the original objectives. When they have been defined, policy decisions are required on which of the affected team roles must be retained in similar or alternative forms, and which can be discarded as superfluous.

The controller using computer assistance interacts with the system primarily through the human–machine interface, which becomes the controller's main tool in the conduct of air traffic control. Most of the changes in team functions are a consequence either of the reduced observability of controller actions and effectiveness, or of the failure to devise forms of computer assistance that can be applied collaboratively by teams to air traffic control. Despite recent developments in computer-supported cooperative work, including consideration of some of its air traffic control applications (Bowers and Benford, 1991), the techniques are still evolving and there are instances of failures (Diaper and Sanger, 1993). Many of the traditional roles of teams become more difficult for controllers to fulfil as team members if the available forms of computer assistance only help their individual tasks. Examples include discerning the controlling style of colleagues, the allocation of responsibilities among team members, and the identification of retraining needs and career and promotion opportunities. Mutual help becomes less practical because it is more difficult both to identify the needs of colleagues and to render effective assistance. The controller turns to computer assistance and human–machine dialogues for help, instead of consulting colleagues and discussing problems with them: the information that can be accessed by everyone who would like to take part in discussions may be curtailed. Controllers may have much less evidence available to detect when a colleague is tired or incapacitated or inexperienced, and far fewer facilities to intervene even when such judgements are still possible with some validity. Each controller's picture of the traffic tends to be confined more within his or her region of immediate responsibility, because the reduced contact with other team members tends to remove opportunities to supplement the picture with a framework of broader information about the traffic in adjacent regions.

Suite designs with controllers seated in a row or a crescent rather than around a large horizontal display or facing wall-mounted displays are associated with computer assistance. These suite designs do not facilitate direct discussion between controllers, and are not intended to. Suites with vertical displays are meant to encourage indirect conversations and consultations rather than direct ones. Speech is routed through the machine, and the substance of signals and gestures has to be inferred from inter-console marking. The main communications are through menus and dialogues between human and machine, and not between the controller and other controllers or pilots. As a consequence of the needs to bring the machine into the loop and to maintain its knowledge of human activities by entering them into the system through the human–machine interface, any team activities that do not include machine participation become external to the system and not an integral part of its functioning. Hence the roles of team activities are reduced, and the residual team activities tend to be marginalized.

The reduced observability of air traffic control and the changes in team roles affect the feasibility of on-the-job training and assessment and the acquisition of shared mental models (Salas *et al.*, 1994). Both training and assessment require a fuller understanding of what is occurring than may be acquired from observations, which must be confined largely to activities through the human–machine interface in more automated systems. This applies particularly where the controller appears to do little but accept the computed recommendations, even though there may be a great deal of detailed thinking and pondering of alternatives on the part of the controller, to which the instructor or assessor has no direct access.

15.6 Restrictions on supervision

Practical supervision of air traffic control is commonplace. Each watch, the group of people who work together in the same air traffic control workspace, usually has a designated supervisor. The person in charge of an air traffic control suite, such as a Chief Sector Controller, may also have some supervisory responsibilities. How active a participant in air traffic control the supervisor is, depends primarily on the type of air traffic control service and the workspace design, and within those constraints depends on the individual's supervisory style. Many current air traffic control systems specify the roles and responsibilities of the supervisor, but allow considerable discretion in how they are exercised. The supervisor who makes broad decisions about staffing levels and the allocation of work among controllers and who receives, briefs, and takes charge of visitors so that they do not disturb the control work may remain quite aloof, for these functions can be almost a full-time job. In air traffic control contexts where decisions are taken collaboratively, the supervisor may join in. A participative supervisor may annotate a controller's flight progress strips, albeit in a distinctive colour to denote that it is the supervisor who has done so. The supervisor may add information that the controller does not know, draw the controller's attention to particular information items, amplify or change data and explain why, and ensure that wider air traffic control interests have a proper influence on the control activities within each suite. With more automated systems, most of these functions become less practical and some are impossible.

The supervisor is likely to have other roles. These can be formally stated and attached to the job rather than the person who does it, or they can be acquired by accretion personally by individual supervisors. Their line management roles may set the standards and ethos of the suite or watch, establish the acceptable norms of performance, determine which circumstances are routine and which require querying or consultation, and support the favoured practices and customs that reflect the characteristic control style of the suite or watch. The supervisor can have a strong influence on the self-confidence of controllers and on their mutual confidence and trust, and hence on the basis for reaching collaborative decisions and on the efficacy and safety of the air traffic control service provided. This is achieved primarily through the nature and the manner rather than the amount of supervisor intervention, through the degree of support from their supervisor that controllers can rely on, and through the timeliness, appropriateness, competence and efficacy of any help for controllers that the supervisor provides, whether on his or her own initiative or at the controller's request. Again, the feasibility of direct participation by the supervisor is often curtailed by computer assistance.

The supervisor sets the tone of the team. The supervisor may have further line management functions, and recommend retraining, transfer, promotion or job appraisal review and assessment. Many supervision practices are encountered in air traffic control. It may be possible to call down at the supervisor's position the displays and spoken dialogues at any controller's position, and thus supervise each controller remotely. The supervisor can be more peripatetic but remain relatively passive by watching the activities in a suite and even plugging into the RT on a suite but not taking part in its activities. The supervisor may intervene more actively in many ways, including participating in control decisions, annotating flight progress strips, contributing to discussions, and consulting controllers elsewhere who would be affected by decisions within the suite. In the extreme cases of exceptional workload or a very unusual situation, the supervisor may for a short time become in effect a full member of the team in the suite. In manual systems, the supervisor can contribute great flexibility, but computer assistance generally makes less provision for supervisor intervention, makes less information available to the supervisor that would justify intervention, and reduces the kinds of intervention that remain feasible.

The traditional supervisory roles depend heavily on the fundamental observability of air traffic control. Most of the roles of supervision have relied on observations. They are needed for independently assessing safety, judging the efficacy of the solutions to problems, tracing the origins and consequences of actions, assessing individual controllers in terms of their competence and retraining needs, and appraising any instructing or training activities within the suite. Active supervision keeps the supervisor well informed about the current air traffic control situation, and brings other benefits which depend on the supervisor being well informed. Manual air traffic control can be understood and appraised by a supervisor with professional air traffic control knowledge who passively watches it in detail, whereas to a naive observer it can seem well nigh meaningless and unintelligible. Without observability, it becomes progressively less intelligible even to the best-informed supervisor.

Many forms of computer assistance that can be justified on their own merits because of the benefits they bring have the incidental effects of changing the kinds of supervisory function that are feasible and the ways in which they can be exercised. They promote the kinds of remote non-interventionist supervision mentioned above, and prevent those that allow active participation by the supervisor. This is a factual and not a judgemental statement. Participatory supervision is not necessarily better, and certainly not under all circumstances. But the changes that prevent it are made for other reasons, and their full consequences should be ascertained beforehand, particularly since ultimately someone called a supervisor could be present in air traffic control environments where any form of effective human supervision is in fact impossible, and this could delude everyone into believing that the safeguards of supervision are present when they are not. Computer assistance can change most of the functions of supervision, including active participation and intervention, support for and identification with team decisions, assistance with overloading, judgements of individual strengths and weaknesses, recognition of training needs, and all roles and judgements based on observability. As more forms of computer assistance are introduced, at some point the option of dispensing with supervision altogether has to be considered.

If all of these potential consequences of computer assistance for supervision are faced, policies can then be formulated and implemented on how they should be handled. It is inevitable that computer assistance changes supervisory roles, and it is therefore necessary to plan what supervisory roles should be retained in the future and how they should be exercised. Completely different means to fulfil some traditional and essential supervisory roles may have to be devised; for example, there may have to be alternative ways to assess the professional knowledge and competence of individual controllers. Some controller attributes that have been important in the past may not remain important in the future. The ability to be a competent, trusted, congenial and predictable member of a collaborative air traffic control team may not matter much when the controller works mainly through the human–machine interface rather than with colleagues and supervisors, because the computer assistance acts through these interfaces and not through team functions. Supervisory roles may persist for legal reasons, safety reasons, efficiency reasons, or management or career development reasons, and forms of supervision that can fulfil such purposes will have to be devised and verified. If supervision no longer exists, other means that are seen to be fair must tackle promotability, retraining, staff allocation, professional norms and standards, and other issues traditionally within the supervisor's bailiwick.

15.7 The development of norms and standards

The relationship between computer assistance and the development of norms and standards in air traffic control is one of its least researched and least understood aspects. While the quality of the air traffic control service is clearly influenced by professional norms and standards, the mechanisms through which they are determined and become entrenched have been studied far more in other contexts than in air traffic control. Yet its professional norms and standards are one of the most striking features of air traffic control to an outsider.

In many parts of the world, the air traffic controller strives to provide the standard of service expected by colleagues as a condition for gaining their professional respect and full acceptance. A hallmark of a profession is that its norms and standards are generated internally, and air traffic control complies with this. There are many human factors implications. The development and perpetuation of norms and standards rely on opportunities for controllers to demonstrate their merit to colleagues by observable actions and conduct, from which each individual's competence and style can be appraised. Each controller requires some discernible feedback from colleagues about the outcome of this quiet appraisal, which is usually in the subtle forms of tacit acceptance, little direct monitoring of activities by colleagues, and leaving the controller to get on with the work, rather than in the forms of overt praise or expressions of trust. Controllers can gain self-esteem from the tacit acknowledgement by colleagues that their work meets professional standards, and can earn the professional respect of colleagues because their work is sufficiently observable for such judgements to be made about it. Any forms of computer assistance that deliberately or incidentally reduce the observability on which such judgements depend can change the mechanisms by which norms and standards are maintained, to the extent ultimately of removing them altogether. The question then arises of whether they can be dispensed with, or must be inculcated by other means such as during training.

Norms and standards tend to be fairly localized. They contribute to loyalties and identification. Watches or small teams of controllers are prone to exaggerate any differences in the norms and standards that prevail. Each becomes certain that their own watch or team is superior to all others because it provides the best service, follows the best practices and exhibits the greatest professionalism, and their certainty depends on the norms and standards of the team rather than on the merits of individual controllers. All members try hard to maintain these norms and standards so that pilots receive an excellent service whichever team member is providing it. Supervisors may influence the prevailing norms and standards by making clear which activities are acceptable and which are not.

Where norms and standards have become entrenched, system changes may have smaller effects than expected. Controllers who have to work with equipment which they think is inadequate may be forever complaining about it to management and to anyone else who might improve it, while they nevertheless strive to provide a good air traffic control service despite the equipment limitations. They try to achieve their own self-imposed norms and standards wherever it is feasible to do so, by ever-greater efforts if necessary. Conversely, computer assistance that makes their jobs easier may not necessarily improve task performance or the air traffic control service, if it is not possible to provide a better service than was provided with the poor equipment or if the controllers seek to maintain the same norms and standards of performance with the new equipment as they regularly achieved with the old. This must not be construed as unprofessional, for indeed it is the opposite. It reflects the strength of the self-generated professional norms and standards that prevail under a multitude of circumstances, which can be quite resistant even to major changes in computer assistance. Norms and standards can also be quite resistant to managerial influences. They may be almost immune to external exhortation, and not influenced much by retraining if controllers still expect the behaviour of their retrained colleagues to remain predictable in accordance with the prevailing norms and standards. Managerial requests for greater efficiency may be construed as ill-conceived, maladroit and even unsafe, if they require any established norms and standards to be abandoned.

Air traffic control is a comparatively young profession, yet its norms and standards have often become entrenched. It seems important to discover the reasons for this and the underlying mechanisms, lest they be removed or undermined inadvertently. Norms and standards exercise further influences. They contribute to the solidarity of air traffic control as a profession and may also help to protect it against some of the most militant forms of external interference. Norms and standards operate at a national level and controllers may take pride in the quality and renown of the air traffic control service that they provide as a nation, as well as in the high standards of their own local watch or team at their centre or tower or other facility.

Some forms of automated assistance introduced for other purposes seem to have significant consequences for norms and standards which may not be anticipated because the major influence of professional norms and standards on air traffic control performance has not been recognized or quantified, and is therefore insufficiently understood. But norms and standards are important enough for controllers to wish to reject any changes that seem to impair the quality of the air traffic control service that they can offer. The implication is that any form of computer assistance that appears to lower existing norms and standards will encounter serious problems of acceptability among controllers.

15.8 Human understanding of system functioning

The main justification for most forms of computer assistance is that they can out-perform humans in some way, for otherwise their additional costs could not be recouped. It follows that the human cannot fulfil the same functions to the same standards in the same way at the same pace, and it is implied that the human cannot understand completely the detailed functioning of the computer assistance as it is taking place. The controller's level of understanding of the computer assistance must meet some clear requirements within other equally clear limitations. Some minimum level of understanding of the computer assistance will be required in order to use it properly and to recognize if it is faulty, and perhaps to trace how far the ramifications of any failure extend within the system. The controller does not need to understand the causes of failures sufficiently to repair them, which is the responsibility of others, but does need to know what remains fully functional despite failures.

A related issue is how the controller's understanding of the computer assistance should be acquired. The two main methods are through formal instruction within a structured training or retraining course, or through adapting the computer assistance to function as a self-teaching aid or directing device, whereby the controller is led through demonstrations and worked examples until sufficient understanding according to performance of the worked examples has been acquired. Learning on the job is not unknown but not recommended. Insufficient human understanding of the functioning of computer assistance can be self-defeating. In the extreme case, a controller with no awareness that a form of computer assistance exists or with no understanding at all of how it functions will not use it. For any form of computer assistance short of full automation, some degree of human understanding is a prerequisite for its usage, with the implications that the fuller and more correct the understanding becomes, the more effective the computer assistance can become, the more often it will be used, and the closer the accord will be between its actual and its intended usage.

The controller may not have a full understanding of the structure of the menus and dialogues that can be accessed within the human–machine interface. Typically the controller is presented only with the options which the machine offers, which depend on the outcome of previous dialogues or of other tasks or events. Different outcomes usually result in different presented options at subsequent stages. This raises the issue of the desirable level of human understanding of the entire dialogue design that would be required for the most effective dialogues. The selected options at any given stage may be chosen solely according to factors that seem immediately pertinent to each choice, and may therefore ignore all the other incidental choices that are also being made unwittingly about the options that will be presented at all subsequent stages of the dialogue. This topic is raised here not because there is a satisfactory answer to it for there currently is none, but because much fuller consideration of it seems essential (Wieringa and Stassen, 1993). Current practice seems to assume that human understanding of the total dialogue structure need not be very comprehensive.

Human understanding of system functioning can become crucial in an emergency with non-standard problems and solutions. The controller may occasionally be forced to adopt solutions that normally would never be entertained, but which offer the safest alternative in the circumstances or the one with fewest cumulative adverse repercussions. The machine may never offer such non-standard solutions as options and may be incapable of doing so, and the controller must be able to override all machine options and implement the preferred human solution. Emergencies are not fitting circumstances for cumbersome or elaborate procedures, and it is desirable that the controller understands in advance how to implement non-standard solutions and receives machine collaboration in doing so. A machine that rules the controller's solution as invalid, and refuses to accept or implement it, is peculiarly unhelpful, since the controller knows its invalidity in normal circumstances already. Ultimately the controller can do only what the system permits, for every controller action has to be implemented through the facilities within the human–machine interface. Routine activities may not demand extensive human understanding of the system functioning, but non-routine activities may rely heavily on human understanding. Wherever departures from routine seem to render the machine uncollaborative, the controller may be placed in the unenviable situation of not only having to devise a safe solution but of thwarting attempts by the machine to prevent its implementation, and the controller will need detailed understanding of the system functioning in order to circumvent its objections to the human solution. Current air traffic control systems have only very limited facilities for the controller to convey human intentionality and objectives to the machine, in the hope of enlisting machine support and collaboration in achieving them.

15.9 Initiation of actions by human or machine

In manual air traffic control systems, most actions are at the behest of the controller. This applies to their initiation, conduct, completion, timing and scheduling. Because the functions of the first forms of computer assistance were to gather, store and compile data, the balance between human and machine initiatives did not change much. Forms of computer assistance that employ the data for computations raise the issue of whether to make available to the controller those computed products that relate to the controller's activities and responsibilities. Generally the initiative for this has to be with the machine, since the human controller could not know what relevant computations there were unless they were presented. As a result, the first machine initiatives intended to prompt human air traffic control actions were mainly in the form of warnings such as a detected potential conflict between aircraft or an aircraft beginning to stray from its planned route by more than a predefined distance. Initially the machine did not indicate any required human response. On the basis of the machine initiative, the controller acted to resolve the conflict or restore the aircraft to its correct route. The machine also determined the timing of its initiative, and the controller might not necessarily have taken any action at all without the machine prompting.

More elaborate forms of computer assistance tilt the balance between human and machine in the initiation of actions further in favour of the machine. To extend the above examples, further computations applied to the data used by the machine to detect a possible conflict can formulate one or more solutions to it and hence provide computer assistance for conflict resolution, or further computations on the data about the straying aircraft can formulate revised headings or other route changes to restore the aircraft to its planned route and to check that the changes proposed by the machine do not engender further problems such as new potential conflicts along the revised route. These kinds of computer assistance based on computation can be elaborated so that the machine provides not only headings to restore the aircraft to its former route but also speed changes to restore it to its former time slot. In these circumstances the computer assistance becomes extensive but the initiation of activity remains with the human controller, who could formulate and implement an alternative solution but is encouraged to accept one of the computed solutions routinely. The human initiation of actions has then become nominal rather than actual in some respects. At a further stage, computer assistance evolves further towards full automation and the initiatives pass to the machine when the solutions are not routed via the controller but implemented directly by the machine. This may be done either without informing the controller at all or by notifying the controller of their occurrence as a matter of record while including no provision for controller intervention. The final stage of full automation would remove the human controllers altogether and leave only machine initiatives.

This last stage seems distant for many predictive, problem-solving and decision-making functions even if it becomes technically feasible. Legal constraints and requirements for the responsibility to remain ultimately human will have to be resolved first. Some of the most beneficial forms of computer assistance in terms of increased efficiency do replace human initiatives with machine ones, and the associated human factors problems focus on what the controller needs to know about them. When machines initiate and complete actions automatically, this clearly reduces the degree of responsibility that the controller can have for them, and if the controller retains the legal responsibility then the controller must at the very least be kept informed about them and be able to exercise sufficient initiatives and interventions to ensure that this legal responsibility can be a reality. In practice the controller requires further knowledge of other machine initiatives to prevent human-initiated actions that are at cross-purposes with machine-initiated ones. A simple example refers to the transmission of information between air and ground. Here dialogues between the pilot and the controller, almost invariably initiated by the human, are being replaced by the machine-initiated transponding of data automatically between air and ground through data links or other devices, and by the automated updating and presentation of the transponded information on displays. Nevertheless, some provision for human initiatives must linger, particularly in non-standard and emergency situations, which poses the problem of how to integrate these human initiatives with machine ones without inducing any hazards or inefficiency.

Decisions about human and machine initiatives can have profound human factors implications, which are seldom recognized at the time and therefore have to be coped with retrospectively. This is not the optimum way to achieve a highly efficient and safe system. Obviously the nature of human initiatives has major consequences for job and task descriptions, for the levels of skill needed, and for feasible human responsibilities. Less obvious are their consequences for the ability to maintain professional knowledge and skills through everyday activities, which may have to be boosted by other expedients such as off-line training and assessment if diminished initiatives result in fewer opportunities for active engagement in tasks, and if certain essential but rarely used human initiatives become insufficiently practised in the course of normal work for them to remain available at once on the rare occasions when they are suddenly needed.

The human factors implications of this balance of initiatives extend to job satisfaction and many team functions, and this balance affects the ways that tasks can and should be grouped into jobs, the division of responsibilities among controllers, the feasible kinds of supervision and assistance, and the recommended methods and content of training. An important means of ensuring that the nature and quality of the air traffic control service provided are broadly uniform, regardless of which individual controllers are providing it, is for all controllers to conform with similar norms regarding the occasions for human or machine initiatives. The tendency in the past to propose various machine initiatives without envisaging their full consequences has to be reversed. The primary reason is not to keep more human initiatives, although there may be legal, technical, air traffic control and human factors benefits in doing so, but to ensure that whenever further initiatives are transferred to the machine the outcome will be effective, efficient and safe improvements to the system, and that these improvements accrue because all their human implications have been correctly identified and allowed for in advance and do not have to be accommodated retrospectively by piecemeal and non-optimum expedients.

15.10 Potentially incompatible human roles

Air traffic control is quite a complex job, as gauged by the time and effort required to train a controller, the variety of roles and tasks, and the multiplicity of the factors considered relevant in controller selection. Because the attributes required in controllers seem to be many and complex, there must be a considerable probability that they might not all be compatible with each other.

Several factors increase this probability. Air traffic control must provide a continuous service. There will be times when controllers are on duty but there is little work for them. At other times, controllers will be extremely busy handling traffic levels at or beyond the planned capacity of the air traffic control systems. Continuously increasing air traffic control demands tend to lengthen the periods of working continuously at maximum capacity until they become the normal condition. It is inherent in air traffic control as presently constituted that there will be times when the task demands are too low and times when they are too high. Much of the controller's work is quite routine. It requires knowledge and skill but does not pose insoluble problems. The controller must be able to deal with emergencies and unusual or unexpected events, but in most air traffic control jobs these are uncommon. The controllers' roles often incorporate periods of relative idleness and periods of very high activity, and include many periods of routine work and a few that require non-standard interventions urgently.

A few further potential role incompatibilities are worth mentioning. The controller is expected to be quick, correct and decisive in individual decision making, and also to function as a collaborative and effective member of a team reaching group decisions. The controller is expected to be independent and self-confident, yet to submit to management directives and wishes that can seem little more than whims to the controller until they have been adequately explained. The controller is expected to concentrate on difficult problems and solve them, yet never to concentrate so exclusively on any single task as to fail to monitor scrupulously the continuing safety of the whole region under control. The controller is expected to maintain all skills and knowledge for instant use, even when they tend to fade because there are so few opportunities to practise them. The controller may be expected to maintain verbal fluency, even though it is needed less often, and to acquire new skills of manual dexterity and data input, yet there may be no inherent connection between such diverse skills and no evidence that possession of some predicts possession of others. The controller may be expected to retain the responsibility for roles, functions and tasks even if the means to exercise that responsibility are whittled away and become largely nominal.

The human role of maintaining an up-to-date and comprehensive picture of the traffic as a frame of reference for interpreting existing information and for incorporating new information is difficult to reconcile with increasingly passive monitoring roles in which the controller relies on computer assistance for prediction, problem solving and decision making, and experiences reduced understanding and knowledge of the situation yet retains full responsibility for it. Some forms of computer assistance seek to direct the controller's attention through a sequential series of events and responses, during which the controller retains the responsibility for attending to everything else that is happening but is not reminded of this automatically.

New forms of computer assistance proposed for air traffic control do not generally start from human roles but require adaptation to them, which increases the potential for mismatches between human and machine roles and for mismatches among the human roles that have become machine assisted. There may be difficulties in matching with the machine different human functions derived from different human roles; for example, the information that the controller needs while computer assistance of a particular human function is being provided may differ greatly from the information that the controller needs if for any reason the computer is not providing that assistance. Consider a more specific example: if computer assistance in the form of conflict resolution presents several viable alternative solutions to the controller, the information that the controller needs to make a choice is very different from the information that the controller needs if the computer assistance can find no solution to the conflict within its rules and so passes the problem back to the controller for a human solution. The controller may then need to know far more than normal to find the best solution. The incompatibilities are not confined to the provision of suitable information but extend to its forms of presentation and to the time available to gather and apply it.

Some potentially irreconcilable team functions have already been mentioned. One concerns the perpetuation of the traditional roles of air traffic control supervision where reduced observability and powers of intervention render most of its original functions impossible or redundant. Analogous considerations affect the feasible roles of human air traffic control assistants wherever the controller is deprived of the means to supervise an assistant effectively. There can be incompatibilities between decisions by one controller that are affected through the human–machine interface by dialogues that are not observable by others, and other decisions implemented manually and observably following team consultations. Incompatibilities may affect some ancillary functions. They are potentially present between roles that involve doing air traffic control and roles that involve watching others do air traffic control. When there is automated assistance, it can become much more difficult to fulfil satisfactorily any other functions in air traffic control that require someone else, whether an assessor, instructor, supervisor or controller acting as mentor, to watch a controller in order to form a judgement either about the controller personally or about the controller's activities.

One kind of potential incompatibility affects workload, stress or boredom more than roles, tasks or observability. Most of the forms of computer assistance, despite a few encouraging exceptions, have not been selective in relation to workload and have not given the controller more means to control workload. Typically if they reduce workload when it is too high, they also reduce it further when it is already too low. They thereby aggravate problems of boredom, inactivity and the maintenance of attention when there is little work, in pursuit of their main objective of relieving the over-burdened controller by some machine assistance. A controller with more control over roles, functions and workload may be more able to resolve incompatibilities between human roles and computer assistance. Whereas stress that originates in excessive demands and time pressures may be alleviated by some forms of computer assistance, the stress that originates from having to accept computer assistance that the controller does not really understand but remains responsible for may be increased by the computer assistance, which may not necessarily remove the sources of stress or resolve incompatibilities within them, but may simply substitute one cause of stress for another. Incompatibilities of this kind have to be resolved by forms of computer assistance that are selective in their application, remain at least partly under the control of the controller, have some capacity to adapt to circumstances, and can provide assistance when it is needed but can be dispensed with when it is not. This has not been the main driving philosophy in the past but there are encouraging signs that it may become so in the future. The technology to implement such a philosophy is available.

16

Effects of the system on the individual controller

16.1 Experience

Through training, the controller learns what to do, but proficiency and tacit skill require experience (Myers and Davids, 1993). The initial experience of controlling real air traffic is usually gained during closely supervised on-the-job training. For a few, being personally responsible for the safety of aircraft proves so daunting and incurs such anxiety that they cannot face the prospect of it as a lifetime career. If they will never have enough self-confidence to become satisfactory controllers, it is best for everyone to discover this as early as possible. For those who do complete training successfully, their first experience of the full responsibilities of air traffic control can still be nerve-racking. This reaction is quite normal. Supervisors and colleagues keep a watchful eye on inexperienced controllers, provide help and guidance where it is needed, give encouragement when it is due, and ensure that inexperience never hazards safety. Gradually the controller acquires self-confidence and begins to earn the trust and respect of colleagues and supervisors, whose watchfulness becomes more relaxed and less unremitting.

Greater experience brings many changes for the controller. Among them are the following. The controller develops air traffic control skills, becomes better at scheduling and prioritizing tasks, and learns to cope with higher task demands. The controller avoids blunders and major errors, adapts professional skills and knowledge to the particular characteristics of the airspace region for which he or she is responsible, recognizes nuances of problems that affect their categorization and solutions, and learns to plan further in advance and to consider strategic as well as tactical implications. The controller's experience is helpful in the early diagnosis of problems, in gaining more control over task demands and work scheduling, in identifying the more extensive consequences of particular actions, and in learning to avoid non-optimum solutions even if their adverse effects are only trivial. The controller makes progressively subtler discriminations in the details and timing of chosen actions, provides a better service in achieving the basic objective of a safe, orderly and expeditious flow of the air traffic, and becomes more aware of the particular needs and expectations of different air users and more flexible in meeting them. Experience brings greater insight into team roles in air traffic control, into individual roles and functions within the team, into acceptable levels of adaptability and conformity, and into the locally preferred practices within each work environment. Controllers gradually develop their own personal styles of controlling, which must not violate any local practices nor clash with the styles of colleagues but be acceptable to them. With experience comes greater depth of knowledge and insight into one's own tasks and those of the colleagues, supervisors and assistants who interface with them, so that teams function efficiently and smoothly as entities. Experience brings the tacit understanding of peers—their roles, practices, preferences, presumptions and attitudes—that is a hallmark of professionalism.

What is taught in training may not overlap much with what is learned from experience, because they impart different kinds of knowledge. New controllers learn from experience that it has been their good fortune to join the best team with the best practices, which they are of course expected to follow gladly and defend against any external criticism or challenge. Training teaches knowledge of air traffic control skills, procedures, rules, work scheduling, standard practices, and the like. Experience teaches ethos and attitudes, professional norms and standards, sources of self-esteem and of the esteem and respect of others, and the criteria applied by controllers to assess the professional competence of their peers. The controller with more experience gains in skill, self-confidence and satisfaction from a job well done. Self-monitoring, though never infallible, becomes more helpful in error prevention because experience teaches when extra care and cross-checking are most needed.

Experience builds skill. Routine tasks require less attention. More functions become overlearned or even automatized so that they can be done concurrently with others. Skills that are appropriate benefit system functioning. Experience builds habits that underpin favoured stratagems and tactics. Habits can also be beneficial, with adequate insight into their occurrence and relevance. Good habits facilitate smooth performance, can engender mental checklists so that no pertinent factor is omitted, render each controller's activities more predictable to colleagues, and may foster thoroughness and consistency. Inappropriate habits can result in excessive rigidity in the choice of data applicable to problems and in their categorization and solutions. Long experience has the inherent potential to build habits that are entrenched, skills that are inflexible, the complacent belief that there can be no more new problems and self-confidence that cannot discriminate between when confidence is justified and when it is not.

Experience is part of the basis for the training and assessment of others. Instructors try to pass on the rules, practices and procedures of air traffic control, together with their own experience of how to interpret and implement them. A human factors problem is that it can be quite difficult to quantify the actual benefits of experience, or validate them in air traffic control terms. The influence of experience tends to diminish with fewer opportunities for human flexibility and if the notion that every problem has one best solution becomes more prevalent. Experience can reduce the controller's anxiety about the work and its responsibilities, and provide a realistic perspective that helps to prevent stress-related symptoms. Experience builds trust among colleagues who learn to rely on it in an emergency. This does not necessarily imply that an experienced controller is necessarily better as a controller or that long experience must always be beneficial, for it can be associated with such human factors problems as complacency, but it does mean that colleagues believe the experienced controller has been tested more often and has proved to be reliable.

It is important to put experience into perspective. On the whole, it is not among the most important determinants of proficiency in the long term over a whole career, although experience is at a premium while the controller is still a relative novice. The most experienced controllers are not necessarily the best when they are also the oldest controllers. Experience is not necessarily a boon when radical changes such as new forms of computer assistance or new data sources are introduced that require the activities of controllers to change. While every effort should be made to continue to utilize entrenched habits and practices and their associated skills, some new procedures will have to be learned and some previous ones discarded. The greater the experience of them, the more difficult it may be to discard them completely. In this role, training is often less helpful than it might be. Much more is known about teaching people to acquire new learning than about teaching them to forget old learning that has become inapplicable. In air traffic control this can lead to real practical difficulties for the experienced controller. The findings of theoretical studies on the extinction of habits, negative reinforcement or forgetting do not seem to transfer well to real-life tasks.

Experience is inevitably confounded with other factors so that its effects are difficult to separate from their effects. In air traffic control, experience and age are often confounded as factors with contrary effects, and many intervening effects of age can make the balance between them difficult to determine (Rinalducci *et al.*, 1993). Whether a more experienced older controller is superior on the whole to a less experienced younger one is a contentious issue, not yet fully resolved (Trites and Cobb, 1964; Warr, 1993). Their combined effects can become obscured by further intervening influences. In air traffic control, as in other professions, people with less experience are often younger and earn less but have more pressing financial responsibilities. They may often be more resilient, particularly in withstanding the effects of shiftwork. As systems evolve, it becomes necessary at some stage to adapt the selection procedures for controllers so that they are a better match with the new requirements. If this adaptation is repeated, eventually the most experienced controllers possess the fewest attributes that have been retained within the current selection procedures, and the retraining requirements may call for aptitudes which the more experienced controllers were not initially selected to possess. New ways of matching human and machine by utilizing the existing skills and knowledge of experienced controllers and circumventing their limitations may offer prospects for progress. The most significant product of experience may prove to be the depth of understanding of air traffic control that results from years of work as a controller without computer assistance, which may never accrue in systems that rely on computer assistance.

16.2 Workload

Workload is probably the human factors concept that is mentioned most frequently in relation to air traffic control. Attempts to assess it are a feature of almost every study. Although workload can be a source of many benefits, most of its accrued connotations in air traffic control are negative, and the quest to reduce it can become a crusade. Occasionally, changes in workload are worthwhile objectives in their own right, but more commonly they are sought for other reasons. Measured changes in workload, which may themselves have suspect validity, may be hailed as evidence of the achievement of broader aims, such as increased traffic-handling capacity. Almost every study of workload in air traffic control has examined high workload. The topic is of legitimate concern, but the preoccupation with it has had two adverse consequences. One is the comparative neglect of the human factors implications of low workload, which are acquiring greater significance since most forms of computer assistance seem likely to aggravate them. The other consequence is a dearth of suitable tools, data and measures to answer the most important practical questions. These do not concern workload that is high but workload that is too high, and must be reduced. Criteria to determine when workload becomes too high, as distinct from merely high, are sparse, but it has been considered as a limiting factor (Jorna, 1991) and attempts have been made to develop workload measurement techniques specifically for air traffic control (Stein, 1985).

Workload must be distinguished from task demands, as they are not the same. For any specified air traffic control scenario, it is usually possible to deduce the full set of controller activities that will be required to perform all the necessary tasks with the available means in order to achieve the air traffic control objectives. This seems measurable, and indeed it is, but not in terms of workload. The reason is that workload depends also on who performs the work. If the same traffic scenario is presented to two controllers, the task demands may be the same but the workload is not. In fact, the task demands also vary more than is often realized (Smolensky and Hitchcock, 1993; Huey and Wickens, 1993). Whether a controller finds the work difficult or easy, complex or simple, familiar or unfamiliar, stressful or unstressful, and satisfying or unsatisfying depends on numerous attributes of that individual controller. These include knowledge, experience, skill and understanding, and, less obviously, professionalism, motivation, tiredness, foresight, patience, confidence, the deployment of resources, trust in the evidence, planning horizons, and colleagues' attitudes. All of these factors, and others, can influence the workload of the controller in meeting the task demands. The consequences of task demands for workload can be mediated by the controller's strategies (Sperandio, 1971), by working methods (Sperandio, 1978), by the development of skill (Bainbridge, 1989), and by various cognitive processes (Tattersall *et al.*, 1991). Workload is a complex and multidimensional concept that can seem unidimensional subjectively, as is implied in the many requests to quantify it, made by people who judge subjectively if their own workload is too high or too low and who expect controllers to provide comparable judgements.

Even a cursory appraisal of the many claimed measures of workload suggests that it must be a more complex notion than it seems, for they cannot all be appropriate, valid and complete measures of the same phenomenon. Measures of air traffic control workload may be of systems, behaviour, performance or error, may be physical, physiological or biochemical, may be individual or social, may be subjective or objective, and may address many cognitive functions (Hopkin, 1979). Different measures may not all yield the same answers; indeed, there would be no point in having so many if they did. Workload measurement must usually employ several measures, chosen not for their availability or ease of administration but because collectively they can provide a satisfactory rationale that includes their relative weightings, their interpretation, and mechanisms to reconcile any incompatibilities between them. Texts emphasize mental workload most (Moray, 1979; Hancock and Meshkati, 1988), and often consider contexts with multiple tasks (Damos, 1991). Not all measures are equally appropriate: subjective measures require, but do not always receive, careful statistical treatment (Nygren, 1991), and test and evaluation environments in particular require measures that are both sensitive and robust (Wierwille and Eggemeier, 1993). Developments in other contexts can be helpful (Farmer, 1994).

Typical complicating factors in assessing workload can be illustrated by two examples. Higher traffic demands may be associated with increased heart rates, recorded as a physiological measure of workload, although this does not always occur (Costa, 1993). Higher traffic demands may also impose on the controller more overt physical activity, greater cognitive demands, greater effort, and greater anxiety, each one of which could also affect heart rate, and it may be impossible to separate their respective influences. As a second example, following the introduction of colour coding on air traffic control displays, controllers may report reduced workload and be certain that their task performance has improved, although objective measures fail totally to demonstrate any improvements in their performance. It is tempting to reconcile such contradictory evidence by presuming that one kind of measure must be wrong, but both might be correct within their limitations as measures, because associated changes in unmeasured factors such as memory or imagery could explain both the subjective and the objective evidence.

Measures of workload tend to exaggerate the influence of some factors such as the ergonomics of the workspace, and underplay the influence of others such as trust. The workload of a controller can be increased greatly by an adjacent inexperienced and untrusted colleague, for the controller does his or her own work but also feels obliged to watch everything that the colleague does. Mistrusted equipment also increases workload significantly. Indices of workload which are insensitive to such influences are at best incomplete and at worst very misleading.

Workload is obviously affected by the amount of traffic, but must not be equated with it. An oceanic planning controller may be dealing with 60 or 70 aircraft and a military controller with a maximum of 4, but no valid deduction can be made from this statement about which has the higher workload. Amount of traffic alone has limited value as an index of task demands, except to compare changes within the same airspace and within the same type of air traffic control. As an index of controller workload, the amount of traffic has even less value, for all other factors would have to be quantifiable or held constant, and they never are. Operational irregularities may not be associated with high workload (Stager and Hameluck, 1990).

It is not that workload is useless as a concept. There are gross differences between light and heavy traffic in the resulting work for the controller, but the non-unitary concept of workload is inadequate to cope with all of them. Comparative subjective assessments that workload has increased or decreased may be reasonably reliable even when the reasons for the reported changes remain obscure, but absolute judgments about workload seldom justify full credence. A controller who suddenly loses the picture of the traffic may have been unable to anticipate that a small increase in workload would prove too much, whereas a controller who reports that the existing workload is so high that it would be impossible to handle more traffic may nevertheless do so efficiently and safely if the necessity arises. The lesson is not that controllers' complaints about excessive workload should go unheeded: on the contrary, they indicate anxiety, and can foretell safety hazards, and should be acted upon. The lesson is that controllers' complaints cannot alone guarantee the detection of excessive workload, which may be present without complaints by them.

If the same number of controllers must handle more air traffic, some increase in the workload of each seems unavoidable without suitable forms of computer assistance or automation. Much effort has been expended to reduce workload by such means, and to devise and prove measures that can quantify the reductions achieved. It was initially assumed that reductions in one kind of workload would permit other designated kinds of workload to be expanded. This fallacy yielded a crop of human factors problems. Reductions in workload may be accompanied by reductions in the controller's understanding if they are achieved by reduced information processing. Time saved on routine data entry cannot necessarily be reassigned to any particular alternative function such as decision making, and, even if it could, there might be no commensurate improvement in decision making. If one kind of cognitive resource is being fully used, successful reductions in that kind of workload may depend on freeing some of the resource or on utilizing alternative resources, but must not impose further demands on a cognitive resource that is already saturated. Until recently, reviews of workload have tended to treat it more as a fixed than as a multiple resource (Lysacht *et al.*, 1989).

High workload brings benefits in job satisfaction and interest, in the maintenance of skills through opportunities to use them, in the absence of boredom, and in impressions that time does not drag during working hours. The reduction of workload for other reasons may introduce such problems unless they have been foreseen and allowed for. Most people, given freedom to work as hard as they wish, choose to be busy. Means to reduce high workload that also increase enforced idleness under low workload are flawed and unpopular. In terms of human–machine matching, greater controllability of workload by the controller is a better objective than the general reduction of workload. Many forms of computer assistance could foster this objective if controllers were trained to use them selectively and flexibly in relation to task demands, so that the controllers remained to that extent in charge of the system. If this could be achieved, many of the traditional workload problems would fade away, since each controller could ameliorate the task demands, to the extent that any excessive workload would no longer be system driven but self-imposed.

16.3 Stress and fatigue

Among the most frequently claimed effects of the air traffic control system on the individual controller is the induction of stress. The topic has seized the public imagination, attracted much research funding and generated many publications. It seems a good time to take stock, to assess the importance of stress as an air traffic control problem, to consider the evidence about its causes and prevention or alleviation, to appraise its effects on safety and efficiency, and to discuss any requirements for further research. The word 'stress' is commonly applied to cover both cause and affect. Stress causes stress. Attempts to confine the concept of stress to causes, and to refer to the effects as either distress or strain have not been widely adopted, and much of the literature retains its popular but ambiguous meanings. A proclivity to impute causal connections to correlations has not helped. Many research studies that purport to have identified causes of stress have merely found statistical correlations which could denote causal connections in the claimed direction but could also denote causal connections in the opposite direction or relationships with no direct causal connection but extraneous common causes.

Some time ago, several published papers on stress in air traffic control provided balanced reviews of the evidence then available (Crump, 1979; Smith, 1980), and Costa (1993) has provided an update of them. Most recent papers describe particular research or approaches, and are less comprehensive. One authoritative review (Melton, 1982) concluded that stress as an air traffic control problem had been exaggerated. The main driving force behind research on stress has often seemed to be political rather than scientific, which is unusual in air traffic control but carries the implication that in retrospect it might have been more productive to divert some of the funding to other topics. Stress shares, with workload and boredom and a few other concepts that seem simple to the layman, the characteristic that the more it is studied the more complicated it seems to become.

The most potentially important issue concerning stress in air traffic control is whether it could hazard safety. A few studies seem to have presumed some connection and a few others seem to have presumed none, but most have not addressed the issue at all. Yet the most compelling evidence of an urgent need for action would be that stress if prevalent among controllers could be hazardous. Evidence that is less scientific but nevertheless much superior to guesswork or speculation comes from accounts of accidents or incidents, which do not seem significantly more likely to occur when controllers are under stress than when they are not. However, great caution must be exercised in this matter because it is very difficult to establish the essential basis for such judgements, either in terms of the actual prevalence of stress and its consequent expected frequency of involvement, or in terms of comparable circumstances that have not led to accidents or incidents. If the question is whether stress is a contributory factor to more incidents than would be expected, the answer appears to be that it is not. It is a separate issue that the occurrence of an incident often generates stress-related symptoms afterwards, as soon as the controller realizes what has happened.

Stress can be a contributory factor in several medical conditions. One kind of evidence about stress in air traffic control can be obtained by comparing the incidences of stress-related illnesses among controllers and among other populations carefully matched according to demographic variables such as age, gender, educational qualifications, and the like (Rose *et al.*, 1978). Data of this kind will not pinpoint the causes of stress symptoms among controllers, but can indicate whether the symptoms are more common among controllers than would be expected. The answer to this question raises a further complication because it seems to depend most on which nation the controller works in. If some factors do make controllers more susceptible to stress-related illnesses, they may not necessarily originate in the air traffic control itself, for several nations with the highest densities of air traffic are among those where controllers do not have more stress-related illnesses than comparable groups. This being so, the complexity of the issues is increased by the need to identify the full range of factors that could induce stress, in order to understand what is happening. Although most research on stress in air traffic control has been conducted in the United States, studies elsewhere (Shouksmith and Burroughs, 1988; Farmer *et al.*, 1990, 1991) caution against extrapolating findings to other nations.

Three main kinds of factor can be distinguished: those directly related to air traffic control; those encountered in air traffic control but related to it indirectly; and those that are totally unrelated to air traffic control but may affect controllers. Examples of each of these three sets of factors are considered below.

Stress-inducing factors related directly to air traffic control include:

1. Continuous high task demands that impose time pressures on controllers in systems functioning at or above their maximum traffic-handling capacity, so that controllers are always busy with no prospect of respite.
2. Numerous minimally equipped aircraft with inexperienced pilots, who may ignore or be unaware of air traffic control procedures and requirements and behave so unpredictably in air traffic control terms as to require unremitting close monitoring.
3. Mismatches between traffic demands and the equipment, particularly when the controller must use old or unreliable equipment or cumbersome procedures never intended for the high traffic demands experienced.
4. The usage of sophisticated equipment that is mistrusted or insufficiently understood or has unrecognizable failure modes, especially if the equipment seems to impede the exercise of the controller's legal responsibilities.
5. Working alongside inexperienced or unpredictable colleagues so that the controller must do his or her own tasks but feels obliged to monitor the activities of colleagues in order to ensure the safety of the system.
6. Divisions and allocations of work, and forms of supervision or of human or machine assistance, that do not match the requirements of the controller who perceives either a lack of essential support or excessive interference in the work.
7. Air traffic control procedures or instructions that are vague or ambiguous or that apparently defy training or normal practice.
8. Insufficient support in emergencies or equipment failures, so that a situation could deteriorate very rapidly in the event of any untoward occurrence.

If any of the above items are suspected of contributing towards stress among controllers, the origins of that stress lie within the air traffic control itself, in that the means for alleviating it require changes in some aspect of the air traffic control system. They could concern almost any aspect of it, including the equipment, the provision of information, the training, the instructions, the workspace, the staffing, the supervision, the procedures, the division of work, and the permissible system capacity, but the problem has to be dealt with as specific to air traffic control.

Among possible stress-inducing factors related indirectly to air traffic control are the following:

1. The extent to which management is supportive of and sympathetic to the controller's needs and requirements at work.

2. The conditions of employment, including pay and financial security, career prospects and development, medical support, and policies on promotion and retirement.
3. Working hours, including rostering, work–rest cycles and rest breaks, work scheduling and planning in advance, night duty, provision for holidays and time off, and the sharing of the most demanding shifts.
4. Attitudes of management towards outside criticism of controllers by the media, politicians, administrators and accountants, particularly when this criticism is unjust or ignorant but controllers have no right of reply.
5. The effectiveness of management in encouraging controllers to identify with and support their profession, workplace, watch and team, and in engendering pride, high morale and camaraderie.
6. The clarity of definition of the controller's responsibilities and their legal status.
7. Policies on recruitment, and the national status of air traffic control as a profession.

The alleviation of stress from any of the above factors is through changes in management or in the conditions of employment of controllers, and not through changes in the air traffic control itself.

A different approach, applied to the controller as an individual, has to be adopted towards the third kind of possible stress-inducing factor, which is totally unrelated to air traffic control but may affect controllers. Typically, individual controllers develop stress symptoms in an air traffic control work environment with no history of prevalent stress symptoms. Examples of factors unrelated to air traffic control include the following:

1. Individual talents and skills that have become insufficient for the work through ageing, burnout, or other individual reasons.
2. Reduced individual ability to cope with stress.
3. Reduced individual ability to cope with and adapt to shiftwork, particularly the associated disruption of sleep patterns.
4. Financial or domestic worries, concerning families, divorce, bereavement, illnesses, drugs or alcoholism, for example.
5. Personal disappointment or disillusionment.

The alleviation of stress from such origins cannot be achieved by changes in the air traffic control, or in the management or work conditions, but requires treatment of the controller as an individual. A permanent or temporary change to less onerous duties may sometimes help, but the root causes of the stress have still to be tackled. A change of job may add to stress rather than alleviate it, and the controller may continue to show stress symptoms in another job. To diagnose correctly the origins of stress-related symptoms in the individual is of crucial importance. Otherwise changes of the wrong kind, made with the best of intentions but irrelevant, can increase the stress that they seek to alleviate.

It is necessary to be clear about the signs and consequences of the successful alleviation of stress. If stress does not hazard safety, its alleviation will not improve safety. If stress is not limiting the traffic-handling capacity of the system, its alleviation will not increase that capacity. Many kinds of cost saving may follow the alleviation of stress, ranging from lower staff attrition rates to reduced human error, and from improved controller–management relations to lower servicing costs because frustration, irritation or anxiety do not lead to heavy-handed use of the equipment. If the alleviation of stress results in gains in safety and performance, so much the better, but for a good employer the alleviation of stress should be a worthwhile objective for human-itarian reasons.

Apart from medical evidence, there are many other signs of possible stress in the individual controller. Some which have affinities with the medical evidence relate to self-medication, excessive drinking, high rates of absenteeism and high attrition rates. In others the cues are mainly social, as the individual becomes more difficult as a colleague, easily irritated, readily tired, more diffi-cult to manage, or depressed for no obvious reason. Sometimes, as with divorce or bereavement, the causes of stress are obvious, and the controller receives the help and sympathy of others until they pass. Subjective data may sometimes illuminate the problem, but should not be relied on. The controller under stress may show decrements in performance and be well aware of them, or make strenuous and successful efforts to prevent or disguise the effects of stress on performance, at the cost of increased tiredness and anxiety. The time-scale of the controller's work planning may shorten, and the consequent anxiety can lead to frequent and obsessive rechecking of details and to rigidity in the formulation and the implementation of solutions to problems. Many kinds of measure may be helpful but none is exclusively related to stress or can diagnose positively that stress is the sole explanation for symptoms. Per-formance decrements and all the other signs mentioned can have other causes as well.

Stress can be a product of lifestyle. Only in the work environment does it have a bad name, and many people impose stress upon themselves deliberately in their favoured leisure pursuits: loving thy neighbour and loving thy neigh-bour's spouse can have quite different effects on stress. Many people nowadays either do work harder than they ever did and experience more pressure at work, or claim that they do although more objective evidence may fail to substantiate their claims. Lifestyle and expectations can be powerful driving forces, and in air traffic control so can unrealistic ambitions. There is human factors evidence of genetic predispositions and personality factors that render individuals more liable to stress because they are over-ambitious, drive them-selves too hard, or have unrealistic expectations about what they can achieve (Cooper and Payne, 1991). However, any pattern of relationships of this kind is always complex: the relationship between the likelihood of stress-related symptoms and the controller's personality has not been demonstrated as con-clusively in air traffic control as in some other contexts. The intensively researched human factors topic of stress in air traffic control has produced much useful evidence from well-conducted studies but insufficient to explain or prevent stress in air traffic control, and many issues remain unresolved.

A related concept is burnout (Shirom, 1989). As with other concepts such as stress and workload, burnout gives the initial impression of having an obvious meaning until its actual usage is examined and the literature on it reviewed, when it emerges as a confused notion with suspect validity (Hopkin, 1993a). In air traffic control, it is generally applied to an individual controller who is suffering from some disillusionment and loss of motivation in combination with a real or imagined reduction in task performance, when these circumstances are neither transient nor readily explicable by external events. The relationship of burnout to stress has not been the subject of specific research in air traffic control but has been studied in other contexts, where increased cognitive demands and low competition within teams appear to make burnout more prevalent (Sonnentag *et al.*, 1994). In relation to air traffic control, the status of such findings is as hypotheses requiring verification. Burnout must be treated as a problem affecting an individual.

A curious phenomenon, which is not common but not unknown among controllers, is night shift paralysis. This usually takes the form of the full retention of consciousness and awareness of the traffic situation, combined with a total immobility and incapacity to take any control actions at all. It can last from a few seconds to a few minutes, and is very worrying to those who experience it. Folkard and Condon (1987) have described it.

Another concept related to stress, and usually deemed to be important enough to have a separate heading, is fatigue (Hockey, 1983), which has been linked to stress in air traffic control for a long time (Grandjean *et al.*, 1971). As a topic, fatigue is not being neglected in human factors, as a recent journal issue devoted entirely to it testifies (Mital and Kumar, 1994). However, fatigue in air traffic control has never received as much attention or funding as stress. A recent paper on examples of fatigue in aviation does not include any air traffic control examples (Rosekind *et al.*, 1994), and although Stokes and Kite (1994) deal with both stress and fatigue their coverage of air traffic control is mainly about stress. A renewal of research on fatigue in air traffic control seems overdue, particularly because so many more air traffic control systems are now functioning at or near their maximum traffic-handling capacity for long continuous periods (Finkelman and Kirschner, 1980), and because computer assistance may compound fatigue by aggravating the conditions of low information processing that can induce it (Finkelman, 1994). Perhaps the effort to counter boredom is itself fatiguing (O'Hanlon, 1981). Such findings reinforce the importance of retaining rest breaks even when there is little work.

Fatigue at work can exemplify the interacting influences of many different kinds of human factor. Motivation is usually relevant, and sometimes of overriding significance, but other influences may include pay, performance, the nature of supervision, the interest or challenge of the work, the task and job demands and the degree of control over them, and the avoidance of the extremes of too much or too little information processing, both of which can be fatiguing (Finkelman, 1994). Again, such findings from other contexts are at best hypotheses for air traffic control, but they do suggest that a broad range of measures is essential to obtain valid findings about fatigue and its prevention. One possible explanation for the comparative lack of interest in fatigue in air traffic control could be the availability of practical means to prevent some of its most extreme forms through changes in rostering and work–rest cycles, coupled with the unacceptability of fatigue as a potential but avoidable safety hazard, though the imputation of a connection between safety and fatigue relies more on commonsense than on scientific support.

16.4 Boredom

The concept of boredom describes a subjective state. Findings that purport to be about boredom must therefore be expressed in subjective terms. If they are not, very persuasive accompanying evidence that they can be converted readily to subjective terms is both essential and difficult to provide. Boredom is a fascinating topic but presents a research challenge. All claims to study it by experimental methods should be treated with suspicion. The slightest hint that the objective of an experiment is to study boredom can rouse such interest among the participants as to dissipate any prospect whatever of boring them. In the unlikely event of success in devising an experimental task that bores everyone who does it, no mean achievement, subjects become uncollaborative and refuse to take further part in it because they are bored. More probably, tasks that bore some fascinate others. Monotony can be a task attribute and sometimes denotes a subjective state, but does not guarantee boredom. Because almost every task eventually becomes boring to some, boredom is an air traffic control human factors problem, but no task reliably bores everyone. Boredom has not been studied nearly as much as stress. Recommendations about it rely mostly on assumptions and anecdotal evidence that have never been tested, and many commonsense assumptions about it appear to be wrong. The main reviews of boredom in air traffic control are not recent (Hopkin, 1980a; Thackray, 1980).

Among the reasons why more human factors resources should be devoted to boredom as a problem are the following:

1. Humanitarian considerations of treating people as well as possible should include attempts to prevent boredom while they are at work.
2. Knowledge of the causes and cures of boredom would provide mechanisms for controlling it.
3. Evidence is needed on the relevance of boredom to attrition rates, absenteeism, industrial unrest, sickness, and complaints about trivia, to aid the formulation and implementation of policies on these matters.

4. Many air traffic control changes associated with computer assistance seem to aggravate the problem of boredom among controllers incidentally, and evidence is needed to predict whether future planned changes will also increase boredom.
5. The effects of boredom on safety and efficiency have been insufficiently defined.
6. The full range of the effects of boredom should be identified, including less obvious effects such as those on equipment serviceability.

There are many assumptions and suppositions about boredom that should be confirmed or disproved. Boredom seems less prevalent during training, particularly during its initial stages. Boredom must not be equated with inactivity: it is probably impossible to discover from measures of activity whether people are bored or not. Some of the origins of boredom usually lie in the work, but most of its consequences are not work related. Although simple tasks may be more boring or become boring sooner, boredom is not confined to simple tasks. Boredom may have some association with certain physiological states or personality attributes, but cannot normally be inferred from such data in any circumstances. Boredom may have no close association with task performance, but any association it has is more likely to be with good task performance than with poor task performance. Efforts to alleviate boredom may therefore be more likely to impair task performance than to enhance it. It is impossible to deduce reliably from observation whether a controller is bored or not.

Much of the evidence about boredom in air traffic control comes from controllers' complaints. Perhaps some boredom is inevitable in any job that requires continuous staffing but occasionally has no task demands and no work, yet where the introduction of other activities to fill the time productively is limited because the controller must be able to resume the main air traffic control tasks again at any time at short notice. Staffing levels cannot match traffic demands exactly and continuously, and must err towards underloading rather than overloading since excessive overloading is known to be potentially hazardous, whereas it is not known whether excessive underloading is potentially hazardous or not. The progressive dissociation of the controller from active tasks within the control room towards tasks that require less overt activity and less direct participation, such as monitoring, the allocation of resources and the scheduling of functions, might aggravate boredom, but this must be proved and not presumed.

A further relevant factor apparently originates in a change in social attitudes. People with the intelligence and education of the typical controller seem less willing than they once were to tolerate protracted boredom at work. They want not just a job, but an interesting and satisfying job, and the latter is difficult to reconcile with protracted boredom. While some people may actually welcome boredom at work, few in air traffic control do. They may escape from it by resigning, which leads to high attrition rates and perhaps recruitment difficulties, by complaining, which leads to problems of management, or by exploratory activities which employ equipment in innovative ways and are seldom advisable in air traffic control.

Aspects of system design that might increase boredom in air traffic control were listed years ago, when their speculative nature was emphasized (Hopkin, 1980a). Their gist is repeated here, as negligible progress in the understanding of boredom has been made in the interim, and none of the issues has been resolved. Aspects of systems that probably increase boredom include the following:

1. Increased passivity of human roles, with less active and participatory involvement and more monitoring.
2. Requirements to maintain alertness which in retrospect have served no useful purpose and have been unnecessary.
3. Fewer opportunities for highly skilled behaviour or for the direct application of professional knowledge.
4. Less need for human intervention or innovation.
5. The curtailment of human flexibility which becomes confined to functions that can be exercised through a human–machine interface.
6. Increased autonomous activity through a human–machine interface by each individual, and reduced activity as a member of a collaborative team.
7. A requirement for the human to adapt to the technology rather than for the technology to adapt to the human.
8. Restrictions on the relevance, applicability, frequency or influence of human cognitive functions such as decision making, problem solving and prediction, especially if accompanied by an increase in routine data entry and retrieval tasks.
9. Reduced interest, challenge, effort or job satisfaction from the job itself.
10. Increased task demands combined with fewer or less flexible means to respond to those demands.
11. Reductions in the esteem in which the job is held by others, and in opportunities to gain the esteem and respect of others from observation of the work.

16.5 Attitudes

Controllers form attitudes towards the following aspects of their work:

1. Air traffic control as a profession, including local, national and international features of it.
2. National and international authorities that devise or enforce air traffic standards and practices.
3. The specific air traffic control system within which they work.
4. Employers and management, particularly in relation to the controllers' conditions of employment.

5. Other professions and disciplines such as engineers and pilots with whom controllers work or with whom they compare their own profession.
6. Peers and colleagues at their own and at other air traffic control locations.
7. Those who plan and design current or future air traffic control systems, facilities and workspaces.
8. Those who service and maintain the air traffic control system.
9. Those in positions of power and influence who make pronouncements about air traffic control, particularly if they seem ill-informed, naive or unjust.
10. Those who provide media coverage of air traffic control or who otherwise influence public attitudes towards it.
11. The aviation community in general.
12. The demands placed on controllers by the air traffic.
13. The equipment and facilities provided to meet those demands, particularly if they seem inadequate or out-of-date.
14. The rules, practices, procedures and instructions within air traffic control.
15. Automation, computer assistance and technological innovations such as data links that influence the nature and satisfactions of the controllers' work.

The strongest attitudes usually have a direct association with an aspect of the controllers' job. For such a young profession, air traffic control has generated firm loyalties that are defended when challenged by outsiders. The controller is well aware that air traffic control can provide others with a tempting excuse to attribute their own failings and problems to it. Pilots informing passengers of a delay may ascribe it to air traffic control when it is the real reason, but also when air traffic control is wholly blameless or when it becomes involved only because airlines plan unrealistic turn-round times at airports or because regulatory authorities sanction published flight schedules that include more departures within a short period than an airport can handle.

Most controllers have very favourable attitudes towards their actual work. They gain great satisfaction from controlling air traffic, and many would not want any other job. The work can possess many of the attributes that promote favourable attitudes (Rajecki, 1990): skills and the opportunities to use them, direct control over the tasks and good feedback of their effects, pride and self-esteem from a job well done, sufficient observability to gain others' respect, high morale and camaraderie in a team environment, clear and significant responsibilities, and the provision of a service to a high professional standard. By contrast, controllers' attitudes are commonly much less favourable towards their management, equipment, and some of their conditions of employment. Many believe that air traffic controllers are undervalued compared with pilots.

In former times when controllers thought that they would benefit from more public awareness, they sought publicity for air traffic control, and public appreciation of their jobs, their professionalism and their high standards. Now they have learned that most publicity is adverse publicity, and that to draw public attention to themselves is not necessarily to their advantage. Nevertheless, concern lingers in some parts of the profession over the widespread ignorance of air traffic control and misconceptions about it among the general public, and even within the aviation community and their own management. Controllers' attitudes towards management are swayed by management decisions that appear to them to reveal woeful ignorance about the conduct of air traffic control, particularly in regard to its funding and equipment. The attitudes of management to controllers have been influenced by a reluctance sometimes by controllers to concede that a failure to provide them with the very best equipment need not denote managerial ignorance, spinelessness or intransigence, but results from stringent financial or technical constraints within which controllers and management alike have to function, for nowhere are the national resources for air traffic control limitless.

Shared attitudes are effective means to develop and sustain professional ethos, norms and standards. To be fully accepted in a team, each controller must not only conform with its ways of behaviour but also adopt its attitudes. The team, the watch, the facility, or the profession has a view on most relevant matters and speaks with a common voice about them. When challenged, controllers are expected by their peers to close ranks and present a united front. This is not an artifice, but a sign of identification with a profession and of belonging to it.

Controllers' attitudes towards their equipment can be quite complex. It is much too sweeping to talk about attitudes to automation, for controllers' attitudes vary according to the perceived value of each form of assistance in the context of the controllers' skills and abilities and the task requirements (Crawley *et al.*, 1980). If there is any general attitude towards computer assistance it is one of caution. Many of its earliest forms offered benefits in such forms as reduced workload which never materialized, and a few instances of this bred attitudes of initial scepticism towards unsubstantiated claims. Attitudes towards computer assistance are therefore selective. Some of its forms that seem helpful are much liked, and most controllers who have tried them would welcome their more widespread introduction: an example would be the more advanced distance-from-touchdown indicators in air traffic control towers. Other forms of computer assistance are viewed with some disquiet: any which may introduce unheralded and apparently arbitrary updating of information while the controller is using it can produce this attitude. Some changes such as the introduction of colour coding on displays may engender more favourable attitudes than their merits justify. Others, such as the replacement of paper flight progress strips with electronic ones, underplay the complexity of their consequences and can result in ambivalent attitudes towards them. New equipment that adapts to air traffic control a technological advance not initially devised for it, can arouse unfavourable attitudes if the reasons for change seem to relate more to the technology than to air traffic control needs. Forms of computer assistance such as conflict detection that fulfil a real air traffic control need are generally received favourably, once their reliability has been established. A guiding influence on attitudes towards computer assistance is that it should be of real help and should not oust traditional controller skills, responsibilities or initiatives.

Attitudes have been neglected in air traffic control studies recently and their significance underestimated, though this was not always so (Smith, 1973). When a new concept, technology, system configuration, or procedure is tried for the first time, usually in an evaluation or real-time simulation, an obvious important intervening influence is the attitudes of the participating controllers towards it, which are often formed quickly and become entrenched rapidly, yet no real effort is made to understand how and why this occurs or to profit from it or circumvent the biases that it introduces. The anecdotal evidence is extensive, but nevertheless requires validation, that if controllers form favourable attitudes towards their equipment they will strive very hard to use it effectively and to learn all they can about it, whereas if they form unfavourable attitudes towards it they can proceed to prove how ineffectual it is in meeting their true needs. If comparable effort to that devoted to optimizing the workspace and ergonomics was expended on promoting favourable attitudes, comparable gains in efficiency and performance might be achieved, provided that the attitudes could be justified by genuine benefits. Full realization of the potential of equipment is probably contingent on favourable attitudes towards it, since these help to secure the vital persistence and thoroughness during initial learning and while any initial deficiencies are corrected. Favourable attitudes both engender and derive from goodwill. A criterion of satisfactory management practices is that they promote favourable attitudes among controllers.

16.6 Trust

Air traffic control depends on trust. Only in good weather and in the vicinity of airports can pilots expect to glimpse the traffic pattern of which they form a part. Pilots have to trust controllers to issue instructions that are safe and efficient. Controllers have to trust pilots to implement those instructions correctly. Both have to trust their equipment, their information sources and displays, their communications, and the safety of their procedures and instructions. The controller has to trust that the computer assistance will not be misapplied as an omniscient tool bringing blame and retribution (Westin, 1992).

A few kinds of air traffic control information contain some evidence, often indirect, about how far they should be trusted. It is the older rather than the newer systems that can still indicate this. A characteristic of more modern systems is to present or make available a great deal of quantitative information, with no provision to indicate an appropriate degree of trust or changes in the degree of trust that are appropriate. Speech, used more widely in past than in future systems, provides an example of this. On the basis of spoken messages, pilots and controllers make judgements about how trustworthy the speaker appears to be. If a controller seems inexperienced or unsure, the pilot may seek extra supporting evidence. If a pilot seems inexperienced or unfamiliar with the airspace, the controller may give the aircraft extra attention to check that it does nothing unexpected. Assessments of trustworthiness are based on the content, manner, phrasing and pace of speech, and on other speech characteristics. Data transponded between air and ground that replace speech are usually of high quality and very trustworthy, but give no indication of this.

Radar, formerly primarily radar but now usually secondary radar, provides a further example. Primary radar contains evidence about signal-to-noise ratio, clutter, echoes, and propensity to fade, from which the controller judges how far and under what circumstances the trustworthiness of the depicted information might be degraded. Secondary radar showing processed and synthesized information contains no comparable evidence. In many circumstances where the degree of trustworthiness is known or could be computed, there are no satisfactory codings to depict it or to guide the controller's choices and decisions accordingly. It is not sufficient merely to present information to the controller without indicating how far it should be trusted, yet such indications are rare. Without this guidance, the controller builds trust in new equipment and new forms of assistance cautiously, and this takes a long time even after the controller fully understands their functioning. The process of building trust is very vulnerable to major setbacks from minor contrary evidence. Even one or two instances of untrustworthy equipment behaviour may prevent permanently the subsequent development of full trust in it, unless prior indications of temporarily reduced trustworthiness have been given.

Computer assistance is not introduced into air traffic control unless it can be trusted. The consequences for safety would be unacceptable, no matter what the benefits for performance or capacity might be. This means that the computer assistance is relevant and reliable, and technically can meet the air traffic control needs. Although the process of building trust can be long, ultimately trust in the computer assistance can therefore be justified as an objective. But when it has been built it leads to a different human factors issue. The acceptability of a proposed computer solution to a problem depends on whether it seems correct, but also on whether the solutions offered in the past have been correct. If they always have been correct so that the controller always accepts them, the controller comes to rely on and trust the computer assistance implicitly, which may be to trust it too much (Muir, 1987). The controller's checking procedures become a vestige of those originally followed, and checking skills are lost because the trusted computer assistance has made them redundant. In this respect, the equipment has begun to overcompensate for potential human inadequacies. It becomes impossible to tell from the controller's behaviour whether the unaided controller could still resolve the problems. Excessive trust therefore carries the seeds of over-confidence in the human, over-dependence on the machine, and over-protectiveness of the computer assistance (Parasuraman *et al.*, 1993).

Successful verification and validation should bolster trust, or lead to realistic degrees of trust when they are less than successful. Air traffic control contains many procedures that should be helpful in this respect. Messages have to be acknowledged, read back or confirmed. Actions are studied to provide feedback and to verify that an instruction has been obeyed. An obvious example is the controller confirming from the radar display that an aircraft is changing its heading in accordance with instructions. Feedback, verification, checking, and compliance with standard procedures are means of helping to build trust. Changes in these means, usually made for other reasons, can thus incidentally undermine judgements about how far the data should be trusted. This is not always acknowledged at the time, and can promote uncertainty in the controller about the trustworthiness of the data presented.

16.7 Job satisfaction

Since controllers like to think that air traffic control is unique, they might be disconcerted to know that the main influences on their job satisfaction seem to accord well with theories of job satisfaction (Arvey *et al.*, 1991). Their main sources of job satisfaction originate from the work itself, and their main sources of dissatisfaction from the conditions of work. To over-simplify, controllers like controlling aircraft but do not always like the conditions under which they have to do it. Job satisfaction is a motherhood concept, self-evidently beneficial for its own sake and beyond criticism, though not always justifiable by its consequences, which can be uncertain. Yet it is far from self-evident what would have to be done to accomplish the simple objective of improving the job satisfaction of the controller, or what the consequences of improved job satisfaction would be when expressed in terms of efficiency, safety, cost-effectiveness or capacity.

Some concepts need disentangling. Job satisfaction, in its general sense and not tied to any specific theory, covers the satisfaction from the content, conduct, organization and conditions of the job. It is not the same as job enrichment with its connotations of extended responsibilities or status, or job enlargement with its greater variety of tasks but not necessarily greater responsibilities, or job rotation with its range of jobs with equivalent pay and conditions that are done in turn. This last concept applies quite commonly to air traffic control in the few nations where it is customary to rotate the jobs within a suite or within a centre. It introduces more flexibility and variety into the work, enables the controller to see the totality of the work from different perspectives, maintains a broader range of skills, and encourages effective teamwork because every controller has direct experience of the jobs of others in the team.

The main human factors problem concerning the controller's job satisfaction in air traffic control is not to improve it but to sustain it. Many of the proposed changes, particularly forms of computer assistance with major cognitive implications, seem to threaten job satisfaction by reducing skills, responsibilities, active interventions and initiatives (Crawley *et al.*, 1980). If the human role seems to have become like a machine role to those who must fulfil it, this is not fundamentally satisfying to people with the education and abilities currently required in controllers, although in some circumstances it may satisfy those with more limited ambitions and abilities. Controllers value and become accustomed to their high levels of responsibility, and wish to keep them.

Theories of job satisfaction suggest that the achievement and maintenance of job satisfaction are quite complex matters. Among the most relevant factors are the following:

1. jobs that are a good match with abilities and with acquired skills and knowledge, and provide frequent opportunities to use them;
2. high and relevant levels of training;
3. work that provides opportunities for adaptability, innovation, flexibility, interest and challenge;
4. work that is capable of meeting the known human needs at work;
5. identification with a profession and full acceptance by those within it;
6. work rewards and conditions of employment that are accepted as fair and just;
7. work that does not impose unreasonable demands on the individual that are beyond human capabilities;
8. work that earns appropriate recognition and status;
9. work in which controllers are consulted by managers, system planners and others who make decisions about the work itself or the working conditions.

Even if all these factors can be accommodated successfully, it does not follow that higher job satisfaction would benefit safety, efficiency, happiness, attrition rates, or other quantifiable features of air traffic control. There is probably a positive relationship between job satisfaction and controller health in air traffic control (Kavanagh *et al.*, 1981). It may be as important to be perceived as trying to improve job satisfaction as to actually improve it (Witt, 1994).

In the past, controllers have not found it easy to transfer to other jobs unconnected with air traffic control if they dislike their work or the conditions under which it has to be done, as their training and skills have been specific to air traffic control. In the future, this will become less true because air traffic control jobs will resemble many other jobs more closely. The skills in using human–machine interfaces to fulfil complex cognitive functions will become more transferable, so that controllers will be more employable elsewhere. Satisfaction with their air traffic control job will make them less likely to look for alternative employment outside air traffic control.

17

Conditions of employment

17.1 Management/controller relations

Relations between air traffic controllers and their management have often been uneasy. In many countries, controllers work in the public sector and are therefore subject to its conditions of employment and its characteristic management styles. The privatization of air traffic control, so that it becomes more self-governing within broad financial targets and within a regulatory framework that meets international and national safety, employment, health and other standards, may become more common in future, and introduce further options and more fluidity in controller–management relations. Many of the general findings about industrial relations seem applicable to air traffic control (Hartley, 1992; Hartley and Stephenson, 1992).

Inadequate communications between management and controllers have most commonly been blamed for any poor relations between them. Occasionally these relations have deteriorated into mistrust and antagonism, which the lack of communication between them has aggravated. The outcome has been some mutual bafflement on the part of each concerning the objectives and intentions of the other. In an atmosphere of mistrust, this bafflement has sometimes been construed as wilful or deliberate obduracy, even when there has been no real evidence to support such negative interpretations. In quite a few countries, relations between management and controllers have at some time become so bad as to lead to some form of strike or work-to-rule practices by controllers seeking to draw attention to their grievances. The most notorious of these was the strike of controllers in the United States, preceded by more than a decade of poor relations there which were the subject of numerous independent enquiries that identified unsatisfactory management/controller relationships and poor communications between them as contributory causes, and recommended improvements. Current relations may not be ideal but they are generally better, and more is now known about the reasons and remedies for poor relationships. Although the following are among the main contributory factors to them, their respective importance differs considerably between nations.

The planning of air traffic control systems, the specification of their facilities and equipment, and the designs of their workspaces have often been completed without the active involvement or participation of controllers. This has caused resentment among controllers, particularly if the resulting systems have conspicuously failed to fulfil promises or expectations, or have apparently revealed a misunderstanding of the functioning of air traffic control, for example by increasing instead of reducing workload. Controllers have been tempted to interpret this lack of consultation as deliberate, even when it has betokened thoughtlessness rather than ill-intent.

Management often failed to communicate to controllers the reasons for management decisions that seemed to affect controllers adversely, so that controllers began to interpret the lack of communication as a sign that management did not want controllers to know the reasons because there was something to hide. Although management decisions might have sound and compelling reasons, controllers to whom the reasons were not communicated might not see any good reasons for the decisions at all. When the reasons lay in political or financial or technical constraints that management might not like but could not avoid, failures of communication led to antagonisms between management and controllers that were particularly unproductive because both of them, having the best interests of air traffic control at heart, were really on the same side but were able to achieve much less singly and divided than they might have achieved jointly by being in agreement and united. Technical unreliabilities, financial targets, costs, and political time-scales may be the main constraints on management decisions, but while they remain unpublicized the associated decisions can seem hasty, neglectful, inept, or parsimonious, implying a low national priority for air traffic control and little appreciation of controllers. Without good management/controller communications, management will be blamed for all unpopular decisions, whether they are within or beyond their sphere of influence. Rightly or wrongly, controllers are quick to believe that management decisions often fail to meet their needs because management does not understand what they do. The reverse criticism is less often voiced but equally apposite: controllers have not always been exemplary in their attempts to understand the constraints under which management must function. In other words, the communications have been poor in both directions, and so the blame for their inadequacy must be shared.

Perhaps air traffic controllers are a difficult workforce to manage. Their selection and training together are intended to produce people who weigh evidence carefully, reach decisions promptly, implement them and abide by their consequences, possess cherished skills and knowledge which they seek to apply, are accustomed to responsibility and some independence, identify closely with their professional norms and standards and professional ethos, defend their profession against outside challenges from any quarter, and do not take kindly to being told peremptorily by others what they should do. Such attributes may be desirable in the air traffic controllers. They are not, however, the attributes of a docile and pliable workforce. Controllers need approaches by management that are in tune with the above attributes, so that they support and accept management decisions because they have been consulted about them in advance. They have not always been approached in this way.

Being closely identified with their profession, controllers are very keen to promote and maintain its high standards of safety and efficiency. They want the best not merely for themselves but for their profession. If they have to use old and unreliable equipment, they interpret this as having to work under unnecessary handicaps that make it more difficult to provide a good service, and as symbolic of lack of appreciation, low status and low financial priorities of air traffic control on the part of management and others. Suitable new equipment is desirable in its own right and as a token of the status of air traffic control and the esteem in which it is held. Where air traffic control has apparently had a low national priority for funding and recognition, controllers have tended to ascribe this to a basic unwillingness on the part of management to promote its interests with sufficient vigour on their behalf.

In judging how they are treated and the quality of their management, controllers make comparisons with other professions, and particularly with pilots. Where pilots have better pay and conditions than controllers, this can rankle as a sign that they are more highly valued. Again there are different national practices, and pilots and controllers differ in so many ways that there may be no universally acceptable and objective criteria with which to judge in which respects they should be treated alike. Controllers who believe that they are as responsible as pilots for the safety of air passengers may expect the same treatment as pilots. Controllers also compare their own conditions with those of controllers in other nations, and blame management if their colleagues in other nations seem to have negotiated a better deal than they enjoy.

For controllers in many nations, their job has become more demanding in recent years because of major increases in air traffic. Current or pending traffic demands often reach or exceed the planned handling capacity of present air traffic control systems. Some lag is inevitable between the acknowledgement of this and the provision of a revised or replacement system capable of handling the increased traffic. To controllers the processes of planning, designing, procuring and commissioning a new system can seem to take far too long. It is tempting for controllers to accuse management of lack of foresight, but not very helpful because most of the management directly concerned will have departed long since, and probably had to contend with a multitude of delaying obstacles, from protracted financial wrangling to public enquiries. Moreover, to increase capacity can be difficult and time-consuming, especially if it involves different route structures, sources of data, communications or applications of computers, especially when there must be some prior evidence that the planned increase in capacity will actually materialize. Controllers who experience more fatigue from working with maximum effort for protracted periods without respite tend to blame management for it. This predicament is not confined to air traffic control. Continuous and protracted high effort to cope with very high task demands and workload without additional resources or reward is a common feature of modern lifestyles, and perhaps an inevitable corollary of any single-minded quest for better value for money. But in air traffic control, safety must remain the top priority.

Controllers enjoy their work, and are aware that much of their satisfaction comes from applying their knowledge, skill and experience. They are wary of any system changes proposed by management that render any sources of satisfaction obsolete, as air traffic control jobs could then become more boring and less professional. This caution, in the context of a history of benefits that have sometimes delivered less than they have promised, naturally induces some scepticism over future benefits from technological advances or computer assistance that offer to assist, to the point of rendering them redundant, cognitive functions that have been sources of satisfaction. Forms of computer assistance are generally welcomed once it is clear that they really do help. Management now makes greater efforts to ensure that the evidence in favour of future changes is more persuasive than it has often been for past changes.

Although major differences in national practices and attitudes remain, in general both management and controllers now realize that they must collaborate more closely in the future. Recent years have seen many attempted improvements in communications between management and controllers, including the formalized participation by controllers in planning future air traffic control systems. These developments are taking place in a context of changing expectations about work, pressures to cut costs and staff turnover and to maximize productivity, and requirements to utilize all facilities and equipment fully. But the era when poor management/controller relations prompted frequent public scrutiny seems to have passed, and it is vital to reinforce the greater goodwill that often prevails now but is still fragile. This relative improvement starts from a low base, and there is much scope to improve management/controller relations further.

17.2 Consultation

In some nations, those in charge of air traffic control are specialist managers whose personal careers remain within management and move in and out of air traffic control as they advance. In other nations, air traffic control managers are recruited mostly from the ranks of controllers who have been recommended for promotion, and who progress from local management to higher management jobs that all deal with air traffic control in some way. The expectations of controllers differ according to which kind of management they have. They accept that professional managers in charge of them may not possess a full understanding of air traffic control, but expect such management to educate themselves about air traffic control sufficiently for controllers to be able to live with the practical consequences of their management decisions. Controllers have much higher expectations about former colleagues promoted into management because they are known to understand air traffic control. These expectations can be dashed when a former colleague apparently deserts them for the world of management, with its political, financial, technical, planning and procedural concerns that are not shared by most controllers. One of the most crucial failures of communications is the inability of managers who have been controllers to explain to their former colleagues the constraints under which they now work, and to maintain constructive dialogues with them. To the workforce, the loyalties of their erstwhile colleagues have changed and they have been "lost to management". When this occurs, a channel for constructive controller/management communications has been lost also.

This problem could be tackled in several ways. These include in-house briefings, frequent public statements on policy, use of internal publications to provide explanations, open-house correspondence on complaints and publicity for the responses to them, widespread dissemination of policy statements and changes and the reasoning behind them, accessibility of management to the workforce, visits by higher management to air traffic control establishments and face-to-face consultations with controllers, and guarantees that constructive criticism will not count against career prospects. But the most productive approach seems to be through consultation (Ganster and Fusilier, 1989). It is particularly important in air traffic control that any changes should be broadly right at the first attempt, since extensive retrospective tinkering and modifications to remove their major deficiencies will at best be protracted, costly and unsatisfactory, and at worst be completely impossible. A process that is becoming more common and more acceptable is to try to ensure that changes will be efficient and acceptable by enlisting the active participation of controllers in the decision making that results in the changes. This practice has become more formalized recently, but in a variety of ways, none of which can yet be identified as the best.

Some of the problems associated with this process have become clear. A tendency for the most forceful and articulate controllers to represent their colleagues makes sense in that they can express their views effectively, but makes less sense if neither the content nor the strength of their expressed views is fully representative of their profession. A parallel issue of representativeness applies when the views of controllers as a profession are conveyed through a trade union or guild or other professional body. Controllers who participate in management and planning activities may have an associated role, to give their colleagues a fair and impartial account of what has taken place, including the reasons for decisions that have been agreed but may not be popular or could be misunderstood. The controllers' representatives in consultations and decision making will have the main role in promulgating the decisions to their colleagues, since they are best placed to understand and explain them. It seems vital for them to be well briefed beforehand about what their colleagues would be willing to accept, so that the processes of consultation do not result in arguments or accusations of bad faith among the controllers themselves. Controllers who represent their colleagues may have to accept two kinds of constraint that they do not like. One is that their terms of reference may seem too narrow so that they wish to raise matters which cannot be negotiated or are beyond the control of management. The other is that management never has a totally free hand, so that both management and controllers must reach agreement within policy guidelines and financial constraints that are imposed by others and with which neither may agree.

The most productive and straightforward consultations between management and controllers deal with quantitative issues on which a decision must be reached. Examples are conditions of employment, rostering and work–rest cycles, pay, physical aspects of the workspaces, and the implementation of national and international health and safety regulations. Numerous policy issues can be resolved, regarding recruitment policies towards minorities, the vexed question of smoking, the health and well-being of controllers, and other matters which can be expressed as agreed policies and recommendations.

More difficult is consultative participation in multidisciplinary activities dealing with policy, planning and design, where the probable outcome is a workable compromise acceptable to all the many points of view represented but ideal for none. Controllers in such consultations must be able to discriminate between the issues on which they can compromise and those on which they must not. The process requires considerable goodwill and tolerance for it puts the objectives of air traffic control as a whole above all sectional interests. To participate successfully in this process, controllers need special training and prior knowledge about how the groups function and what sort of product is expected from them. The participating controllers share the collective responsibility of the group for the decisions taken, and difficult issues arise for them when there is a dearth of factual evidence so that decisions rely heavily on opinion and belief. This applies to many sensitive topics such as morale, attitudes, self-esteem, job satisfaction, pride in the work and the factors that influence such intangibles. Many people in the consultation process may hold entrenched views on such matters, refuse to accept all evidence that does not support them, and seek to denigrate all contrary views. There is always some risk that consultations between management and controllers on such topics, which can be highly productive, may entrench biases and misunderstandings or raise false hopes. Successful consultation depends on a realistic appraisal of what might be attainable. It must not imply that everything is negotiable, and can never be used as a delaying tactic without forfeiting goodwill. It is advisable for consultations to begin with issues that require quantitative conclusions. Improved relationships built on such agreed conclusions may then assist the resolution of more qualitative and contentious issues.

17.3 Careers

In most countries air traffic control as a profession has limited career prospects. This applies particularly where management is recruited separately, but even where management and other specialists within air traffic control, such as system planners and computer specialists, are recruited mainly from controllers the number of vacancies will be only a small proportion of the controller population. The prospects of advancement to posts above the main journeyman/controller grades in which most controllers are employed are not good. Controllers are ill-advised to treat their entry to their profession as the first step in a glittering career. However, some recruitment literature, their selection from the very many applicants for air traffic control jobs, aspects of their training, and a professional ethos that controllers are unique, can combine to engender grandiose expectations among some young controllers. The outcome can be unrealistically optimistic career expectations, and the disillusionment in mid-career when these have to be scaled down may be compounded by increased awareness that their skills as controllers are not readily transferable elsewhere and that experience as an air traffic controller is not a major asset to most other prospective employers.

Some of the promotions into supervisory and allied roles still within the air traffic control environment can be more nominal than substantial as advancements, for they may couple greater responsibilities with marginally increased rewards and benefits. Another factor is that many of those in air traffic control do not actually want to do anything else but control air traffic, and they may have to sacrifice career advancement for this if there is no provision to reward special merit as a controller with additional status and pay. A complication is that high assessments early in a controller's career may not be good predictors of subsequent career progress or of eventual merit when fully experienced. Those who do well in training or in their initial on-the-job assessments may thereby hope for better career prospects than they actually have, since a promising start may not be durable throughout a career.

Most of the main influences on career prospects in the past seem likely to remain, with one exception that could prove to be crucial. As controllers work more through a human–machine interface, many of their skills will become less specific to air traffic control and more transferable to the many other contexts requiring skill and experience with human–machine interfaces. In the future, experience as an air traffic controller may be perceived as more relevant than it is now to many other jobs, in aviation itself, nuclear power plants, chemical processing plants, transportation, air defence systems, public utilities, and offices, wherever the human operator fulfils complex roles within large systems by working primarily through a human–machine interface. There will be many such jobs in the future, and the requisite skills could be in short supply. The skills may relate more to software packages than to the context of their application, and thus may become more transferable across occupations. This could improve the career prospects of any controllers who feel trapped within their profession or disillusioned by some aspect of it, by enhancing their prospects of obtaining a satisfactory job elsewhere in which there is a use for some of their existing skills.

Therefore air traffic control in the future may have to tackle with renewed urgency the rather neglected problem of building satisfactory careers within air traffic control and rewarding those with the highest levels of skill, knowledge and experience in order to keep them. If others view air traffic controllers, with their proven professionalism and sense of responsibility, as attractive prospective employees, air traffic control will have to compete with the conditions offered elsewhere to retain its workforce.

17.4 Human needs and aspirations

Numerous aspects of the jobs and conditions of employment in air traffic control determine whether the work is likely to satisfy human needs and aspirations. An extensive and rapidly expanding literature reflects increasing concern with social attributes of work. Many theories have been propounded, and a great diversity of factors have been claimed to relate to or shed light on human work needs (Quintanilla, 1991; Kanfer, 1992). Much of the supporting evidence is of the kinds described as soft in scientific parlance, which is never intended as a compliment but may be either pejorative or neutral in tone. In the latter case, the implication is that the data are not from controlled experimentation but from techniques such as surveys and structured discussions, and may have uncertain validity. Findings may not transfer across cultures, nations, jobs or workspaces, even across those which seem superficially similar, but at all these levels there are influences on human needs and aspirations in the workspace. The relative importance of various kinds of influence is far from settled. The extent to which evidence from other contexts is applicable to air traffic control is generally unknown. There is considerable debate about the extent to which it is possible, ethically permissible, or desirable to modify human needs and aspirations by attempting to manipulate the main formative influences on them.

No single theory or perspective is accepted widely enough to serve as the obvious starting point for considering controllers' needs and aspirations. The influences on human needs and aspirations seem to become more numerous, complex and interacting the more they are examined, and the extent to which they are controllable at all is far from clear. The reasons for trying to satisfy human needs and aspirations are not clear either. They range from humanitarian reasons, through a wish to cut costs by lowering attrition rates, to an intuitive belief that a more satisfied workforce must be a safer workforce. Even if there were a policy of trying to satisfy controllers' needs and aspirations wherever possible, it could not be readily converted into practical steps that would achieve such an objective. It is difficult to quantify the consequences of any changes on needs and aspirations as they are not made for such reasons, and measures that claim to assess needs and aspirations or changes in them can seem contentious and unvalidated, and generally weak on quantification. Some of the most compelling evidence is negative evidence, where a workforce has resigned in droves, turned militant and uncollaborative, or gone on strike. Such evidence certainly can reveal a problem, but is not very good at providing constructive solutions.

In an earlier publication (Hopkin, 1982a), an attempt was made to compile from the literature a list of those identifiable human needs at work that seemed most relevant and applicable to air traffic control. This yielded a list containing 29 items, which revealed the complexity and variety of the postulated influences, but lacked theoretical underpinnings. None of the items seemed trivial, but there were no criteria that could be applied to them all to judge their relative importance. Extensive subsequent research, much of it theoretical, has not brought any simplification. Perhaps the most important need for humans at work, which is so simple and obvious that it is easy to overlook it, is that there must be some work to do. One of the most significant practical differences between humans and machines is that people cannot tolerate protracted enforced idleness. Circumstances in which the controller must be at work but there is no work because there is no air traffic, or in which there is traffic but it is controlled by machines and the human remains passive and never intervenes, are counterproductive in satisfying human needs at work. The human controller needs work to do that draws on professional knowledge, skills, abilities and experience, and that seems worthwhile to the person who does it in terms of a product made or a service provided. Human needs are not satisfied by work without apparent purpose or value. Tasks that lack objectives are not really work. This is not mere theory but an immediate practical problem in any attempts to enhance safety by the independent parallel performance of the same functions by the human and by the machine.

Humans need knowledge of the results of their work. In air traffic control, the safe and efficient control of the traffic constitutes the primary source of that knowledge, coupled with some recognition that the means by which an excellent outcome was achieved were also themselves excellent in their choice and employment of professional knowledge, practices and skill. Human needs at work are satisfied best when the competence of the achievement can be observed, judged and where appropriate admired by professional colleagues whose judgement is respected and whose acclaim is prized. As computer assistance reduces the observability of the work by others, it becomes more difficult to retain the visible sources of colleagues' esteem based on observing the controller's competence. A need at work is for some comparative appraisal of it, and for reassurance that it meets external criteria, not only those imposed by statute and regulation, but also those derived from the professional ethos, norms and standards. The provision of enough evidence to make such judgements has implications for efficiency and safety, and for the satisfaction of the needs of people at work. Controllers need a degree of acceptance by their colleagues that does not exclude them from the camaraderie of teams. Changes that lessen the opportunities for this have corresponding effects on the satisfaction of human needs at work.

It is easy to confirm that some of the most obvious human needs have been met. Human needs are matched with the physical environment so that the work does not induce any problems of vision, posture or health, or otherwise harm the controller. A human need is not to be exposed to work demands that are unremittingly excessive, to problems that have no solution, or to demands that are grossly at variance with individual competence and abilities, either by requiring skills and knowledge that the controller does not have or by consistently failing to utilize those that the controller does have. Ideally, needs are met best if the controller has some influence over task demands and workload and some autonomy over work scheduling. Humans need appreciation at work. Their work and its value must be acknowledged, and reflected reasonably in pay, conditions of employment, managerial attitudes towards them, and the wider image of air traffic control. An index of the extent to which human needs can be fulfilled is found by comparing the extent to which the system drives the controller or is driven by the controller; that is, controller responses versus controller initiatives. This is not an argument against computer assistance but a reminder that the satisfaction of human needs and aspirations at work is more readily achieved where the individual can guide, influence and initiate work to some extent and is not solely a recipient of work imposed by the traffic or the system.

Needs are also influenced by expectations and aspirations, which is one of the reasons why the latter should be realistic. Controllers' aspirations can be thwarted by the level of their achievements or because their expectations always were unrealistic. Air traffic control has one of the hallmarks of a profession, in that its standards of individual performance are mainly generated and maintained internally. These standards affect the effort and motivation of each controller, which are also influenced by personal needs and aspirations. A controller with no further expectations of career advancement may not care much what the opinions of management about the controller are as long as they do not put the controller's job at risk, but will still care greatly that the opinions of colleagues remain high enough for the controller to retain their professional respect as a colleague and full team member. Work which isolates individual controllers tends to make it more difficult to satisfy all their needs and aspirations, particularly if such considerations have had no influence on the design of tasks and jobs. All significant changes in the tasks and jobs of controllers, especially those involving computer assistance, should be presumed to have significant consequences for the satisfaction of human needs and aspirations at work, which should therefore inform and influence the nature of the changes and their method of introduction.

17.5 Morale

Morale has also been the subject of much research, not always leading to definitive conclusions. Findings from other contexts may not transfer to air traffic control but require verification. Morale can have such significant local variations that at least some of the preponderant influences on it must be local ones. Influences on the controller's morale are discussed here as a hierarchy, starting with individual influences, progressing to influences from the immediate small group, the team, the workspace and the facility, and then considering influences at regional, national and international level.

The morale of each individual controller depends partly but not entirely on influences within air traffic control, for it is also affected by personal attributes of the individual and by other life events. The extent to which the morale of each individual can be sustained by immediate colleagues varies considerably, depending on individual attributes, on whether its main determinants at the time are internal or external to air traffic control, and on how far any commonality of morale extends. The influence of individuals on the morale of others in air traffic control depends on the flexibility of the system, on the correct interpretation of the origins of the main causes of high or low morale, and on the extent to which others sympathize with the causes of low morale. Individual morale can be influenced strongly by air traffic control events that concern the individual directly. An example is personal agonizing in the aftermath of an incident in which the controller has been involved, over how and why it happened and what could have prevented it. The morale of the individual is reduced by burnout, itself a contentious topic, mentioned previously in relation to stress. Loss of morale may also be associated with certain illnesses, and such cases have to be treated clinically. Thus a cluster of influences on morale can be personal to an individual, in that colleagues in the same workspace are not also affected by them.

At the next level are influences on morale associated with small groups, teams or watches within an air traffic control workspace. These influences may emanate from a single individual, or from factors that affect everyone in the group but not others within the same facility. A team may have particularly high or low morale because of exceptional competence or ineptitude on the part of its supervisor or leader, whose influence pervades the whole team for good or ill. Team morale is affected by the inclusion in the team of a controller of renowned competence or of a controller who seems weak enough to worry colleagues. Such an individual's influence extends throughout the team who may emulate the practices of the admired individual and watch overtly or surreptitiously those of the inept one. The morale of the team is also affected by what it is capable of doing in relation to what is required of it, as determined by the task demands, the equipment, the feedback, the procedures, the supervision, and the safety record. With a good match between capabilities and demands, an efficient team can sustain high morale even if the equipment is poor, particularly if its competence is widely acknowledged. However, the provision of good equipment not only aids performance but can improve morale if it is seen as a token that the air traffic control service is highly regarded and should be properly funded. A further determinant of the morale of teams is the general belief among the members of each team that they belong to the best one, because it provides the best service, is the most efficient and safe, is the most competent and should be the most admired, and really sets the standards for all others. Where there are any minor divergencies, in the practices of different watches for example, each watch takes it for granted, if morale is high, that their own practices are superior and that the other watches should copy them. Computer assistance may reduce morale and increase stress if it is perceived to be a tool to watch the controller's actions more closely (Westin, 1992; Carayon, 1993).

The morale at a facility or workplace such as a centre or tower is at the next level in the hierarchy. Management style becomes an important influence, particularly in procuring modern equipment appropriate for the work, providing good conditions of employment, consulting the workforce and reflecting their views fairly, defending them against unfair external criticism or disparagement, and setting high standards of competence and conscientiousness by example. For high morale, management must be well informed, sensitive to controllers' needs at work, and willing and able to communicate effectively with the workforce on all matters of mutual concern. The true reasons for any known failures to meet the needs of the workforce should be explained, whether they are insufficient political or financial priority, technical difficulties, or other reasons. What is important is that management is seen to be trying to provide good work conditions and to promote efficiency, safety and good relations. Most workforces know that management cannot accede to all their requests and may lack the power or the resources to implement even those they agree with, but local management should generally be perceived as understanding controllers' requests and in sympathy with the reasons for them, if local morale is to be high. The resultant positive attitudes towards management can be supportive of management when it in turn is challenged and put under pressure. High morale fosters close identification with the workplace, and pride in the work done there. A sign of this can be positive statements about air traffic control as a profession that encourage young people to consider it as a career.

The national or regional authorities in charge of air traffic control can affect morale through their guidelines for its conditions of employment and their provision and allocation of air traffic control funding. If they have argued successfully for the allocation to air traffic control of a fair share of national resources, this can raise morale because the requirements of air traffic control are not being passed over. Controllers feel entitled to a fair and impartial hearing of any general complaints and grievances, and morale is reduced if there is no forum for this. A management responsibility is to defend air traffic control and its professionalism and standards against ill-informed or malevolent comment, by providing informed refutations and by forestalling the ignorant speculations that are rife following any incident about which the evidence is incomplete and on which those most directly concerned must not make public pronouncements ahead of the official inquiries. If management is effective in ensuring high professional standards and status among controllers, this benefits morale.

Political considerations can affect the morale of controllers in several ways. The most general concerns the level of national prosperity, which determines whether there are resources of which air traffic control may have a share. Another influence depends on whether politicians shoulder the blame for the previous neglect of air traffic control that has led to traffic demands that exceed system capacity and to consequent highly publicized delays, or try to blame the controllers who are handling more traffic than the system was designed for, or blame the air traffic control management which is attempting to recover from the consequences of previous underfunding. This is not a matter of finding scapegoats. Morale ebbs away if controllers are blamed for delays which are not their fault, particularly when their forewarnings and those of their management went unheeded. It can boost morale if those with the responsibility accept the blame, and do not attempt to pass it on to controllers or anyone else who cannot answer back.

Air traffic control often receives criticism from the media. This is of two kinds. If it is well informed and can be justified, controllers are normally willing to accept it and learn from it, and to try to improve accordingly. However, this kind of criticism is rather exceptional. More usually the criticism is surprisingly ill-informed and tendentious. In terms of morale, such criticism is counterproductive. If controllers or their representatives try to correct biased and ill-informed criticisms, they may find that their critics are unreceptive to any evidence that refutes their biases, even when the evidence is overwhelming. Constant disparagement of a profession whose members are doing their best does no good for its morale. Air traffic control is not uniquely subject to such criticisms, for many other professions have had the same experience, but in air traffic control they have become far more prevalent in recent years because the essential infrastructure for handling all the extra traffic has not always been planned and financed sufficiently far in advance.

Finally there are some international considerations where the procedures and practices of air traffic control have been agreed internationally and are subject to international regulatory authorities. Attempts are made to reach international agreements about the minimum standards and conditions of employment for controllers to ensure air traffic control does not harm those who work in it. Low morale among controllers can be engendered in nations which do not comply with these regulations but require controllers to work excessive hours without adequate rest breaks for poor pay, using old and often unserviceable equipment with deficiencies which controllers must compensate for. Controllers are generally aware of how their treatment compares with that of their professional colleagues in other nations, and pilots flying between nations become very conscious of gross disparities in the air traffic control services offered. The morale of controllers in nations where they are treated poorly is naturally lower than their morale where they are treated well.

17.6 Rostering

In air traffic control, there are sometimes strongly held views on rostering and the related topic of work–rest cycles. Considerable research, much of it inconclusive, has tended to entrench rather than resolve different views on the aspects of rostering that are most important, and to exaggerate the significance of rostering and work–rest cycles for performance and safety while underplaying their significance in determining attitudes towards conditions of employment. Many, though not all, controllers work shift patterns, including some night duty. If they are public servants they will have a statutory length of working week. Discrepancies between their statutory and actual hours can arise from overtime working or from difficulties in converting statutory hours into practical shift lengths. Further complications include intermediate stand-by conditions, where controllers must be at their workplace and available for work but are not actually working and which apply mostly to shifts at night, and on-call conditions where the controller can remain at home but may be required for work at any time at short notice and is not free to pursue normal activities. Many nations have enacted legislation covering permissible continuous working hours and shift lengths.

In many countries, controllers are assigned to watches. The members of a watch normally work together and become familiar with each others' strengths and weaknesses. In other countries, rostering is more flexible, and a controller may often work with colleagues whom he or she does not know well. Each arrangement has its benefits and drawbacks. Controllers who share the same watch tend to identify closely with it, whereas others may identify more with the facility or the whole profession. Watches which include the same people all the time may gradually adopt divergent practices which may have to be redressed by management or supervisors if they cannot be accommodated as differences in watch style but begin to affect safety or performance or the quality or nature of the air traffic control service provided. Supervisors may instigate or encourage these different practices if they function as leaders in the watch. Friendly rivalries between watches are common and can benefit air traffic control provided they remain in perspective. The same rostering has to be agreed among all watches.

Recommended rostering practices do not always accord with controllers' wishes. Where shifts are worked, the normal recommendation is for them to become progressively later during the day (Folkard and Monk, 1985); for example, consecutive days with morning shifts, then days with afternoon shifts and then days with night shifts may be recommended, though this is not a universal practice (Costa, 1993). On the whole, air traffic control is a sufficiently cognitive activity for frequent shift rotation to be recommended rather than rosters that repeat the same shift for long periods. No more than two consecutive night shifts should be worked, followed by a period of days away from work. In this context, controllers and their management often clash. Controllers often prefer rosters with the maximum continuous periods away from work, and they may be willing to accept long shifts or arduous work schedules to achieve this. However, this pattern may not be recommended either for well-being or for safety since it can incur tiredness and greater efforts to sustain alertness. As always in air traffic control, considerations of safety must be paramount, and if necessary must prevail over both the wishes of controllers and administrative convenience (Melton and Bartanowics, 1986).

If rosters are published well in advance, controllers can plan their social activities accordingly and have time to make alternative arrangements when essential commitments clash with rostered hours of work. Where there is a shortage of controllers, the pressures to work overtime can become extreme because an air traffic control service must be provided. The longer-term solution of such difficulties must rely on making air traffic control attractive enough to build and keep sufficient competent controllers to provide a continuous air traffic control service to the required standard without overtime. Excessive hours may help to alleviate any financial worries, but at the cost of fatigue, loss of sleep, disrupted meal patterns or domestic friction caused by long and antisocial working hours. Goodwill may be needed to reconcile cost pressures to roster the work so that all the statutory hours are actually worked productively with the requirements of safety, efficiency and occupational health. Despite the efforts of such bodies as the International Labour Organization to define good rostering in air traffic control, major cultural and traditional differences remain in the rostering practices and working hours that are accepted as reasonable. This problem becomes more serious as more controllers handle traffic levels at or beyond the planned system capacities for long continuous periods without respite. With consistently high task demands, shift rostering and total working hours may become more significant for safety and well-being than the detailed manipulation of work–rest cycles. The adverse consequences of poor rostering may not appear immediately as reductions in performance or safety, but gradually as poorer health among controllers, higher attrition rates, increased equipment unserviceability due to frustration and tiredness, and greater stress from the extra task demands and from domestic disharmonies exacerbated by the controller's exhaustion and disinterest in them.

A main influence on rostering practices is their acceptability to controllers. Although the interests of safety must prevail, evidence on the effects of rostering practices on safety is sparse and largely equivocal. If it could be shown that rostering practices influence the probability of occurrence of an incident or accident, such compelling evidence could not be ignored, but there is a remarkable lack of it. Despite many studies on the effects of work patterns on safety, the main conclusion must be that unless the work patterns become extreme any effects on safety appear to be very small. Since rostering is such an emotive issue, this is a disappointment to all concerned. The fact remains that relatively minor changes in rostering or work–rest cycles do not seem to affect performance or safety much, and that other factors are always implicated. For example there could be some tenuous evidence that a controller might be most vulnerable while still building the traffic picture after coming on duty, but this trend is too weak to provide a basis for executive action. Its apparent corollary, that incidents could be minimized by long shifts that reduce the frequency of handovers and the need to rebuild the picture, seems false because the resultant fatigue would act in the opposite direction.

Most people negotiate rosters from an entrenched position. It is important that the human factors evidence is scrupulously impartial, and cannot be construed as biased towards management, controllers, or any other viewpoint. The human factors specialist giving advice on rostering can expect to become unpopular with everyone, because people do not want to know that an issue that is of vital significance to them is actually quite unimportant in terms of most of its known consequences.

17.7 Work–rest cycles

Work–rest cycles are also a contentious issue in air traffic control. Some recommended guidelines may be offered. The maximum working week should be of the order of 40 hours. The maximum recommended shift length, including short breaks and a longer break for a meal, lies between eight and ten hours, though night shifts may be somewhat longer if they include periods on standby but not actually working. The maximum period of continuous work without any break at all should not normally exceed two hours. All rest breaks should be away from the air traffic control work environment so that the controller can relax and cannot continue to watch the displays. It should never be necessary to resume control duties immediately and unexpectedly during a break. Normally a rest break should allow time for a drink, for the use of toilet facilities, and for brief conversations without undue haste. Its actual length depends on how near these facilities are to the place of work. Rest breaks should always be taken, whether the controller has been busy or not. Obviously a busy controller needs a break after two hours. Less obviously a controller who has had little to do while on duty also needs a break as the time will have dragged. Light traffic loading is not an adequate reason for dispensing with rest breaks.

The scheduling of shifts and breaks should accommodate relevant factors external to air traffic control. One of the most important is the traffic conditions while travelling to and from work, particularly when going home after a busy shift. Driving home in the morning rush hour after working all night is not a good idea. Another factor is the scheduling of work so that it does not cause more domestic difficulties than it need do, particularly if a partner has a day job with normal hours. Work–rest cycles should treat particular individual difficulties sympathetically wherever this can be done without unfairness or special privileges. For example, older people returning to night shifts after some time away from them may need extra time to re-adapt to them, and may experience initial difficulties in remaining alert all night (Davies *et al.*, 1991). A few may find it impossible to re-adapt, but most can, given time and patience. The personality dimension of morningness, which may have a physiological basis and makes some livelier in the mornings and others livelier in the evenings (Horne and Ostberg, 1976), has interacting effects with work–rest cycles among controllers (Costa, 1993). A further factor is the feasible extent of job rotation within the watch. A rest break of 20–30 minutes may imply that the controller does not return to the same seat which is now being occupied by another controller for two hours, but replaces a further controller who is almost due for a rest break.

Practices vary greatly on whether rest breaks are scheduled in detail by management or are left for controllers to organize themselves within set guidelines. The latter arrangement works well in some locations and supervisors may oversee the details, but it probably would not work well everywhere. An alternative is for each controller to take a break when he or she feels in need of it rather than at statutory intervals. This can succeed if the arrangement is not abused, but those who are over-dedicated might have to be prevented from working too long without a break, in their own and others' interests. It is not always feasible to plan the traffic and workload for a handover at a particular time and there is considerable merit in handing over responsibility when it can be handed over rather than when it must be, with the proviso that the flexibility does not become excessive.

Recommendations about work–rest cycles take account of evidence about circadian rhythms, the numerous bodily rhythms which follow an approximately daily cycle (Costa, 1991). Sleep patterns are the most obvious of these (Hopkin, 1982b). A criterion for adjusting work–rest cycles and rostering has always been to avoid disrupting the controller's sleep, so that controllers do not come to rely on aids to go to sleep followed by aids to stay awake. The earliest studies relied on subjective records of sleep, until this evidence was shown to be fallible because many people have short periods of sleep which they remain unaware of and therefore deny. Nowadays the basic data are records of sleep based on instrumentation rather than subjective reports. Most studies have concluded that once people have adapted to the unusual sleep patterns typically imposed by air traffic control they can have about their normal amount of sleep. Of course the individual differences are considerable, and carefully controlled experimental designs are essential to gain a true estimate of the effects. Age is a confounding factor, and has to be controlled with particular care. The comprehensive evidence on the cognitive effects of ageing should be applied (Stuart-Hamilton, 1994). On the whole, the effects of work–rest cycles on controllers' sleep appear to be smaller than is usually claimed. Circadian diurnal rhythms have other effects, notably on eating. An interaction is that some sleep patterns interfere with regular meals more than others do, which can be particularly significant for controllers who combine a sedentary occupation with the ready availability of coffee with its stimulant caffeine that can also affect sleep patterns.

Some principles for judging and recommending work–rest cycles emerge. One is that their effects on efficiency and safety tend to be less than expected until they become extreme. Another is that the most sensitive measures of the effects that they do have may not relate to performance or efficiency, but to equipment serviceability, domestic factors, age, or congeniality as a colleague. Within set guidelines, it may be sensible to let controllers arrange their own work–rest cycles and assume the responsibility for the proper staffing of the air traffic control system at all times. On the whole, the balance of evidence favours rotating work–rest cycles and rosters rather than fixed ones over a continuous period, but this evidence is neither particularly strong nor unanimous. Work–rest cycles should not depend on how busy the controller is; the controller who has not been busy still needs a break. The effects of work–rest cycles on diurnal rhythms can largely be compensated for by human adaptability. The effects of work–rest cycles on sleep patterns and on the amount of sleep are less than is commonly supposed after a controller has adapted to them, although the ability to adapt depends partly on age. Once work–rest cycles and sleep patterns have become established they should not be changed for frivolous reasons. Work–rest cycles should be varied only within the recommended guidelines.

17.8 Occupational health

Controllers have regular medical checks, on which their licences may depend. These have salutary and beneficial effects on their health, and prompt positive steps to improve their lifestyle and health habits at least temporarily in advance of their annual medical, when they try to lose weight, stop smoking and drink less alcohol. Some of the health problems of controllers may have more to do with the lifestyle that they share with many other professions, than with air traffic control itself (Rose *et al.*, 1978). Controllers might be expected to be healthier than many comparable groups for two reasons. The first is their compulsory annual medical examinations, which force them to become fitter, and which allow some illnesses to be detected earlier. The second is that some illnesses must be less common among controllers either because people with the symptoms of them are not initially selected as controllers or because they cannot remain as controllers if the symptoms develop. Examples of the latter include extreme obesity, and alcohol or drug dependence.

In contrast, there are other factors that might be expected to make controllers less healthy than comparable groups. Some of the older air traffic control workspaces, particularly any that still have horizontal radar displays, are difficult to sit at comfortably, and may induce postural problems in a minority of controllers. Some air traffic control workspaces require frequent scanning between wall-mounted and workstation displays or between either of these and the outside view; these changes can require frequent head and eye movements, refocusing, and changes in visual accommodation, which, while not directly harmful for the eye, may induce complaints of visual fatigue and irritation. Every effort is made to allow a controller to continue working by prescribing spectacles or other visual corrections that relate to the visual content and viewing distances of the particular air traffic control workspace, but some air traffic control environments can nevertheless lead to visual problems in a few controllers. Some occupational health problems originate from or are aggravated by physical environmental characteristics (Evans *et al.*, 1994), of which inadequate ventilation, stuffiness, and humidity that is too high or too low are probably the commonest in air traffic control. Most of these potential occupational health problems should become less prevalent as their causes become better known, as the relevant human factors principles are more widely applied, and as modern technology removes many of the constraints that have led to the problems and curtailed the solutions. On balance, shift-work, particularly night work, would be expected to have detrimental effects on occupational health, with its disrupted sleep and meal patterns. Shiftwork may also lead to domestic problems which further disrupt sleep and add to stress (Bohle and Tilley, 1989).

Over the years, air traffic control has acquired a reputation for generating stress. Stress is held to be intrinsic to the nature of the work, caused by air traffic control rather than conditions of employment or shiftwork or management methods. Data have been collected in many nations, but have not been consistent. Most evidence denies that air traffic control is a potent generator of stress-related symptoms, and the residual evidence implicates other factors such as shiftwork, conditions of employment, and individual clinical influences as much as air traffic control itself. These findings do not mean that stress is never a problem in air traffic control, but that it is not a more serious problem there than in many other professions which make great demands on the work-force. Aspects of contemporary lifestyles which may aggravate the problem include lack of control over task demands, few acceptable ways of relieving frustration, unsatisfactory diets and drinking habits, inadequate fresh air and exercise, and a tendency to try to work on despite minor illnesses so that colleagues are not let down. Occupational health may be affected also by controllers' attitudes to their work, that is, by the context in which the work is done as distinct from the work itself. The development or perpetuation of favourable attitudes to air traffic control and to its conditions may improve health, whereas conditions that undermine favourable attitudes may impair it. Controllers' health is also associated with their job satisfaction (Kavanagh *et al.*, 1981).

Some occupational health problems are self-correcting. Those who find the responsibilities of air traffic control intolerable will eventually leave the profession. Those who develop problems of alcohol or drug dependence will be forced to leave it. Those who develop postural or visual problems may remain in the profession, but in jobs that do not involve the active control of traffic. Those who suffer burnout may have to retire early or be transferred to less onerous duties. Those with serious problems in their personal life, such as divorce, bereavement or major financial crises, may have to be temporarily relieved of their air traffic control duties while their problems are being re-solved. Those who cannot adapt their sleep patterns to shiftwork effectively as they age may have to be transferred off shiftwork.

Occupational health problems which are genuine but not major can become exaggerated if they catch the fashion of the moment or become enmeshed in semi-political crusades. An example concerns radiation hazards from air traffic control displays. All marketed displays have to meet international standards for radiation, and are tested to check that they do. The regulations are strictly enforced, and actual display emissions are generally far below the permitted levels. For controllers at work, radiation emissions should never be a problem. Any possible hazards from this source in air traffic control environments would primarily affect maintenance personnel and not controllers. Because the radiation levels are so low, the debate over the possible effects of radiation on pregnant women seemed magnified, but air traffic control workspaces have not been designed to accommodate comfortably women in advanced stages of pregnancy, who may then encounter some postural difficulties in sitting comfortably and reaching all the input devices.

A further issue applies to visual displays. Where people experience visual difficulties in air traffic control, it is easy to attrubute these to deficiencies in the displays or the visual environment, and to dismiss other possible causes. For example, people in late middle age commonly experience eyesight changes and need a visual correction for the first time. This will apply to a substantial portion of controllers when they reach that age, regardless of what displays they are using or what tasks they have been doing, but they tend to blame their displays. Visual fatigue may originate outside the workspaces. A controller may spend all working hours watching cathode ray tubes in the forms of radar displays and tabular information displays in an ergonomically specified visual environment that satisfies human factors requirements, and then spend many leisure hours watching a cathode ray tube display in a non-optimized visual environment at home. Visual fatigue in the controller may originate in the non-optimized environment or from the cumulative interacting effects of both environments, but is least likely to be caused solely by the optimized work environment, and therefore least likely to be preventable or cured by changes to it. Both postural and visual problems can be compounded because the main postures and visual activities are so similar in the work and leisure environments. There are bound to be interactions; problems in either are bound to affect the other. It is important not to fritter away resources on the work environment if the problem does not originate there. When influences have many causes and there are significant statistical relationships between them, it is easy to infer causal connections that do not exist.

One occupational health problem in air traffic control commonly requires clinical intervention. After a busy shift, an alcoholic drink or two can help the controller to relax and wind down, particularly if the controller keeps wondering if a particular problem could have been solved better. The drinking that began as an option may become a necessity, and the drinks may have to be stronger to have the same effect. If sleep becomes disrupted, strong coffee may be needed before going on duty, and then stronger stimulants. A cycle of sedatives or depressants and stimulants becomes associated with the work–rest cycles, and worries about dependence on them may add to the original anxieties. Clinical intervention may be necessary to break this cycle of dependence.

Although the traditional problems of posture and vision may recede in air traffic control, other occupational health problems may be aggravated by pending developments. Forms of computer assistance that leave the legal responsibility with the controller but reduce the means to exercise it can increase controller anxiety. The reduced observability of air traffic control may make controllers realize that they are more on their own because others are less likely to notice their problems or to be able to assist them. Collective responsibility and mutual assistance both become less feasible. Pressures to work overtime, to work for longer continuous periods or to prolong working hours, may appeal to the controllers' professionalism and dedication and are acceptable occasionally, but they can lead to stress and anxiety as they become more frequent. Insufficient acknowledgement of occupational health implications in job design, task design, environment design and workspace design may incur occupational health problems which are very difficult to resolve retrospectively without major costs and modifications. In dealing with occupational health problems derived from air traffic control, prevention really is better than cure, for there may be no effective cure. In setting conditions of employment it is important to consider their occupational health implications and problems, and whether the cure for them will be by treating the individual, or by changing the conditions of employment, features of the workspace, or aspects of the air traffic control itself (Mohler, 1983).

17.9 Retirement

The need for active preparations for retirement to ensure a successful transition to it has been accepted as a desirable feature of air traffic control conditions of employment. Nations have different retirement ages for controllers. In some, controllers retire at the same age as other public employees. In others they retire after a fixed number of years as a controller, so that retirement age depends on age at entry to the profession. Some states have an inflexible policy on retirement age; others are flexible subject to retained proficiency and health. Sometimes minimum and maximum retirement ages are specified. Where relatively early retirement has been negotiated by controllers' representatives on their behalf, this can seem beneficial to controllers who are still young but much less beneficial as retirement nears, especially if their plans have assumed that alternative work will be available after retirement, but none is.

In some professions, such as medicine, a gradual winding down of activities instead of a sudden retirement may be a feasible option with some locum work following retirement. Such opportunities are rare in air traffic control where retirement is usually a clean break, with normal work almost up to the retirement date and none thereafter. Air traffic control skills do not readily transfer elsewhere, and therefore work opportunities outside air traffic control that benefit from air traffic control experience are correspondingly few.

In common with many professions, air traffic control assigns its most senior and responsible posts to its older and more experienced controllers who then retire from a highly responsible job. This can be a sensible application of their accumulated experience, but counterproductive as a preparation for retirement as the change of lifestyle associated with retirement becomes even larger. A more appropriate preparation would scale down duties and shiftwork and try to transfer the controller to the retirement locality during the final years of work, so that any traumas of retirement are not compounded by moving house and leaving friends and neighbours as well as colleagues. Because of shiftwork, air traffic control has the problem that many controllers share their leisure activities as well as their work activities with their professional colleagues because they are off duty at the same unusual times. This helps morale and camaraderie at work, but means that retirement may end many social and leisure activities and friendships as well as work relationships. A recently retired controller can become sad and miserable if retirement is based on continued association with a group of former colleagues who have not retired, but whose conversation and mutual interests continually re-echo air traffic control events in which the retired controller can have no active part. This prolongs the controller's adjustment to the inevitable severance of many of the links with air traffic control that normally accompanies retirement. For a controller who has identified wholly and exclusively with the profession, this adjustment is painful enough without prolonging it.

Provision for early retirement in individual cases for clinical reasons or factors such as burnout must always be made. Retirement plans must be based on realistic expectations, particularly about the prospects of further employment related or unrelated to air traffic control. While every problem cannot be foreseen, many could be prevented by practical counselling about the real prospects and opportunities likely to be available in the retirement locality. This is particularly important if the controller has some flexibility over the timing of retirement. Ideally, the plans for retirement should have been made well in advance, so that the last few years of work can assist the retirement process. While older controllers may find the air traffic control work more tiring and may have greater difficulty in adapting successfully to new equipment or procedures, there is not really sufficient evidence in terms of performance or safety alone to justify policies of a fixed retirement age, because individuals differ so much in their retention of skills and abilities, in their health and in their wishes as they age. However, administratively and in terms of future recruitment and training requirements, it is much easier and may seem more equitable to have a fixed compulsory retirement age for air traffic controllers. If this is the policy, human factors evidence can be applied to help to make it successful by tracing the general quantitative relationships between ageing and other relevant attributes (Davies *et al.*, 1991), and by providing human factors guidance on the preparations for retirement that are most effective (Talaga and Beehr, 1989). But the main tenor of human factors evidence is and always will be against a single retirement age because individual controllers differ so much.

18

Measurement

18.1 *Purposes of measurement*

The basic reason for measuring air traffic control systems or the controllers within them is to provide surer foundations of knowledge on which to base decisions. A multiplicity of more specific objectives falls under this rubric. None of the lists of measurement objectives that have been compiled (Hopkin, 1980b, 1982a) is fully comprehensive. Almost every facet of human factors in air traffic control can be measured somehow. Some principles have become clear. No single objective ever requires every measure that has been devised or could be available. No single measure is suitable for every objective. The efficacy of measurement is crucially dependent on the correct choice of measures and tools. Different kinds of measure are required for different kinds of objective. There are scarcely any objectives for which only one measure will suffice. The need to employ several measures for most purposes leads to problems of interpretation whenever different measures yield sets of evidence that are not obviously compatible and may even seem contradictory.

The more clearly the objectives of measurement are defined, the more likely they are to be realized. Measurement for its own sake is pointless. The many measures mentioned in this chapter are all pertinent to air traffic control in certain circumstances and most have been employed at some time, but measures that are inappropriate for the objectives are wasteful and can mislead. The reliability and validity of measures must be proved rather than be assumed. Measures should be chosen because, according to all the evidence available, they are the best to meet the objectives, and not because they are cheap, quick, available, familiar, expedient, or have served other objectives well in the past. Claims that the same kinds of evidence can legitimately be interpreted in many ways, for example in terms of performance, error, workload, capacity or stress, suggest some muddled thinking behind the choice of measures, weak rationales, bias in interpretations, or some arbitrariness in the explanatory concepts.

The main broad purposes of human factors measurement in air traffic control are suggested below. In practice these have to be subdivided into more specific purposes before the most appropriate measures can be identified.

1. Descriptions of actual or attainable system functioning, including levels of safety, efficiency and performance, in relation to the main influences on what is achievable and the main factors that can change those influences.

2. The quantification and verification of plans, procedures and proposed changes, in terms of system functioning or the functioning of controllers within the system.

3. Consideration of human and machine roles within systems, their matching, and the interactions between them.

4. The quantitative comparison of viable alternatives within air traffic control systems.

5. The aetiology, prevention and circumvention of any hazards to safety, breakdowns, failures, omissions and errors, whether of the system or of the human, and the identification of their full range of consequences.

6. Consideration of all the aspects of the system that can affect the well-being of those who work within it.

7. Examination of the ways in which the system can affect or satisfy human needs and aspirations at work.

8. Consideration of individual differences between controllers, both their effects on the air traffic control system and the effects of the system on them, with reference to recruitment, selection, training, job allocation and conditions of employment.

9. The specification of what is achievable in relation to the resources required to achieve it.

10. The identification and proving of appropriate measurement techniques in relation to air traffic control tasks and objectives.

Measurement is not simply data gathering. The assiduous collection of data that are never used and have no influence on events is a waste of time and resources. It is essential to choose measures that can answer a question validly, and also highly desirable not to choose any measures that show nothing, are unproven, or yield data that cannot be interpreted. Piles of data set aside until there is the time and opportunity to analyse them properly, which there never is, represent misguided efforts that could have been applied more profitably to other objectives. There has been no consistent policy on the verification and validation of measures for air traffic control (Wise *et al.*, 1993). All measures require this to some degree but few have been validated adequately, and a reluctance to do this kind of work has led to findings that have been difficult to sustain or interpret.

It will be correctly deduced from the above remarks that air traffic control lacks a proven and comprehensive theory of measurement. Measures have often been adopted because they have been helpful in the past, rather than because they possess a sound theoretical rationale which explains why they have been helpful and the conditions to be satisfied before they can be. Measures in vogue elsewhere have been copied, without close scrutiny of their applicability. The choice of psychological constructs, cognitively based activities, records of events, organizational descriptions, physiological indices, social measures, error taxonomies, and many other kinds of measurement has often seemed to depend as much on current popularity as on proven relevance, and on a willingness to settle for descriptions of findings rather than to seek explanations of them. Currently ergonomic measures seem to be on the wane, cognitive measures in the ascendant, and organizational measures on the horizon. Such trends owe more to fashions outside air traffic control than to needs arising from within it. In choosing measures, it is important to recognize relevant developments elsewhere, particularly when they cover factors that have previously been neglected, but what seems important now may prove transient, and in choosing measures appositeness should always be put before popularity. This apparently commonsense advice can be difficult to apply because those who fund work requiring measurements may prefer the new exciting measures which they have recently heard so much about, to the orthodox, validated, worthy but dull measures that seem less likely to discover anything new. Reviews of measures emphasize scientific methodology (McGuigan, 1989), the practical evaluation of work (Wilson and Corlett, 1994), or interdisciplinary measures of performance (Ilgen and Schneider, 1991).

18.2 System behaviour

The form in which many of the questions about air traffic control measurement are posed expects an answer couched in terms of system behaviour. Questions mention inputs to or outputs from the system, functions occurring within the system between the inputs and the outputs, traffic flows and capacities, or failures. They may be about the whole air traffic control system or subsystems within it. System measures do not normally distinguish between human and machine functions and may apply to both or to either, but they are expressed in system and not human concepts. If manual functions are automated, this may constitute a major change in the technology and in the human tasks and responsibilities, but if the function does not change much, neither will its description or measurement as system behaviour.

Air traffic control has multiple aims as a system. Often quoted are safety, orderliness and expedition, but there are many others: impartiality, cost-effectiveness, noise abatement, fuel conservation and granting users' requests. Because the purpose is to handle air traffic, measures of the system behaviour relate directly or indirectly to this. System inputs deal with information about the traffic, derived from sensors, aids and computations. Other inputs deal with conditions and constraints relevant to traffic handling, such as applicable rules and regulations and procedures, and serviceability states and weather.

General system measures concern traffic demands and system capacities, traffic patterns and flow rates, navigation aids, and bunching and separation minima. There are data about each aircraft under control. Some kinds of data relate to effects of the system on the aircraft, such as re-routeings, delays, restrictions, permitted manoeuvres, and fuel usage. Other kinds relate to effects of the aircraft on the system, and include its type, the equipment carried, and its performance capabilities. Some of the system information sensed about specific aircraft is categorized into measures of flight level, speed and heading, for example. Further information about rules, procedures, instructions and responsibilities defines how air traffic control is conducted and the parameters within which the measures of system behaviour apply. System-based computations derive capacities, check separation standards, and compile evidence on delays and diversions. From all this evidence, broad system parameters can be derived: examples are safety levels, the incidence of delays, information flows and constrictions, channel occupancy times and distributions, the derivation and application of serviceability and maintenance schedules, and system reliability expressed as failure rates or down times. This last concept of down times is normally applied only to system factors that affect system behaviour, such as power failures, radar unserviceability, non-functioning communications channels, and exceptionally severe weather, but not to human factors that affect system behaviour such as controllers on strike, even though factors of the latter kind may be responsible for most of the down time from the point of view of the system users. Partly as a result of this narrow interpretation of categorizations of system behaviour, the amount of effort devoted to the various causes of down times may not relate much to their relative prevalence.

Many indices of system performance and efficiency rely entirely on measures of system behaviour. This applies to operational research and analysis, to quality control, to indices of component reliability, and to planned system capacities and traffic flow rates. Measures of system behaviour serve many vital purposes in their own right, but human factors purposes are not usually among them. For example, system data about voice channel occupancy times can yield very little human factors information until they have been converted into human factors data by the addition of information about the purpose, nature, content, pace and relevance of the voice messages.

Measures of the behaviour of systems and subsystems are not the same as measures of human performance (Taylor, 1957). Appropriate system measures can show that the performance of a subsystem has improved in system functioning terms but cannot alone discriminate between better equipment, revised procedures, or a better trained workforce as contributory causes. Measures of system behaviour often provide the framework within which other kinds of measure, including human factors ones, have to find the causes, the explanations and the interpretations for systems effects, or within which other measures must be interpreted. In air traffic control, measures of system behaviour are essential for many purposes, but cannot suffice for measuring the controller.

18.3 Human activity

Some human factors measures can be based, at least in part, on measures of human activity (Meister, 1985). To be valid, such measures must be totally passive in that the processes of measurement must not interfere in any way with the activity to which they are applied. This really implies that such measures must be indirect, and preferably automated. Failing this, they must be so unobtrusive as to be accepted completely as a normal and integral aspect of the work environment.

The most easily recorded human activities are timed human actions. The recorded usage of a keyboard can exemplify this. Time and event recording is a well-established form of measurement. Another kind of human activity can be recorded through videos or photographs, to provide a continuous, intermittent, selective or randomized record of events, depending on their objectives and how they are set up. For example, a continuous recording of controllers' head movements taken from behind has been used successfully in an evaluation to prove that when controllers become very busy they no longer look at what their immediate colleague is doing. Head movement recordings, though lacking the fine detail and discriminability of eye movement recordings, are simpler to analyze and can also reveal whether controllers consult wall-mounted displays. If film or video cameras are used, it can be sensible not to load them at first, until controllers have become so accustomed to them as to ignore them. Even then, their presence may inhibit controllers' activities sufficiently to invalidate some of the kinds of data they gather; controllers may never stretch out and relax so much when there is little to do, they will edit their conversations particularly on non-work topics, their conduct will be more orthodox and less abrasive, and supervisors and others may not stand in their usual positions if they would thereby cut off the camera's view of a workstation.

One technique for measuring human activity is through random sampling of it during normal work. In one study, a camera showing the workspace was triggered remotely from another room whenever anyone in the other room walked past the triggering mechanism in the course of their normal job there. They had no form of feedback and could not know what would be photographed. Quite a large random sample of the controllers' activities was photographed. The purpose of the particular study was to look for any incipient postural problems associated with the suite designs, though this measurement technique could have much wider applications. Analysis was of two kinds. The main one sorted the photographs according to postural categories specified in advance. The subsidiary analysis subjected all evidence of unusual postures to the critical scrutiny of a medical specialist to rule on any occupational health implications. At all costs the trap of scientific invalidity must be avoided, whereby the data are gathered first, used to derive hypotheses, and then used again to test the hypotheses derived from them. This constitutes scientific sharp practice. Hypotheses derived from the original data can be tested validly only by new data. It is essential to formulate each hypothesis before gathering the data, and indeed the hypothesis should determine the data to be gathered. It is recommended that photographic techniques should be used only to test hypotheses and not to generate them.

Records of human activity are factual and not judgemental. They can quantify the times spent by controllers on collaborative tasks, show the normal sequences and patterns of activities and the occupancy times of communication channels, classify the contents of transmitted messages, and reveal relationships between scanning displays, inputting data, and using communications channels. They can show when the controller is inactive in the sense of being non-interventionist, and may sometimes distinguish between a controller doing nothing and a controller who is busy looking at displays or pondering decisions without being overtly active. Activity analysis techniques can reveal when unorthodox sources are used and what controllers fail to do—which inputs they never make and which displays they never look at. In one interesting example, controllers ostensibly working in teams of two often obtained information not from the displays of their colleague in the team but from the displays of another colleague in an adjacent team, because the alternative source of information seemed more relevant. This kind of evidence demonstrates mismatches between the task requirements and the facilities provided.

Some measures of human activity can show how busy a controller is. Some of the strongest evidence can be negative in that the controller never sits back and relaxes. Human activity measures may cope well with special circumstances such as a watch handover, if they record what the incoming controller needs to do and needs to be told, and identify the functions that the outgoing controller completes before handover. Activity analysis may serve as background information to interpret other measures. In so far as changes in physiological and biochemical states are a function of physical activity as distinct from other factors such as stress or workload, an activity analysis can show if there were changes in physical activity, and hence can help to prevent the misattribution of causes to observed events. Similarly, errors or omissions may be partly explained by an activity analysis which shows continuous high activity without respite.

Measures of human activity are not a sufficient basis by themselves for judgements about workload, capacity, errors or omissions, but in conjunction with other kinds of data they can strengthen the foundations of such judgements. To some extent air traffic control activities can expand to fill the time available. If a controller is constantly active, not all of the activity is necessarily being imposed by the task demands; some may be self-generated. An activity analysis will record such activities, but may not be capable of distinguishing them from activities directly derived from operational demands. The kinds of data gathered to describe human activities are often construed as human factors data by people in other disciplines, who try to apply them to answer human factors questions. Among their beliefs are that such data can establish how long human functions should take, or can quantify human functions mathematically for modelling purposes. The reason is that comparable data about non-human system components can often provide satisfactory models or mathematical descriptions of the functioning of those non-human components. It is a hard but essential lesson that apparently equivalent data are not sufficient to model human functioning or to predict human performance, for several reasons. The technique cannot distinguish between functions which are superficially similar according to an activity analysis but very different in human terms, and so it tends to lump them all together. It also tends to group activities that are correct or incorrect, appropriate or inappropriate, or essential or redundant, for it cannot distinguish between them. The technique cannot cope adequately with omissions, any occurrence of which is a vital aspect of human performance, and cannot provide an adequate classification of the influences on human activity that any satisfactory model must include. Thus a human activity analysis is a useful feature of measures of the controller within the system, provided that its very considerable limitations as a technique are recognized.

18.4 *Task performance*

Air traffic control is complex, and comprehensive measures of task performance have to be correspondingly complex (Ilgen and Schneider, 1991). A distinguishing feature of most performance measures is that they are not confined to descriptions of performance but include directly, or can be adapted for, judgements of the performance in such terms as efficiency, safety, correctness, and timeliness. The notion of performance usually implies some judgemental attributes. Measures of task performance focus on the controller or on teams rather than on the system, and on human rather than system functions. They imply that some form of job or task analysis or alternative procedure has already been applied to identify tasks and functions as separately measurable entities. This seems straightforward, but often is not in air traffic control. For example, the handover of control responsibility for an aircraft as it leaves a sector and enters an adjacent one may be an active task with measurable events or a tacit task relying on a silent handover with no measurable human events. Successful measures of task performance therefore have to accommodate the same task with or without overt activities.

Most recordings of air traffic control and evaluations of controllers assume that some measures of task performance are essential for nearly all the purposes of measurement mentioned at the beginning of this chapter, and therefore include some. Two particular issues arise. One is the level of detail required for the particular purposes of measurement. There are at least five levels of detail at which air traffic control tasks can be described and measured

(Crawley *et al.*, 1980). If the wrong level of detail is chosen, the data may fail to answer the questions posed. The second issue is that similar measures of task performance have been interpreted as providing valid evidence of many different kinds, for example on safety, capacity, efficiency, workload or stress. The facts from measures of air traffic control task performance can be clear and quantitative, but the concepts, constructs and contexts that can legitimately be applied to interpret their meaning and implications may be hotly disputed.

If air traffic control is primarily a human activity and any computer assistance manipulates data rather than supports human tasks directly, task performance measures can in principle provide relatively comprehensive and complete descriptions of the controller's activities, although they can become complex where there are extensive team functions. As computer assistance supports human tasks more integrally and directly, any adequate description of them must include more machine events, whether they initiate human tasks, signal the completion of tasks or of stages within them, denote transitions between tasks, or constitute essential task elements as in human–machine menus and dialogues. A perennial problem in the measurement of task performance is the inclusion of the controller's mental processes where these have no immediate counterpart as recordable events. A controller may muse over a complex air traffic control problem and decide to take no immediate action. No measures of task performance may be capable of discriminating between this and failure to notice the problem, and other kinds of measure may be required to supplement or interpret the data on task performance.

Task performance in air traffic control normally includes speech, which therefore has to be measured. It is easy to record it and to derive system measures such as channel occupancy times from the recordings. It is time consuming but not particularly difficult to classify speech under a variety of headings related to its information content or roles. It is much more work and far more difficult to relate each item of speech to its context in order to judge its safety and appositeness, because such judgements must refer to its objectives, the traffic scenario, the choice and timing of tasks, the results achieved, and the feasible alternatives. The judgements must rely on manual individual assessments rather than general automated ones, with a considerable element of subjective interpretation. It is vital to define rigorously beforehand the purposes of analysing speech and to adhere strictly to those purposes. Broad analyses of speech content with no clear objectives are likely to induce more problems than they solve.

Measures of task performance cover human achievements. They include measures of what was done, when it was done, with what frequency and how quickly, and also measures of how well it was done, whether it was the right thing to do, whether it was done but should not have been, and whether it was not done but should have been. The data obtained from measures of human performance may provide simple factual descriptions, may be scored against theoretical or actual maxima or criteria, or may be compared with corresponding data from other places, other conditions, other times or other people. Task performance measures are the basis for assessing numerous human air traffic control functions. For each of them, the task performance measures that seem most appropriate and practical have to be chosen beforehand, which is still in some respects an art rather than a science. Task performance measures normally must be supplemented with other measures, most commonly of system behaviour, subjective assessments, and errors if these are not included in the tasks. There is no single optimum package of task performance measures. A combination of them that fits the objectives and can provide comprehensive, fair, complete and unbiased measurements of the human achievements should be chosen afresh each time. Other kinds of measure may be needed to indicate at what human cost the measured task performance is being achieved.

An adequate description of the following human characteristics normally must include some measures of task performance, in conjunction with other measures. The characteristics include efficiency, skill, learning, experience, capacity, workload, decision making, problem solving, prediction, understanding, attending, remembering, judging, task sharing, and the application of tactics and strategies. There is some debate over whether situational awareness can be included or must be measured subjectively (Garland and Hopkin, 1994; Hopkin, 1994b). Some task performance measures are usually applied to study speech intelligibility, the scheduling and division of work and responsibilities, and levels of human activity.

Although task performance measures are applied to cover all the above functions, all measures of task performance are not equally apposite, reliable or valid in meeting objectives or in their use of resources. A common but false presumption has been that every kind of human factors question needs to include task performance measures in its answer and is therefore amenable to study by experimental methods. Some years ago, Hopkin (1982a) compiled a list of 19 broad human factors task categories that normally occurred in air traffic control, on which further evidence could with benefit be gathered, and suggested appropriate task performance measures for each. Little progress has been made in the meantime, and more evidence on most of the listed items is still required. Among the issues mentioned then were handovers as indices of workload, the effects on performance of including selective information on the quality of data presented, the causes and effects of boredom, the main influences on the timescales of the controller's decisions, and training requirements to enable the controller to benefit from greater control over workload.

18.5 Inconsistency

The interpretation of measures of task performance and of all other air traffic control measures depends on the consistency of the data. The consequences of inconsistency depend on its nature, on the measures affected, on the measurement objectives, and on the context of measurement. Some of the main kinds of inconsistency are discussed below.

An individual controller may perform inconsistently. Possible reasons include inexperience, or unfamiliarity with some aspect of the work, and individual worries originating at work or away from it. Because the measured effects in the form of inconsistency of performance do not necessarily have common causes, it can be quite difficult to ascertain their origins clearly enough to provide effective remedies, but the approach exemplified by recent work on the unifying concept of readiness to perform (Gilliland and Schlegel, 1993) could perhaps be extended to individual inconsistency by defining the problem in terms of its common effects rather than its diverse causes. Alternative explanations of individual inconsistency include deficiencies in the task performance measures, and unidentified and therefore uncontrolled critical influences on performance that are actually more important than the known independent variables.

Inconsistencies between controllers in their performance are widely recognized to be a consequence of stable differences between individuals in their capabilities, training, knowledge, experience, attitudes or motivation, which affect the individual's understanding of the traffic and preferred tactics and strategies. Differences between controllers are also influenced by the degree of local tolerance or acceptance of individual innovation, idiosyncracy or variability. A factor that can stabilize inconsistencies between controllers is that air traffic control is complex enough for the same objectives to be achievable in more than one way with equivalent safety and efficiency, which removes some of the normal grounds for reducing measured inconsistencies between individuals.

Teams often evolve their own preferred practices, which may characterize the inconsistencies between teams while serving as marks of team membership and team conformity. Since each team favours its own practices and believes them to be the best, and may be somewhat vague about how far they differ from the practices of other teams, attempts to reduce inconsistencies between teams will meet opposition unless there are compelling reasons for them, especially if the inconsistencies have become tacit indices of team cohesion, loyalty and solidarity. The extent of inconsistencies between teams depends on the diversity of permitted acceptable practices, the degree of commonality in training, and the collective memory and experience of the team in handling previous incidents or problems. The above sources of inconsistencies between teams may extend to watches, centres and towers. At this level, inconsistencies may also be attributable to other factors, such as different management styles and traffic scenarios.

Some inconsistency between nations is inevitable in the measurable conduct of their air traffic control. Obviously there are geographical, traffic and equipment differences, but more specific human factors differences are in the calibre of the controllers employed, their conditions of employment, the satisfaction and motivation of the workforce, the status of the profession, and the norms and standards that have evolved. These can result in differences between nations that can be construed as inconsistencies at a more general level.

Training differences are a further source of inconsistencies. Those trained initially as procedural air traffic controllers may perform some tasks differently from those trained initially as radar controllers. Those who have had to adapt to various forms of computer assistance may perform their tasks differently from those who have never conducted air traffic control without such assistance. Despite international standards, some differences between schools of air traffic control are inevitable in the content of their courses and in their teaching methods, with resultant differences in how controllers apply what they have been taught to the control of air traffic.

A different kind of inconsistency, but one that can have significant human factors implications, is between the more flexible and adaptable real-life air traffic control data and the more rigid and orthodox data from any surrogate of air traffic control such as an evaluation, simulation, model, or training environment. In real life, unorthodox procedures never covered in training or evaluations may be unavoidable to maintain safety in exceptional circumstances. Accepted local practices arise to meet local requirements and produce local inconsistencies that evaluations do not cover. Occasionally there is misplaced denigration of training or evaluation practices as too conventional, cautious or inflexible for local needs, but such orthodoxy is usually based on safety and should not be discarded lightly. Wherever it is, inconsistency can result. Inconsistencies between evaluations and real life can be because all the circumstances of real life were not foreseen or could not be replicated in the evaluation, which cannot possibly cover every eventuality. The omission of crucial real-life factors, however inevitable, can become characteristic of all evaluation findings wherever the typical discrepancies between evaluation findings and real-life experience are not identified and fed back routinely to prevent their recurrence in subsequent evaluations.

There may be inconsistencies between human and machine. Some of these are obvious and refer to their respective capabilities. The point of computer assistance is that the machine excels the human in many functions, especially of data gathering and computation. The human can excel in devising unorthodox solutions and at implementing them if given the means to do so, and the human carries the legal responsibilities. Many problems at the human–machine interface concern inconsistencies between human and machine: data that can be assimilated readily by one usually have to undergo some transformation to be assimilated readily by the other. This may result in inconsistencies within the data. The information presented to the human may be a much simplified version of the data used for computations. Too much simplification leads to discrepancies and inconsistencies, to the extent that the human and machine can disagree about whether a separation standard has been infringed. One source of inconsistency is the practical difficulties of building some commonsense into computer assistance, a human concept that needs very considerable transformation before it can be expressed in machine terms. There are also inconsistencies between human and machine in initiating mutual help. The notion of user friendliness implies inconsistencies between human and machine requirements in systems where such friendliness is lacking.

Air traffic control has multiple objectives. Some inconsistencies between objectives are inherent wherever they clash, and these extend to the weighting of the objectives and to deciding which should prevail when they are irreconcilable. Safety is always paramount, but expedition may be achieved at the cost of orderliness or vice versa, and both may be downgraded in the interests of noise abatement, for example. If there are disagreements about the correct weighting of air traffic control objectives, inconsistencies in the measures will remain until these primary disagreements have been resolved.

Some inconsistencies within measures originate in the nature of the air traffic and the local procedures for handling it. Identical traffic patterns in different regions would be handled quite differently because of differences in route structure, danger areas, restricted zones, local equipment and practices, and numerous other factors such as the equipment in aircraft and the experience of pilots. Measures have to be able to reveal and explain such differences, and must be interpretable in terms of them. No measure can be completely reliable, especially when chosen for reasons of applicability, relevance, and the minimization of effort and resources, rather than for its reliability, which often remains unknown. Even with known poor reliability, a measure may be the most feasible in the circumstances, and better than no measure at all. But any measure is only as good as its internal reliability.

There are also inconsistencies between measures. These may be spurious, an artefact of inadequate reliability, if a measure is irrelevant or does not in fact measure what it claims to measure. If the inconsistency is not spurious but genuine, this is a key justification for using multiple measures. If controllers do one thing but believe that they are doing something else, this can be revealed only by measuring both what they do and what they believe they do. This does not mean that one of the measures must be wrong or that the inconsistency must be resolved. Inconsistency is an important finding in this context. Controllers may believe that their task performance has been improved by the introduction of colour coding on radar displays when it has not, that a form of computer assistance has increased the incidence of conflicts when it has not, or that the air traffic control service at one air traffic control facility is better than that at others when it is not. The point is that the beliefs are sincere and that controllers act on their beliefs. Multiple measures are the key to understanding what is happening. Inconsistency as a concept seems to have acquired some slightly perjorative connotations, as unpredictability has and adaptibility and flexibility have not. But one person's inconsistency may be another's flexibility. Performance dubbed as inconsistent may in fact be successful and adaptive, and optimum in the particular circumstances.

Inconsistencies are almost taken for granted when evidence about the circumstances of an incident or accident is being gathered. A total lack of inconsistency between accounts of an occurrence is strong prima facie evidence of collusion. Inconsistencies between accounts of the same event are always to be expected, and hasty judgements must be avoided about which account is right or which is wrong, for none will have a monopoly of correctness and none may conform closely with the actual events. Inconsistency does not imply human deviousness or insincerity. What it does imply is human fallibility in perceiving, understanding, interpreting, remembering, and describing what has occurred.

There can be inconsistencies between findings and expectations. Plans may envisage that air traffic control will be conducted in one way, and in practice it is conducted in another way. An expectation may be that computer assistance will be used in accordance with the designer's intentions whereas the controllers who actually use it may discover alternative applications of it that suit them better. If this is not realized, measures of the usage of the computer assistance will be inconsistent or misleading, and certainly incomplete.

All of the above sources of inconsistency can contribute towards variance that is unaccounted for in measures of air traffic control. The greater their extent, the less satisfactory the controlled and measured variables seem in providing a rationale for what occurs. The more the sources of inconsistency that can be identified, the greater the potential to control them becomes, and the better the prospects of measures where the variables studied seem preponderant and important influences rather than trivial ones, because they explain much of the variability instead of failing to account for most of it.

18.6 Errors

Normally it is essential in actual air traffic control and in air traffic control evaluations and simulations to measure human errors, in order to identify those that are possible and prevent any that could be hazardous. There are two distinct approaches to error measurement in air traffic control, and both are needed. One approach categorizes all errors, and assigns probabilities to their occurrence. Whenever alternatives of any kind are compared, the one with the lower incidence of errors would usually be preferred. The other approach to error measurement is quite different. It examines single errors without categorizing them, at least initially. The ultimate intention is still to prevent all similar errors, but only a single example of a particular error suffices to establish that the system did not prevent it and that it can occur. Through error categorization, and especially pre-categorization, much of the significant detail about each error can be lost. The approach to the investigation of incidents and accidents examines individual errors in detail (Baker, 1993), to establish the circumstances that gave rise to them and make recommendations to prevent their recurrence without being bound by preexisting error categories, although the errors will be classified subsequently in relation to existing keywords and familiar error taxonomies where these fit, whether the taxonomies are of human error generally (Reason, 1990; Senders and Moray, 1991) or of air traffic control errors in particular (Stager, 1991).

Simply to record that an error has occurred will not achieve much. Counting or classifying errors does not usually lead to practical action. Errors are not necessarily similar, and to force them into predetermined categories can be counterproductive in trying to explain them. Evidence about the nature and circumstances of each error requires the gathering of detailed information about it. Extrapolation to other circumstances is suspect unless there is firm evidence about which of the many factors were the crucial ones. The successful investigation of error must deploy extensive human factors resources to identify and weigh all the relevant factors as a mainspring for recommendations. Errors are heterogeneous in their origins and in the diversity of their consequences (Reason and Zapf, 1994). It is easier to classify them in terms of their consequences than their origins. Errors that could be dangerous must have priority over errors that are inconvenient or inefficient. However, it is important to consider if a particular error that has merely been inefficient could be dangerous in different circumstances. Human factors evidence is currently incomplete on how well controllers can monitor and correct their own errors as they make them.

From single instances of errors, practical error taxonomies can be built, but they are most successful when they can be linked to more fundamental principles. An example concerns the possibility of confusions between callsigns. Incidents record actual callsigns that have been confused. Principles provide classifications and explanations of the incidents because the design decisions about the methods of entry of the callsign into the system predetermined the kinds of human error that are feasible in entering callsigns into the system. Alternative principles for entering a callsign could include: spoken alphanumerics, where most human errors will exemplify known kinds of phonetic confusion; or visual reading of the callsign, where most human errors will involve the misreading of the alphanumeric characters known to look most visually similar; or premature or delayed channel switching, where cutting off part of the spoken callsign results in errors that can be predicted because it is known how they arose; or the use of keyboards, the layouts of which will influence the typical errors of data entry that result from striking a key adjacent to the correct one; or moving a marker through an electronic list of callsigns, where the human errors will mostly refer to callsigns next to the correct one in the list; or placing a radar cursor over a callsign label on a radar display, where human errors will often be associated with physically adjacent or overlapping labels; and so on. Although single instances of errors are not generally predictable, types of human error are, and known evidence and principles allow them to be deduced from design decisions. Because they can be deduced in principle, many can be detected, and some may be prevented altogether, whereas the potential hazards from the others can be minimized.

There are no design decisions involving human activity that prevent all human errors, and so the complete prevention of human error by optimum design decisions is not a practical option. The main human factors contributions are to deduce, as scrupulously as possible, all the sources of human error inherent in the design decisions, and to advise on practical means to minimize their occurrence and their consequences. First attempts will never be perfect, and it is therefore important to be able to detect and learn from any residual human errors that do occur, to prevent their recurrence and also to reduce still further the incidence of uncorrected human errors in the system.

18.7 Omissions

Omissions are difficult to measure in task performance terms because they refer to events that do not take place. Therefore any form of activity analysis and most performance measures are useless for studying them. External criteria are usually needed as evidence of an omission, because internal criteria can only cover a few specific kinds of omission such as the exclusion of a single item within a fixed sequence of items which collectively comprise a single task or subtask. Most omissions in air traffic control are not of that kind. They arise because the controller did not do something, did not notice something, did not understand something, or did not follow a prescribed standard routine. Omissions are measured only insofar as they are sought. An omission may be the crucial cause of an incident, but it will remain undiscovered and unknown unless someone thinks of it or notices its absence.

The simplest omissions are corrected because they become apparent at once. They may be irritating, but their obviousness ensures that they are not dangerous. The omission of a single keying within a series of data entries will often result in nonsense that the machine cannot or will not accept. If the omission is because the controller has made a keying error it may simply be corrected with no other consequences. But if the controller did actually press the correct key and the omission is because of a machine error, which could be anything from a faulty key to an overloaded communications channel, the consequences for task performance can become much more significant, particularly if the same fault recurs and it is not an isolated instance. This can interfere with the pace of keying, with the rate of learning, and with the forms of feedback required. The controller may want to check that each key entry has been made before making the next one, and the data entry will never become fast and fluent under such conditions. Much more is known about the human as a monitor of errors than as a monitor of omissions.

An attribute of many omissions is that there may not be agreement that they are omissions. If an event that is vital to one controller seems redundant or superfluous to another, its non-occurrence is an omission to one but not to the other. Without standardization, any event that is not mandatory could be

construed in this way. When the controller is busy, activities such as acknowledgements may be dispensed with but recorded as omissions. It is ironic that
the easiest recordable categories of omissions are those which the controller
would often not concede to be omissions, since their absence was deliberate
and not the result of ignorance, distraction or forgetfulness.

Omissions are generally defined in relation to task performance by other
events to which they refer and which did occur. This may allow omissions to
be identified and classified but does not explain why they happen and is not
sensitive to different omissions with different consequences. To some extent,
difficulties in measuring omissions are an artefact of modern automated
recording methods. Records of timed events and measures of task performance
may be unable to deal with omissions because there is no timed event and no
task performed, whereas subjective methods of measurement may find omissions no more difficult to deal with than errors are. If mandatory activities are
defined in advance, their non-occurrence in applicable circumstances, which
must also be defined in advance, can be recorded automatically, and in principle omissions from any standard activities can be detected in this way provided that there are always recordable events when there is no omission.
Omissions are a common consequence of forgetting, but this may also help to
explain why they have not been studied very much, because forgetting has not
been studied much either, though memory has (Hopkin, 1988).

When non-essential events are omitted because the workload or task
demands are high, this should not necessarily be deplored as it can be a sensible stratagem to adopt. A consequence is that some of the most sensitive
measures of high workload or high task demands are omissions of non-
essential functions, whether they are omitted entirely or merely postponed.
They can be better measures than secondary tasks at tracing the effects of
different degrees of high workload, because they avoid arguments about competing or compatible mental resources, because they can be more sensitive
than any alternative measures, and because the order in which different kinds
of omission are introduced is an index of their relative significance to the
controller, in reverse order of importance.

18.8 *Physiological and biochemical indices*

Physiological and biochemical indices have been used to show the effects of
the air traffic control system or air traffic control work on the human controller. Their most common human factors applications have been as measures of
workload, effort, fatigue or stress (Melton, 1982; Costa, 1993). They have also
been used to study aspects of controllers' health, and sometimes to amplify
performance measures taken in conjunction with them, as in the case of
records of head or eye movements (Stein, 1988).

Sometimes apparently similar data have been interpreted in different ways. The feasibility of using physiological or biochemical indices obviously depends on the availability of suitable resources. Automated recording and analysis of the data that promote the gathering of data non-intrusively, also allow more sophisticated forms of analysis. Not only may heart rates be averaged over a larger range of chosen time periods, but measures of much greater subtlety can be developed if they are more relevant or sensitive, or better in other respects as measures. Examples based on heart rates are change of heart rate, rate of change of heart rate, and indices of heart rate variability.

Early attempts to employ physiological or biochemical measures in air traffic control were cumbersome and difficult to justify because the measures were intrusive and very demanding of resources. Crump (1979) and Melton (1982) review them. Physiological and biochemical indices are specific to each individual and that is how they usually have to be interpreted. Their significance depends on comparisons between equivalent data from the same people under different conditions or at different times. Recent technical and electronic advances enable some kinds of data to be transponded instead of being gathered by the attachment of special equipment to the body or by obtaining samples of body fluids. Biochemical measures require suitable laboratory facilities and equipment for analysis. Physiological measures need computer analysis as they typically produce a lot of data. Measures of brain potentials require specialized instrumentation and analysis, but have not yet progressed sufficiently to warrant their practical application (Humphrey and Kramer, 1994).

As the gathering and analysis of physiological and biochemical data have become more tractable, their popularity as measures has waned. A common reason was not that they were unsuccessful, but that much of the information that they yielded could be obtained more simply and parsimoniously with alternative measures. To show that the controller's heart rate increases when the controller is handling more traffic is a finding that is useful in its own right and to validate other measures, but it is not necessarily easy to interpret and indeed does not always occur (Costa, 1993). Usually the handling of more air traffic also entails more physical movement on the part of the controller, which could also affect heart rate regardless of the mental loading of the tasks. This complication, and other associated effects such as feeling warmer because of the increased activity, have not always been taken into account.

Even in the case of individual controllers, but especially if the physiological data from several controllers are averaged or otherwise combined, the physiological and biochemical data might not provide a sufficient basis to justify executive action. They could show, for example, that heart rate increased or became high but this raises the issue of what heart rate or what increase in heart rate is acceptable. The evidence required is not that the heart rate is high, but that it is toŏ high. The basis for intervention would have to come from associated measures, of impairment of performance, of potential hazards to safety, of increased errors, of reduced efficiency, of poorer health, or of increased susceptibility to occupational illnesses. Evidence of these kinds has generally been elusive. Increased heart rate might also reflect that the work was challenging, interesting, rewarding or satisfying as well as being more demanding, and increased the controller's motivation and effort and determination. If such changes according to physiological measures can indicate a medley of benefits and disadvantages, their interpretation can become ambiguous or contentious. Ironically, some of the most useful physiological indicants of the need for action might reveal interactions between the controller's work and non-work environments (Tattersall *et al.*, 1991).

Most of the reasonably standard physiological and biochemical measures have been tried in air traffic control at some time or other. These include heart rate, heart rate variability, rate of change of heart rate, skin resistance, breathing rates and breathing volumes, blink rates, eye movements, head movements, body temperature and blood pressure. The commonest biochemical indices have related to the sympathetic nervous system and measures of epinephrine and non-epinephrine excretions. Many measures have also been taken for clinical and occupational health reasons, as distinct from human factors purposes, and these are considered as measures of health. For many measures, there has been avid discussion of the required quantities of data, the best sampling methods, and the appropriate analysis techniques. General guidelines on this cannot be supplied since the answers depend on the questions and the objectives. Clearly, a question about the physiological sensitivity of an individual controller to minor changes in task demands, and a question about the potential health consequences for controllers of working at a centre with high prevailing task demands, require very different data to answer.

Attitudes towards physiological and biochemical measures, which were formerly favourable, now seem to have swung too far in the opposite direction, although there are recent signs of renewed interest in eye movements and blink rates in particular (Stern *et al.*, 1994). Physiological or biochemical data can be very useful if the results of a change are not reflected directly in any performance measures but in the changed effort required by the controller to sustain the same performance, for they can reveal this. They can also show if the same changes in task demands have disproportionately larger effects on some controllers than on others. They may even suggest when an individual apparently fails to realize the significance of events and treats them too complacently. Their independence from other measures may give them a vital role in attempts at validation. The fact that the same measures have been interpreted differently in the past, in relation to workload or stress or task demands or effort or motivation, for example, indicates that the purpose of using physiological or biochemical measures has to be very clearly defined, the rationale for choosing the most appropriate measures has to be impeccable, and the interpretation of the findings must not be speculative, otherwise the suspicion lingers that the data bolster a favoured interpretation but actually prove little. The authors of the review of blink rate cited above note that fatigue and continuous time on task are often equated in blink rate studies, and mention several factors known to influence blink rate: *ergo*, blink rate can be a measure of all of them. They do not fall into this trap, but it is easy to see how others do and then claim that blink rate is measuring the factor of their choice.

Most measured physiological or biochemical effects may be attributable to factors within the work environment, to factors outside it, or to individual attributes, and often to a multiplicity of factors of each kind with unknown interactions between them. If the data warrant action, they must make clear whether the action concerns the air traffic control system or the individual controller. This issue is not specific to air traffic control, but the major applications of physiological and biochemical measures to air traffic control occurred when it was believed, wrongly, that air traffic control had unique problems, such as stress. This is no longer a sufficient reason for employing physiological and biochemical indices. In some instances physiological and biochemical indices may reflect the individual controller's adaptability. Detectable but moderate changes in response to changed circumstances may denote balanced assessments of their consequences as a result of long experience, whereas lack of changes in response to circumstances may denote inexperience in the case of uniformly excessive reactions, or complacency in the case of uniformly low reactions. In human factors studies of air traffic control, physiological and biochemical measures always have to be used in conjunction with other measures and interpreted in relation to them.

18.9 Subjective assessments

Subjective assessments are among the most popular measures of air traffic control. They are convenient, inexpensive and always available as an option. They may in some circumstances yield data which can be obtained in no other way. They can be very helpful in supplementing and explaining other measures, but are never wholly adequate as a substitute for other measures and are usually best treated as complementary to them. Subjective assessments should not be employed in default of anything better, but should be one of a group of mutually compatible measures that collectively are most apposite to meet the objectives. Among their most popular applications are in subjective workload assessments, employing tools such as the NASA Task Load Index or the Subjective Workload Assessment Technique (SWAT) (Hendy *et al.*, 1993), the Instantaneous Subjective Assessment (ISA) (Evans, 1994), or more empirical improvisations. The strengths and weakness of subjective workload assessments (Tsang and Vidulich, 1994) and their psychometric properties (Nygren, 1991) must be understood if they are to be applied and interpreted correctly.

All subjective measurements are not equivalent. Each is capable of yielding different kinds of information, and the particular subjective assessment should be chosen carefully in relation to the questions to be answered. One possible categorization of subjective measures is the following:

1. Direct assessments by individuals using pre-structured measures. Examples are questionnaires, rating scales, and checklists. It is possible to construe many standard psychological tests as subjective assessments of this kind, but they are treated as a separate topic here.
2. Direct assessments by individuals using unstructured measures. Examples are narratives, verbal protocols, case histories, and review and replay techniques.
3. Direct assessments by individuals or by groups using unstructured measures. Examples are diaries and log books.
4. Direct assessments based on dialogues or interchanges using structured or unstructured measures. Examples are debriefings, discussions and interviews.
5. Indirect assessments using structured or unstructured measures. Whereas direct assessments are made by the individual concerned, indirect assessments refer to subjective assessments of individuals by others. Examples are observations, commentaries, expert opinions, peer ratings, and over-the-shoulder assessments.

The comparative strengths and weaknesses of each of these subjective assessment techniques have been reviewed extensively in relation to their applications to air traffic control (Hopkin, 1982c). In any given investigation or for any given purpose it is seldom sufficient to rely on only one subjective technique. Equally, it is never necessary to use them all. General characteristics of subjective assessments are discussed below.

The first point to emphasize is their very subjectivity. They reflect beliefs, opinions, attitudes and impressions. They are at the mercy of bias and prejudice, distortions of individual emphasis, and the fallibility of human memory. They are not right or wrong. If they are in agreement with other measures, this may support and help to validate both the subjective and the other measures. If they are in disagreement with other measures, this does not justify discarding either the subjective measures or the other measures, and need not imply that one or other is wrong (Muckler and Seven, 1992). Subjective assessments are often most valuable when they seem at odds with data obtained by other means. A frequent example has been the belief of controllers, measured subjectively, that the colour coding of essential information displayed to them has improved their performance, whereas no improvement in performance, measured objectively, has been shown. This usually means that both kinds of data are not based on the same evidence. The controllers' belief that their performance has been improved is genuine as a belief. Perhaps the colour coding has made the tasks seem easier, or has made improvements that objective measures do not cover, or has changed an intervening variable such as motivation, or has changed tasks or task requirements in unidentified ways, or has removed some former recognized sources of error or difficulty without introducing any new ones that have become obvious. People act on their beliefs, even when they are mistaken. Subjective measures may therefore explain controllers' actions.

Subjective assessments can be an effective means to extrapolate beyond the evidence gathered, especially in evaluations. By drawing on the professional expertise of controllers, they can gather informed opinions on circumstances which are not covered by other measures but might nevertheless pose difficulties, be potential sources of weakness or vulnerability, or hazard safety. Such evidence can be speculative, of uncertain validity and sometimes wrong, but it is almost invariably better than the alternative of no evidence at all. It can generate hypotheses for testing, and alert planners to critical combinations of circumstances. If the design of the evaluation and the choice of other measures take proper account of the supplementary subjective assessments, the measured conditions in the evaluation can be spread so that the gaps to be filled in by subjective evidence are not too large and are evenly distributed within the independent variables.

Evidence from subjective assessments made by many individuals is often compiled into representations of controller opinion which denote that controllers were consulted. Some subjective measures, though not all, are capable of revealing issues on which controller opinion is unanimous and where it is split, and the strength of the opinions held. It can be helpful to know the issues on which there is a consensus of views, particularly on broad issues such as whether a workspace is a satisfactory air traffic control environment. However, some of the most important evidence from subjective assessments can be that from single individuals. Although all the participants in an evaluation may have experienced the same traffic samples, or all the controllers in a centre have been working the same traffic, they have been doing different jobs and therefore have different perspectives. If only one or two controllers at particular workstations mention certain influences or flaws or errors in their subjective comments, these should not be ignored or swamped by a compiled consensus. The subjective judgements of the controllers in a centre or during an evaluation do not rely on identical evidence but have different bases, and even when they have the same bases it is often impossible to prove this. Whereas the amalgamation of subjective assessments given individually is always to some extent suspect as it rests on assumptions which may be ill-founded, individual remarks represent genuine beliefs and should be treated on their merits. On the one hand, subjective data suffer from all the well-known sources of bias in evidence and recall. On the other hand, people act on their beliefs and subjective assessments tap the basis of their actions. It is inconsistent to dismiss the subjective views of an individual controller in one context such as an evaluation because they are fallible, but to accept them in another context such as confidential incident reporting because they deserve serious consideration and are presumed to possess some validity. In relation to air traffic control and aviation safety, no source of potentially relevant information can ever be discarded on principle, just as no source is axiomatically always valid.

If subjective assessments are gathered, there is an obligation to treat them seriously. The process of gathering them conveys the implication to the person whose opinions are sought that the subjective evidence gathered will be heeded and can have some influence on the issues covered. It is a mistake to obtain subjective assessments on issues which are closed and cannot be changed. At best it is misleading to mention the issue and thereby raise false hopes for change, and at worst it is highly counterproductive, for the lack of subsequent activities in response to the views expressed gives the impression that they were ignored because they were disparaged or unpopular. A disadvantage of subjective assessments is that they can imply more options than there are. They should never be used lightly. They are often an irreplaceable and indispensable source of valuable evidence, but seldom sufficient alone.

18.10 Social factors

In almost every air traffic control workspace, several controllers are working together. The extent to which they function as teams varies considerably. Computer assistance can change team functions incidentally, and also some supervisory functions and forms of assistance. Organizational and management policies affect the relationships between management and controllers. The professional bodies that represent controllers' interests assume new roles in response to changed political and economic climates and revised forms of international collaboration and coordination. Measures of social factors apply to interactions between the controllers in a workspace, between controllers and others in the same work environment, and between controllers and management. They refer to aspects of the controller and the controller's job that can be mediated by social influences, such as norms and standards, motivation, conformity, and acceptance by colleagues. They cover any disparities between the nominal and actual social structures.

Social measures of air traffic control functioning are common, but their purpose has seldom been to study social factors. For example, the recording and analysis of air traffic control speech is standard in simulations and evaluations, in operational research and analysis dealing with the capacity of communications channels, and in the investigation of incidents and accidents. Speech recordings contribute to the retrospective investigation of incidents, and to the legal and financial records of the flight.

Social measures related to team functioning include the occupancy times of speech channels, the information content of speech, classifications of speech by equipment (face-to-face, telephone, headset, etc.), classifications of speech by the speaker and the listener, classifications by its form (question, information, confirmation, instruction, etc.), and classification by content, sequences, degree of standardization, and other dimensions. Speech is supplemented or replaced by indications, gestures, head and eye movements, and other means of conveying information between controllers within the workspace. For example, the workspace layout may allow others to see from the flight progress strips in the pending bay that a controller is about to be particularly busy. Computer assistance that removes this information source may result in signals, speech or other indications that the controller is too busy to be interrupted.

Many orthodox measures of social factors, such as the above classifications of speech, can be adapted to air traffic control. Relevant movements within the workspace, such as those of a supervisor, can be recorded and related to other actions and spoken messages, to describe the extent and nature of the supervision and its relationship to the functioning of the controllers as a supervised team. The pattern of the supervisor's physical movements, including their durations and pauses between them can trace the supervisor's social interactions. Records of the body movements, head movements and eye movements of individual controllers, combined with information on speech and gestures, establish common patterns of interaction among controllers, and can, if necessary, be integrated with many of the other measures considered in this chapter. Much is possible, but lavish resources are required for social measures since the accumulation of sufficient social data to permit statistical analyses is time-consuming, and it is especially important with social data to know beforehand the objectives and to confine the data gathering to satisfy their specific needs.

The feasibility of performing air traffic control tasks by a team of controllers can be restricted by the physical layout of equipment. Displays can only be shared effectively if all can see them, and input devices if all can reach them and if more than one controller never need to use the same input device for different purposes at the same time. Although this is obvious, it can be difficult to achieve by task planning and scheduling, and by the coordination of the tasks and responsibilities of different controllers. The extent to which controllers actually function as a team can be assessed in several ways. One is to look at the degree of flexibility, and particularly at circumstances when a controller helps or is helped by others, or does some of the work for others with their tacit or overt agreement. Another concerns the extent of reliance on colleagues, compared with active verification of colleagues' activities. This can be partly a matter of policy and of local customs. Verification can obviously be a safeguard, but constant verification denotes some mistrust and constitutes major additional workload for the controller who checks.

One measure of team functioning concerns the extent to which the team functions have precedence over individual ones. Where air traffic control allows controllers to support a colleague who is inexperienced or facing particular but temporary difficulties, traffic can be handed to that colleague with few problems in it, and colleagues and supervisors may add effective support and encouragement in other ways. The curtailment of the opportunities for such help with the planned increase in computer assistance lends more urgency to

the expression of these consequences in terms of social measures. Other orthodox team measures include the patterns of conversation and consultation in
problem solving and decision making, and the balance between individuals
and teams in fulfilling these functions. A sign of team cohesiveness is the adoption of common practices and standards in choosing which matters are drawn
to the attention of colleagues and which are not. Each member of an integrated team may not watch colleagues much directly but may rely heavily on
tacit understanding and glean the information indirectly from what is perceived to have happened. Controllers may address numerous enquiries to a
colleague whose behaviour does not conform closely with team norms and
standards.

The examination of how ideas and practices are formed and disseminated
within a workspace or team can indicate the processes by which norms and
standards evolve and the means by which team members form views on the
professionalism of others within the team. Those with most influence on the
standards and practices, and the ways in which they exert their influence, can
be discovered from such measures. The true influences and the formal ones
may not correspond. One index of the strength and cohesiveness of a team can
be obtained by studying the mechanisms by which a newly joined member of
the group comes to agree with the group consensus and to support it publicly,
and the pace at which this occurs.

Social factors can be measured through beliefs, attitudes and opinions, for
which standard measurement techniques have evolved. The levels at which the
strongest group pressures are applied can be identified. These can be at
national level originating within professional groups or organizations, at management level, at the level of the facility or workspace and set by the local
management or with the consensus of all peers, or at the level of a watch or
smaller team within the workspace. Social measures of the nature and magnitudes of differences between teams, watches, facilities or nations can establish
the origins of the strongest social pressures, and can therefore locate the main
influences for change. Data about individuals are also needed to ascertain the
respective influence of team and individual differences, the conformity of individual and team opinions and attitudes, and the extent to which official representatives, supervisors or line managers actually voice the consensus and
diversity of the views held by those they represent. The magnitude of differences in the views of various groups and teams provides a perspective for
assessing the significance of the magnitude of the differences between individual controllers. Social factors cover the defence of the air traffic control profession and its practices when they are challenged. Individual controllers may
not identify strongly with an opinion widely held in their profession until it is
challenged from outside, whereupon they feel obliged to defend it even if they
do not fully share it.

The importance of social influences on air traffic control has recently gained
increased acknowledgement. One aspect of this relates to training. There has
always been some team training of controllers in addition to individual train-

ing, but team training requirements have expanded, though not yet to the extent of team training for aircrew. Some aspects of the training of teams are specific to teams and are not merely the sum of the training of individual team members. The effectiveness of the team as a whole requires extensive training of teams as a whole.

Individual differences can have a profound influence on the social characteristics of teams. Some controllers habitually complain over trivia whereas others never complain at all. Some seek supplementary information or liaise with colleagues on flimsy pretexts whereas others only do so in rare and dire circumstances. Team members recognize and accept these and other individual differences and accommodate them. Indications of stress or overloading by a controller who customarily reacts in this way when lightly overloaded do not produce the same responses from colleagues as similar signs by another controller who very rarely seems overburdened. Team membership is characterized by selective adjustment by team members to each controller's habits, strengths and weaknesses as an individual, and this selectivity strengthens the team as a cohesive unit. The mainsprings for this cohesiveness and team solidarity and loyalty may be weakened by any forms of computer assistance that reduce the amount or observability of the evidence on which the selective judgements about each team member relies. It is therefore important to ascertain what is gained from team functioning in efficiency, in safety, in job satisfaction, in motivation, in identification with the profession and in any other important ways, so that these gains are not lost inadvertently through changes made for other reasons.

18.11 Biographical data

Biographical measures are routinely gathered for most purposes, and the employer's record of the controller is based on biographical information, starting with the data in the initial application to become a controller. The familar curriculum vitae consists mostly of biographical data, such as name, address, age, gender and education. A factual record of progress as a controller is built up, beginning with all the test scores and assessments from the selection procedures, continuing with a comparable record of scores and assessments during and at the end of training, supplemented by additional data of similar kinds from any retraining, and completed by recorded experience and career progress to date as a controller. The controller's safety record in terms of incidents that had some official mention will be included. Any particular honours, achievements and accomplishments will also be noted, and other professional activities such as representing colleagues on professional matters. Medical history, discussed separately below, is a further example of biographical data. In comparison with other measures, biographical data have the merits of being quick and easy to gather, and are generally reliable.

The above kinds of biographical data are generally unexceptionable, but two further kinds of biographical data can be more controversial. Individual employment records are likely to include some comments, judgements and predictions about the individual that are not strictly factual but single out controllers with the best promotion prospects or controllers who are being considered for retraining, for example. Concerns about safety that have no official substantiation may be expressed obliquely. These are matters that the individual may not agree with, which is why they may be controversial. Another kind of biographical data may be controversial because the individual interprets these data categories as personal and private so that they should not be the legitimate concern either of an employer or of a profession. The categories of biographical data construed in this way seem to be increasing quite rapidly. The facts may not be disputed, only their pertinence. Examples of sensitive data can include ethnic group, marital status and history, family responsibilities and obligations, previous criminal convictions, substance abuses, and financial status and solvency. There are some delicate ethical issues on where to draw the line between the rights to privacy and confidentiality, and the ability of line managers and colleagues to provide practical support to a controller experiencing serious difficulties outside work which are affecting the controller's performance.

Selection procedures depend heavily on biographical data, and the initial paper sift of candidates may rely entirely on them. It often includes expressions of previous interest in and knowledge of air traffic control or aviation, though positive replies to such questions may reflect motivation rather than aptitude to become a controller. Biographical data are the ones most liable to become political tools, if they show that identifiable categories such as women or ethnic minorities or the disabled are under- or over-represented among controllers. Where air traffic control is in the public sector and the policy is to distribute public sector jobs equitably among all identified population categories, air traffic control must implement this policy unless it could hazard safety to do so. Sensitive data may have to be gathered to prove conformity with such policies. Targeted recruitment drives may be considered, to resolve the problem that some groups within the eligible population never seem to consider air traffic control as a career.

Biographical data collectively also fulfil an important and very different function. When the biographical data from all controllers are added together they describe air traffic controllers as a demographic population. Biographical data are therefore applied to identify demographic trends within the controller population. The applicants for jobs as controllers can be compared with the population from which they are drawn, to discover which categories are over- and under-represented, and to formulate policies accordingly. As a population, controllers can be compared with any other population for which equivalent data have been gathered. For example, the personality characteristics of controllers and pilots can be compared if there are data for both of them from the same personality tests. Data on the age distribution of controllers can be applied to discover whether they are ageing as a population, to predict future recruitment needs for controllers, to formulate policies on retirement, and for many other purposes. The controllers at a particular location, or the controllers who participate in an evaluation, or any other identified subgroup of controllers can be compared with the whole population of controllers on the basis of their biographical data to ascertain how representative of controllers as a whole the particular subgroup is. Alternatively, a group of controllers may be selected as a representative sample of all their peers by proving that the means and distributions of the sample and of the whole population are the same according to all the biographical measures on which data are available.

18.12 Health

Controllers have to pass a medical examination as part of their selection procedure. This covers medical history, various medical conditions, standards of fitness, and tests of hearing, eyesight, colour vision, and other measurable attributes with minimum requirements for acceptance as controllers. Controllers have annual medical examinations to test their continued fitness and freedom from serious illnesses. Their licence as controllers depends on passing these medicals. As a population, controllers receive more regular medical checks than most of the general population. It might be presumed that as a population they should therefore be fitter than average, and often they are. However, air traffic control has acquired a reputation, usually misplaced, as a source of certain occupational health hazards. The most probable human factors influences on any occupational health hazards are poor workstation design leading to problems of posture, and poor display and lighting designs leading to visual difficulties. Visual corrections tailored to the precise visual conditions of the work environment have to be prescribed for a few controllers. As more air traffic control workspaces benefit from the application of ergonomic principles during their design and specification, occupational health hazards originating in aspects of the workspace design are becoming less common. From time to time there are particular anxieties about possible health hazards in air traffic control environments. A recent example has been radiation from displays, although the enforced standards are stringent and most emissions are so low that it can be difficult to find equipment that can measure them reliably.

Air traffic control has been blamed for stress-related illnesses. Most research and medical evidence have failed to support this view. It is not that stress-related illnesses are unknown in air traffic control, merely that it has no more than its fair share of them. However, interest in the topic and the association in the public mind of air traffic control with stress persist. The situation is actually quite complicated. In most nations, air traffic controllers do not have a significantly higher incidence of stress-related illnesses than matched populations from other professions. Stress seems more likely to be the result of modern lifestyles than of air traffic control. Causality has been ascribed too hastily to statistical correlations. There are two broad origins of stress. One origin could be in the air traffic control itself, from time pressures, task demands, working hours, inadequate equipment or insufficient training, and the alleviation of such stress must also lie within air traffic control through changes in the above factors. The other origin is in the individual, from personal, domestic, financial or other problems unconnected with air traffic control, and such stress must be treated as an individual's problem in order to alleviate it. Even this distinction is an over-simplification. Many air traffic controllers work shifts, which can disrupt normal social and family life, and are associated with other stress-related factors ranging from higher divorce rates to difficulties in sleeping during the day. In a sense these are consequences of being an air traffic controller but are not caused directly by the air traffic control.

Air traffic control is not characterized by exceptional occupational health problems. The morbidity and mortality rates of controllers tend to be normal in that they are similar to those of comparable professions. It could be argued that the life of an air traffic controller is not inherently very healthy. Shifts disrupt sleep patterns and lead to erratic meal times. An essentially sedentary occupation carries high responsibilities. Positive physical fitness is often encouraged and may be beneficial. Many controllers who should have a regular regime of physical fitness only embark on it a few weeks before their annual medical is due when the motivation to become fitter and reduce weight becomes urgent enough to demand action. This typifies the health problems of modern lifestyles which are shared by air traffic controllers and many other professional people. To attribute them immediately and uncritically to air traffic control may be to seek inappropriate remedies for them.

18.13 Standardized tests

Standardized psychological tests provide an objective and impartially scored measure of actual or potential individual human aptitudes or abilities. Test scores rarely mean much in absolute terms but can be very meaningful when interpreted in relation to the appropriate population norms. These norms constitute the most important feature of standardized tests.

Tests cover a large variety of human characteristics. Some can be quite general, such as tests of intelligence. Some relate to skills or to the potential to acquire skills. Some are measures of particular attributes, such as manual dexterity or aspects of memory. Some measure personality, either in general terms or in terms of a few specified dimensions. Some attempt to express social or cognitive human attributes quantitatively. Test scores may be interpreted in relation to norms for a national population or for a more narrowly defined population such as controllers, or in relation to any other dimensions for which test norms have been derived.

Most tests have a theoretical basis. What they are intended to measure is defined first. The test items are then devised, tested and validated, to show that each item contributes towards the valid and stable measurement of whatever the test purports to measure. The test is then applied to a large population to derive norms for it, and often the results from applying the test are fed back to update and refine the norms for interpreting it. The rationale for applying standardized tests to air traffic control therefore rests on a demonstrable or putative connection between the dimension which the test claims to measure and its relevance to some aspect of air traffic control. The establishment of the relevance of the dimension to air traffic control depends on deductions from task analyses of controller functions, the choice of a test that really does measure the deduced dimension, and its validation by demonstrating a relationship between test scores and some aspect of air traffic control performance. Several air traffic control tests devised on a more speculative and intuitive basis have been disappointing as measures. This applies particularly to tests which have high face validity because the test items appear to represent simplified air traffic control problems, but which often predict subsequent ability as a controller less well than expected.

Standardized tests have many air traffic control applications. These include selection, training, allocation, evaluation, retraining, career development and guidance, individual counselling, clinical diagnosis of health and well-being, redeployment and preparations for retirement. Tests may be used as a diagnostic tool to tackle problems experienced by an individual controller, or as an impartial objective measure of a dimension as part of the explanation for an occurrence. The scores from standardized tests can be accepted as independent and impartial provided that they have been administered correctly. Applied on the completion of training, for example, they can be seen to be independent of those who supply the training, of the training location, and of others who could be accused of divided loyalties if they became directly involved. Standardized tests can also be of factual knowledge to check progress or of performance to measure skill and competence.

Some precautions should be taken when tests are used. There may be no test norms for air traffic controllers, and before the test scores can be interpreted correctly it may be necessary to check that controllers as a population comply with the norms available, and to derive norms for them if they do not. Test scores are culturally dependent; despite strenuous efforts to adapt standard tests across cultures, only the most frequently used tests have separate norms for different countries. These are sufficient to prove the need for separate norms for different countries and to discourage cross-cultural test applications.

The theory behind tests is that they tap innate abilities and are therefore less susceptible than most measures to external influences. They give a direct comparative quantified measure of individuals. Unfortunately, theory does not accord fully with practical experience. People who are allowed to practise the same or similar tests indefinitely continue to improve their test scores. Coaching and practice on tests which are similar to but not the same as the standardized tests pose difficult ethical and human factors issues. This has been an issue in educational psychology for many years where there has been much research on the susceptibility of various tests to influence by coaching and practice, but the issue has now spread to occupational and industrial psychology in general and to air traffic control. In the United States in particular, courses have recently proliferated which offer coaching and practice to prospective air traffic control candidates on material similar to that used in the selection procedure. There may be social implications, such as that those most able to afford a course are most likely to succeed, but the potential implications for the validity of the selection procedures and for longer-term training and safety are the main human factors concern. Repeated applications for acceptance for controller training also lead to difficulties if test scores improve with practice, and it may be more valid and fair to prefer the test scores from the initial application to the improved scores from subsequent applications after unknown coaching and practice. These are current issues regarding standardized tests in air traffic control.

18.14 Modelling and related techniques

The concept of modelling has now acquired a very large range of connotations when applied to human factors in air traffic control. A model can be a cognitive structure, a series of mathematical or pseudo-mathematical formulae or equations, a paradigm, a surrogate, a miniature, a replica, or a description. One kind of model recreates incidents in air traffic control (Rodgers and Duke, 1993, 1994). Models fulfil many functions not directly related to human factors, such as determining the effects on airspace capacity of the pre-flight planning of whole flight trajectories (Ratcliffe, 1994). The primary concern with models and related techniques in the present context of human factors measurement deals with them as human factors measurement tools rather than as techniques for the broader description or explanation of air traffic control functioning. There is a history of attempts to model workload, for example (Robertson *et al.*, 1979).

Human factors as a discipline has shown great persistence in trying to build models to describe, represent, predict or explain humans as system components, interactions between the human and the system, the human consequences of system changes, and the consequences for the system of human changes. The earliest attempts at modelling were encouraging. The behaviour of an experienced human operator performing a task such as radar tracking could be represented well in mathematical or engineering terms, which could be incorporated into descriptive models of the wider system and thus include aspects of human behaviour in them. However, the greater the task complexity, the more difficult it was to model it mathematically or in other ways. The practical difficulties of representing problem solving, decision making, prediction, innovation, adaptibility, flexibility and other human functions adequately in models of the air traffic control system have proved daunting, and the resultant models have not succeeded in capturing a sufficient proportion of human variance for most modelling purposes.

Lind (1991) treats models of human information-processing tasks as resources for achieving control over the system, drawing on Rasmussen's (1986) models of human decision making. Kirlik (1993) uses models of strategic human behaviour to suggest that computer assistance should be employed selectively. The commonest approach to modelling in relation to human factors aspects of air traffic control has defined and categorized human functions by task and job analysis, sought formulations that cover the identified functions most adequately, refined them according to context and required level of detail, and attempted to represent the human as a system component. In practice, even quite complex formulae have been unable to represent the functionality of all or most experienced controllers reasonably well under most circumstances. Controllers do not function like other components and any descriptive formulae for them do not integrate well with those of other components. The level of complexity and integrality of air traffic control cannot be captured adequately by formulae based solely on tasks, as the interactions and combinations of tasks have too crucial an influence on human and system functions. Any method which partitions human functionality must be supplemented by means to re-integrate the separated functions.

Descriptions of system functions can be derived from operational research and analysis, fast-time simulation, control theories and other techniques that seek to represent human activity and variability as the functioning of system components, but descriptions which best represent the human are of different kinds, affected by different variables, conditions and influences. Whereas the model of the system deals with factors such as traffic capacity, permitted separation standards, and changes in route patterns or structures, the model of the human deals with such factors as whether a colleague is trusted, whether the manner and content of a pilot's speech suggests uncertainty over procedures, and whether a supervisor practises interventionist or remote supervision. Some attempts to integrate models of the human and the system have foundered because human factors and other disciplines do not gather the same data, do not classify data in the same categories, do not have the same objectives, and do not have sufficient common ground for mutual adaptation. Specialists in human factors and in operational analysis each find that their own primary concerns are of little interest or relevance to the other. To resolve this impasse, it is probably necessary to agree common linking measures across several disciplines wherever the objective is to integrate evidence from several disciplines into a single model.

There have been many instances of successful interdisciplinary collaboration in the resolution of human factors problems, but model building is not among them. Yet success could have a big payoff. If models could be built, their predictions of human performance could be compared with actual performance. Measures could thus validate models, and models could prescribe what should be measured. Combined models of human behaviour and system functioning could predict the functioning of the whole system in circumstances that are not or cannot be measured. These models could identify where there is greatest uncertainty which should be resolved by appropriate real-time or fast-time simulation or evaluation. Discrepancies resolved in this way could tune the model to become a more powerful predictor. Sadly, although this integration of measures seems to offer so much, hitherto it has delivered relatively little.

18.15 Organizational measures

The mention of organizational measures at all in connection with human factors in air traffic control is quite a recent development. Organizational concerns are now a major part of human factors, with many applications (Bradley and Hendrick, 1994). Organizational measures cover the organizational structures that influence human factors aspects of air traffic control. Measures of this kind differ in their purposes, their terms of reference, the kinds of function that they do exercise, and the kinds of function that they can exercise.

At the highest organizational level are international bodies with terms of reference related to air traffic control. Insofar as they deal with standards and regulations, professional air traffic control practices and working conditions, legal aspects of the controllers' tasks and responsibilities, and the commonality of facilities and instructions, they have an influence on measurement in air traffic control and on its prescription and standardization. Examples are the International Civil Aviation Organization and the International Federation of Air Traffic Controllers' Associations. Various national authorities support, fund, seek to influence, are represented on, and determine the policies of other bodies at national level, which must work collaboratively with the international bodies. Nations have different requirements, policies, interests, and management in regard to air traffic control, and reaching international agreement tends to be slow, but where new navigation aids such as those based on satellites, or new forms of transponded data such as datalinks, or new formats or usages of air traffic control language, are proposed, international coordination and agreement are essential and must include agreed measures.

Human factors can be applied to study the functioning and decision-making processes of organizations to check whether their human factors responsibilities can be fulfilled within their current structures (Westrum, 1991). Organizations are often criticized for not making decisions which they are incapable of making, either because the decisions are beyond their terms of reference or because there are no mechanisms within the organization through which they could be made. A common effect of organizational structure on air traffic control human factors concerns integration. It is quite common for those concerned with planning systems, with selection, with training, with task and workspace design, with incident investigation, with conditions of employment, and with personnel matters to work independently within the same organization so that no mechanism exists for tackling collaboratively a human factors issue that violates these organizational compartments. Attempts to achieve collaboration are interpreted defensively by others, as interference, poaching, or blatant empire-building on behalf of human factors. Yet no solution confined to any of the above domains can wholly succeed, and some human factors issues, such as the optimum allocation of the limited human factors resources available within air traffic control, cannot even be tackled in the absence of an organizational structure that permits all the possible applications to be appraised before the allocation of resources is made. Many organizational structures rule this out. Often human factors resources are under-utilized because there is no mechanism within an organization for putting questioners in touch with those most able to answer their questions.

Other measures of organizations, in addition to those that describe their structure, deal with their rigidity, their delegation of responsibility, their flexibility in interpreting their terms of reference, the promulgation of ideas within them, their response to originality, their methods and speed of implementing actions, their forms of internal communication, the exercise of leadership, and the extent and mechanisms for providing feedback and knowledge of results. Further measures cover conformity between nominal and actual hierarchies of responsibility, attitudes towards competitors, policies on training and career management, conditions of employment, criteria by which internal status is judged, the organization's public and private image, the loyalty of the workforce, and its degree of identification with the organization. There are also measures of the extent to which organizations try to satisfy employees' needs at work and succeed in doing so.

A large and extensive literature exists on classifications of organizations, and on their functioning, their influences and their measurement (Golembiewski, 1993). Much of this is applicable to air traffic control organizations but has not been applied to them because the acknowledgement of the importance of organizational factors in air traffic control is still so recent.

18.16 Verification and validation

Interest in verification and validation has recently revived, with the publication of the proceedings of a NATO Advanced Study Institute on the topic that drew most of its examples from air traffic control (Wise *et al.*, 1993), and a further volume of papers (Wise *et al.*, 1994). Originally it was taken for granted that human factors findings required validation (Hopkin, 1993b). The preferred demonstration of validity was compatibility between the obtained findings and a more general psychological theory of human behaviour. Although the desirability of verification and validation was not disputed, they received lower priority in the allocation of funding and human factors resources, until they became exceptional rather than routine. Many human factors recommendations relied on data from handbooks and similar sources which had not always been adequately validated. In selection procedures and in the construction of psychological tests, the practical usefulness of the products continued to depend on their proven validity, so that validation remained intrinsic to their development (Altink, 1991). In the development of software the need for its validation and independent approval before its installation in safety-critical systems resulted in widespread attempts at validation, even in early systems (Hausen, 1984). However, these two examples illustrate the lack of validation of other human factors system aspects and the disparities among the validation practices that remain.

Of immediate concern is that none of the disparate validation techniques that have evolved for particular aspects of systems is applicable to entire systems that have become complex and highly integrated, because none can deal with all the possible interactions, options and scenarios. Formal logic may be applied to the verification of hardware (Yoeli, 1991), and software tools to the validation of simulation (Knepell and Arangno, 1993). This raises the issues of the need for validation and its possible forms. Some require specialist knowledge not generally available. Others require resources which are scarce and cannot be spared. All require funding which is usually directed towards other objectives. Many validation attempts employ criteria independent of and external to the system. Examples of such criteria have included theories, comparable evidence from other applications, and the findings from incident investigations (Baker, 1993). Their essential characteristic was their independence of the system which they were used to validate. This characteristic fails with complex systems since nothing of comparable complexity can serve as an independent criterion. Questions that arise are how the validation of complex systems could be achieved, and whether validation as a process could still be applied to a system externally or must be internalized to become a series of self-validating processes that are intrinsic to the system design and to its iterative evolution (Bangen, 1993). Techniques being developed seem to render the latter option more practical. Stager (1993) has reviewed many of the human factors issues that arise in relation to validation.

When validation studies are conducted, it is quite common for them to reveal that validity is quite low and that the evidence being applied is not very trustworthy. While it is vital to discover this, those who fund validation studies may not construe such results as good value for money since their funding seems to have reduced rather than increased knowledge (Jorna, 1993). The reasons for low validity are numerous. The data on which the recommendations were originally based may be far weaker than subsequent impressions suggest. The conditions attached to the original data and recommendations may have been insufficiently publicized and may not be met, so that the findings do not transfer. The original theoretical interpretation of the findings may subsequently have been discredited. The validation may have relied on the generalizability of the evidence, but the original findings may have been too context-specific to generalize. Other possible reasons include deficiencies in the original studies or misinterpretations of the data.

It is important to validate evidence, and essential to publicize evidence, indicating whether or not it has been validated. Fully validated recommendations should always be preferred over unvalidated ones. The degree of validation should be an influential factor wherever there is a choice, for example between competitive tenders, between system designs, or between alternative equipments. Validated, proven, established, generalizable recommendations are always preferable to those that lack these qualities. The problem is that sources of evidence far too rarely provide sufficient evidence to judge their validity. This even applies to some of the most reputable sources. The human factors specialist is advised, whenever it is possible, to trace the origins of evidence in order to establish its validity before basing human factors recommendations upon it.

18.17 Interactions between measures

Hardly any human factors questions in air traffic control can be answered completely with only a single measure. Several measures are normal. The need to interpret the data from the chosen measures in relation to each other should guide the initial choice of measures. As a set, they must cover all aspects of the human factors issues to be resolved. The exclusion of an essential measure for any reason renders the findings incomplete, inadequate, misleading or wrong. If there are too many measures, some of the data will be duplicative or redundant and more resources than necessary will be expended on analysing and interpreting them. Therefore crucial human factors judgements concern the choice, number, mix and compatibility of measures, so that the data from each measure and the interactions between measures can be interpreted as a coherent set.

Some measures are intrusive, such as those that sample body fluids. Some measures interfere too much with what they purport to measure. Any measure that seems novel can wreck a study of boredom, for example. Some measures are too similar to the task itself: verbal protocols or running commentaries by a controller interfere with the speech that is a strong feature of air traffic control. Some measures which have fared quite well in other contexts are inapplicable to air traffic control. Secondary tasks were unsatisfactory years ago (Hopkin, 1982a), before the notion of spare mental capacity with its implication of a fixed quantifiable mental capacity was acknowledged to be an oversimplification. Any extra task introduced for assessment purposes interferes with what is being measured, and should be eschewed. In air traffic control it is essential to maintain awareness of the primary task all the time in order to know when to return to it.

Because the nature of human factors is interactive, some measures must be interactive to capture it. Because there is a multiplicity of objectives, the set of measures must capture this multiplicity. Any disagreement between measures need not betoken faulty measures since the disagreement may be genuine. Some contradictions between measures are inevitable. It is important to agree beforehand how measures will be integrated and reconciled, rather than to attempt to reconcile the irreconcilable after the event, which can look suspiciously like attempting to fabricate the data. The correct interpretation of data must explain all the findings and any anomalies, and not only those that support the favoured hypothesis.

Measures of air traffic controllers may be chosen empirically because they accord with the expertise or predilections of the investigator or can be accommodated within the available resources, or for similar practical reasons. While this is sensible and perhaps unavoidable, it does not guarantee that the measures are adequate. This requires careful appraisal of the objectives and of the evidence required to meet them. Alternative methods of gathering the evidence must be weighed, and a set of measures selected that can gather all the required evidence with a minimum of redundancy. The methods of interpretation of the data are agreed in advance so that sufficient data on each measure will be gathered for that kind of interpretation, including the whole range of possible interactions with other measures that need to be covered. Measures are still chosen too often for empirical reasons and not because they belong to an integrated comprehensive set of measures. The measures of the human controller have to be integrated with the measures of other aspects of the whole functioning system, and the set of human measures has to be capable of dealing with all the human factors that are relevant to the issue being addressed, and not just some of them.

The nature of human factors requires multiple measures even when the questions seem simple. There will always be interactions between measures which must be discovered and accounted for in the interpretation and explanation of findings. Endsley (1994b), in considering the measurement of situation awareness, has to deal with many of the kinds of measure mentioned in this chapter, and a recent comprehensive study (Hilton Systems, 1994) of the human factors issues relevant to the retention or abandonment of 60 as the mandatory retirement age for pilots also includes many kinds of human factors measure and evidence. Such studies typify the appropriate human factors approach to measurement. Hopkin's (1980b) review of human factors measures in air traffic control is still generally applicable, despite some increase in the number, range and sophistication of measures in the interim.

19

Research and developoment

19.1 The role of human factors

The funding and resources for human factors contributions to air traffic control research and development have never seemed sufficient. Many human factors problems that could have been anticipated and resolved more optimally beforehand with better resources have not been recognized in time, and have had to be resolved retrospectively with less satisfactory but more costly palliatives. There is nevertheless an argument that the limited resources for air traffic control work have not been deployed to best advantage, because too many resources have been expended on research and development to obtain new evidence and too few expended on the direct application of existing human factors evidence to resolve real problems in evolving or operational systems, where the payoffs would often have been more immediate, tangible and significant. Research findings have consistently been under-utilized because they have lacked appropriate mechanisms for applying them (Heller, 1991). Human factors specialists in air traffic control should perhaps concentrate more on solving current urgent human factors problems in real air traffic control workspaces. This is not to decry research, which is absolutely indispensable and must continue, but it is to deplore the chronic under-application of what is already known and attempt to redress it by evolving better mechanisms for incorporating existing human factors knowledge into air traffic control systems. It is disheartening to see elementary mistakes being made, to possess the knowledge to prevent them, yet to lack any means to apply that knowledge.

Most research and development in air traffic control is system oriented. Questions about equipment, aids, facilities, procedures and system functioning are posed in terms of system issues such as safety, capacity and operational practicality. Answers in similar terms are expected. Recently there has been increased recognition of the relevance of human factors to these system issues, and of the need to consider human factors implications at all research and development stages. This implies that human factors is normally one of several disciplines contributing collaboratively to air traffic control research and development programmes in which all the essential requirements of each discipline have to be satisfied. Wherever requirements conflict, compromises must be devised because to leave incompatibilities unreconciled is not a practical option. However, some of the most crucial influences are almost exclusively human factors ones. The human factors specialist must therefore know or be able to find out the optimum recommended human factors solution of every issue, and to know how far and in what ways it may be compromised.

The purpose of research is to acquire evidence that is not already known. Such an obvious statement would be trite, were it not forgotten so often. Research is not to confirm choices already made or decisions already taken, nor to justify dropping administratively inconvenient options. Properly conducted research may yield such results serendipitously, but its main purpose is to gather new facts and knowledge. Inevitably, some findings will be unexpected and even unwelcome, but they must still be applied. It is a waste of money and resources to conduct valid research and then ignore the findings because they are inconvenient. Such findings are often the most beneficial ones, because they identify problems when they can still be resolved. The most difficult human factors findings to deal with are inconclusive ones. They can be appraised rigorously to verify them but nevertheless be correct, either because the human factors problem has no solution or because the compared options have equivalent human factors effects. It is helpful to have a research philosophy for human factors work on air traffic control (Hancock, 1991).

A technological innovation may be considered for introduction into many air traffic control systems in different parts of the world. Its human factors problems are intrinsic to it. Without international coordination of the associated human factors research, duplicate human factors studies will probably be conducted in different nations at about the same time. Human factors resources for air traffic control are a scarce commodity everywhere. Collaboration to prevent duplication is in the interests of everyone. But it is not common. This does not imply that the optimum human factors solution of a problem is the same everywhere, for numerous local influences may intervene to invalidate universal recommendations, but all adequate solutions must share some commonality to be compatible with human cognitive capabilities and limitations and with human health and well-being. Research should be planned in programmes (Stein and Garland, 1993b), and be organized to achieve known objectives defined in advance (Furze, 1993). The magnitude of some research and development projects has recently encouraged the planning of the associated human factors work as a whole, followed by its partitioning among several agencies that can be in different nations, since the work programme would over-extend the resources of any single agency.

Although the human factors research and development specialist must know the ideal ways to conduct experiments and to contribute human factors expertise throughout the system evolution, opportunities to make ideal contributions will be rare or non-existent, and compromises are vital. It can be presumed that there will never be, in the eyes of human factors specialists, enough time, money, resources and effort for all the work that should be done. The most crucial human factors decisions are therefore about the optimum deployment of the limited resources available. Research and development work in air traffic control can be expensive, and there are pressures to maximize the productivity of specialized research and development facilities.

The initial identification of issues and the judgement of whether and how research and development should be applied to them is a vital human factors role. An example is the US National Plan for Aviation Human Factors (FAA, 1990). As the outlines and concepts of future air traffic control systems begin to evolve, usually between 10 and 20 years before the systems will become operational, and as they become progressively more detailed and specific, it is possible as a parallel exercise to begin to specify their human factors implications in greater detail. These implications having been identified, a human factors role is to advise on relevant available research findings, and to identify gaps in knowledge in time to commission appropriate human factors research to fill them. Such exercises in problem identification, if competently performed, invariably raise more human factors issues and problems than can be tackled with the resources available. The most fruitful and productive research and development issues are chosen by considering the value of their prospective findings in relation to the resources deployed to obtain them.

The human factors research and development for an air traffic control system should initially be planned broadly since the most efficient allocation of resources will depend on the merits of each issue rather than on any research category to which it may happen to belong. For example, to prejudge the proportion of human factors effort devoted to selection or to training or to workspace and human–machine interface design may be administratively neater but is not the way to obtain the best value from the whole research programme. This approach requires an informed centralized authority to steer the human factors research as a whole, and to avoid rigid prior compartmentalization of effort.

Of great importance is to distinguish from the outset between specific air traffic control issues that will be ignored unless some resources are devoted to them, and issues which recur elsewhere in aviation or in other complex human–machine systems, from which relevant human factors evidence may be gathered. Correct judgement on this can conserve scarce human factors resources and direct searches towards existing sources of evidence on which recommendations might be based without additional research and development. For example, broad human factors experience of introducing colour coding onto displays may accrue from many sources, whereas the potential value of ghosting as a form of computer assistance is specific to air traffic control. Some human factors problems in air traffic control may have military equivalents (Gal and Mangelsdorff, 1991).

Sources of relevant information include the general human factors literature, the literature of psychology, computer science, physiology and ergonomics, findings from other applied fields such as aviation, engineering, and large human–machine systems, and evidence about methods, measures, equipment, tests, tools and techniques from many contexts. Innovative research pioneering new methods may be rewarding for the researcher, but those who fund it will be less enthusiastic unless it yields practical results. The researcher should know about alternative experimental designs and methods, and statistical techniques including their relative power and validity (McGuigan, 1989), and be aware of the directions in which relevant research is evolving (Megaw, 1991). Common problems encountered in air traffic control human factors research include small numbers of subjects, confounded variables, insufficient control over variables, unknown familiarity with experimental material, equipment unserviceabilities, incomplete experimental designs, fixed and inflexible time schedules, lack of balance across compared alternatives because they are not all equally practical, and other woes to undermine the validity of findings. It is a mistake to gloss over such difficulties. Others need to know of them to judge if the findings apply to their own problems. It can be discouraging when such problems are so common as to be normal but remain unmentioned in accounts of work. The human factors specialist has to be prepared to justify research. Human factors procedures to curb unwanted variance can seem too restrictive to others until they are explained. In interdisciplinary work, others should understand the constraints on human factors recommendations, just as the human factors specialist should understand the constraints affecting colleagues in other disciplines.

A broad list of human factors contributions to air traffic control research and development would include the following:

1. The provision of basic psychological knowledge about human capabilities and limitations.
2. The application of human factors knowledge derived from air traffic control and other relevant contexts.
3. The statement and interpretation of all the human factors problems encountered in each research and development programme.
4. Ensuring that all fundamental human factors requirements are met throughout the research and development programme.
5. Recommendations on the most appropriate human factors tools, methods and techniques.
6. The appropriate measurement of individuals and of groups.
7. Statements of the human factors implications of research and development findings.
8. The interpretation of the research findings in relation to other human factors knowledge.
9. Relating the full range of the implications of human factors research and development to the air traffic control system in human factors terms.

19.2 Real-time simulation

Real-time simulation is the commonest tool for human factors research and development studies in air traffic control (Hopkin, 1978). As a technique it has been judged to be both indispensable and over-used, indispensable because some kinds of evidence can be obtained in no other way, but over-used whenever it is applied to objectives for which it is unfitted or invalid or for which better techniques are available (Hopkin and McClumpha, 1980).

The roles of real-time simulation are not the same in research and in development. Research may be more open-ended and exploratory, yet more rigid in its methods. It may seek to establish the feasibility or practicality of an option, to trace and identify all its implications, to develop measurement tools and techniques, or to quantify effects and interactions through orthodox experimental procedures. When a new technology is considered for air traffic control, research may employ quite a rudimentary air traffic control simulation to discover its potential strengths and weaknesses and to ascertain whether it is a serious option. Research may also indicate where, when and how findings could be applied. Development occurs within a putative context and application which the real-time simulation represents and to which the simulation findings refer. It is not open-ended, nor widely generalizable. It usually simulates a particular air traffic control region with a particular traffic density and mix, and often compares alternative means of realizing the same objectives.

There can be two kinds of approach to real-time simulation for human factors air traffic control research and development. The first replicates the functioning of major air traffic control regions in a purpose-built air traffic control simulation facility. As many as 20 or 30 staffed control positions may be studied concurrently, together with much of the interaction and communications between them and some simulation of pilots and of adjacent airspace regions. The purpose is to test and quantify the viability of forms of air traffic control proposed for the region simulated or for other regions of airspace which it typifies. Many system measures as well as human factors ones are normally taken, for the simulation is not exclusively or even primarily a human factors evaluation. Sometimes the purpose is exploratory, to establish feasibility and to quantify capacities. Sometimes it compares options by simulating them. Measures include system functioning, behaviour, performance, interactions, errors, channel occupancies, and information flows, the human influences on these and the human costs in achieving them in terms of workload, effort, motivation, attitudes, and team roles and relationships. Such simulations are seldom continuous, but each exercise typically lasts for an hour or two and is followed by a pause before the next one. Questions that would require continuous running, concerning gross fluctuations in traffic demands, handover procedures and watch lengths for example, are rarely tackled.

The second kind of approach to real-time simulation is much simpler and often deals more exclusively with human factors. Typically it identifies the questions first, and simulates whatever is thought to be necessary within the resources and timescales available to answer them validly. If the question is simple, so may be the simulation. Comprehensive simulation of every aspect of air traffic control is not sought; the aim is to impose sufficient control to disentangle the effects of the main variables and to ascertain the most sensitive measures. The choice of appropriate traffic samples and scenarios can be crucial for both kinds of approach, to the extent of being the main determinant of the answers obtained.

There has been a tendency in real-time simulation to equate fidelity with validity. Strenuous efforts are made to replicate faithfully air traffic control workspaces, tasks, equipment and communications so that even an informed visitor may not recognize at once that the air traffic control is simulated. Three different kinds of air traffic control are illustrated in Figures 26, 27 and 28, but what they have in common is that they are all simulated and not real air traffic control. There is no adequate theoretical or empirical basis for prescribing which aspects of air traffic control must be simulated with what fidelity in order to yield findings with a specified validity. In the absence of such guidance the commonsense practice has been to simulate as faithfully as possible. This is expensive in time, money and resources. Those who fund such simulations expect, not unreasonably, valid answers to all their questions. The

Figure 26.   Simulation of an en route suite

Figure 27. Simulation for evaluating a new sector suite

Figure 28. A realistic air traffic control simulation

reluctance to concede that real-time simulation cannot answer some kinds of question validly has been accompanied by a tendency to answer simple questions with unnecessarily sophisticated and elaborate simulations. Some of the earliest real-time simulations of air traffic control were ambitious attempts to convert detailed plans into functional equivalents of operational systems (Parsons, 1972). The plans were compiled first. The processes required to simulate them increased understanding of their functionality even before the simulation commenced. The real-time simulation, being independent of the plans, could validate them.

Real-time simulation is a representation of a representation of air traffic. Recognition of this would lead to more realistic expectations about it. Reality is aircraft in the sky. Actual operational air traffic control utilizes a few categories of information about these aircraft to control them as traffic. To do this, it represents them in forms that are very different from aircraft in the sky—as plan views of labelled symbols on a radar display or as paper flight progress strips containing written or printed alphanumerics. Simulation starts from this representation of the traffic and selects from it the more limited features that must apparently be represented to meet the more limited simulation objectives, in so far as it is feasible to represent these limited features at all. Normally some features such as communications with real pilots are not feasible despite their desirability, although some communications issues can be studied such as the effects of short message delays corresponding to their transmission times from satellites (Nadler *et al.*, 1990). As a consequence, some kinds of human factors problem cannot be addressed validly by real-time simulation: examples are the emotional responsibility for real air traffic, the full flexibility of communications, and the longer-term adaptability and tacit understanding of workspaces, procedures, equipment and colleagues that accrues from years of familiarity. The inevitable contrast between the novelty of a simulation for a few days or weeks and real air traffic control as an everyday job means that neither the excitement, challenge, and interest of real air traffic control, nor its boredom, tedium and routine can be simulated.

Most simulations impose restrictions on their usage, and the methods and measures employed must take account of this (Buckley *et al.*, 1983). All participants in simulations receive specific instructions. Initiatives, non-standard practices and short-cuts, the development of professional norms and standards, and team and supervisory roles are typically curtailed in simulation. Many organizational, managerial, scheduling and rostering features of air traffic control, its work–rest cycles, its working conditions and unsocial hours, and the interactions between work and domestic life, are absent altogether from simulated air traffic control. The consequences of errors, failures, incidents, violations, and infringements of the rules differ between simulation and real life. Given these and further examples, the applicability of real-time simulation to human factors problems in air traffic control is obviously limited. Much of a comprehensive listing of valid and invalid applications of operational evaluations as a technique is also valid for real-time simulation

(Hopkin, 1990). Findings from real-time simulation generally require some form of verification (Wise *et al.*, 1993) and are seldom either wholly useless or fully valid. The problem is to establish the appropriate degree of credence for each finding. Whereas comparative findings are often valid because the constraints on their validity affect the compared alternatives equally, claimed absolute capacities, workloads, strategies and error rates are suspect because the conditions for confidence in such findings cannot be met. Real-time simulations tend to underplay individual differences between controllers as unwanted sources of variance in relation to the simulation objectives.

Despite all of the above, real-time simulation as a technique has many strengths (Sandiford, 1991). It can show if an option is feasible. It can indicate how easy it is to learn, predict training requirements, and suggest if controllers should be specially selected. It can identify causes of potential error or failure, and may indicate remedies. It can show whether envisaged benefits could possibly accrue, even where it fails to demonstrate that they have done. It can validate planning decisions (Beevis and St. Denis, 1992). It can confirm that all the necessary information for tasks is present, or specify what must be added. It can confirm the ergonomics of the workspace, and assist the formulation of improvements. It can discover if controllers will be able to integrate their tasks and responsibilities and function as teams, and can gauge the efficacy and intended forms of supervision. It can show the main human factors limitations on system performance, though not quantitatively. It can verify the design of the physical environment. It can indicate the feasibility and acceptability of proposed changes. It can reveal how procedures and instructions could be revised and improved, and suggest optimum sequences in which functions should be performed. It can explore interactions between human and machine roles, and mutual promptings and directives (Hollnagel, 1993). It can identify functions that controllers perform well or that need extra support, and suggest which functions interfere with each other and which do not (Stager and Paine, 1980). But claims to express these many achievements in quantitative terms that remain valid for real air traffic control should always be treated with suspicion.

19.3 Fast-time simulation

Fast-time simulation refers to a cluster of techniques, primarily mathematical or software-based, applied in air traffic control research to represent some functioning aspects of an air traffic control system. Inputs to the fast-time simulation, or interactions between inputs, are varied systematically, and the resultant outputs are measured. Relations between the inputs and outputs are stated as formulae or other expressions that describe changes in accordance with the input variables studied. It is therefore a tool to describe the most important variables, how they exercise their influence, relevant interactions, and the range of their effects. Fast-time simulation is not much used in human factors studies, because its formulae are not suited to deal with human variability. Except for very simple tasks, human functions tend to appear as constants or sources of random variance in fast-time simulations. Humans represent delays, sources of error, and limitations on concurrent functionality.

In fast-time simulation, the normal variables are system ones and not human factors ones. Some of the difficulties in integrating fast-time simulation and real-time simulation findings occur because the former have excluded human variability and the latter have included it. Sources of human variance and of system variance are not usually expressed in similar terms, and therefore resist integration. Practitioners of both real-time and fast-time simulation harbour some mutual incomprehension and mystification about the activities of the other. Each seems to ignore some variables that to the other are important. This is regrettable because in much air traffic control research a combination of fast-time and real-time simulation could be much stronger than either is alone.

Fast-time simulation can identify combinations of circumstances with crucial effects on system behaviour, which are often the most productive combinations of circumstances for real-time simulation where human influences would be most critical, but this kind of integration of techniques is not common. As a technique, fast-time simulation has the advantage of being capable of examining many alternative scenarios quickly and systematically to trace first- and second-order effects and other interactions. Combined with real-time simulation, fast-time simulation can reveal the most influential factors, and hence point towards human factors measures that are likely to be sensitive, but this approach has not been adopted much either. The benefits of closer links between real-time and fast-time simulation have been apparent for a long time, but they have not been integrated into mutually supportive techniques as often as they could have been. Air traffic control research, including its human factors aspects, has been the loser.

19.4 Evaluations

Evaluations for research and development purposes in air traffic control most commonly employ real-time simulation, but alternatives can range from subjective assessments to measures of real air traffic (Hopkin, 1990), and include prototypes (Fassert and Pichancourt, 1993) and field testing (Harwood and Sanford, 1994). They may help to bridge the gulf between laboratory and field research by obtaining many of the benefits of both (Dipboye, 1990). Usually an evaluation has to adapt current technology to represent future developments, and it costs too much for the controllers to continue to participate in it until they attain full proficiency. Therefore the objectives of valid evaluations for research and development may have to be limited to feasibility and viability rather than aspire to capacities and quantification. The concept of evaluation also has broader meanings, in which the worth of work is studied in ergonomic terms (Wilson and Corlett, 1994), or the efficacy of alternative methods to validate human task performance is examined (Sanders and Roelofsma, 1993).

An initial human factors contribution is to judge whether an evaluation could yield valid evidence (Narborough-Hall and Hopkin, 1988). Evaluations are not a universally applicable technique. Their findings are subject to numerous sources of unwanted error and variance that limit their validity, generalizability and contexts of application. Only a part of an air traffic control system can ever be evaluated since the whole system is far too big and complex. Every aspect of its complexity and integrality cannot be captured. Some coordination and communications aspects are particularly difficult to represent adequately. Evaluations normally have to employ specific air traffic control scenarios, samples and regions of airspace which can be either real or imaginary, yet the findings must not be an artefact of them but be generalizable. A rationale to prove or make plausible this generalization has to be built in. Wherever pilots, equipment, communications or other system attributes differ between the evaluation and real life, the evaluation findings are degraded, usually to an indeterminate extent.

All human factors problems that can be prevented before the evaluation begins should be prevented. The workspace, furniture, human–machine interfaces, displays, input devices, communications and features of the physical environment should be free from gross human factors faults. They should not, for example, induce postural or visual difficulties for the participants. The detailed layout of each workstation in the evaluation should meet the human factors requirements stated in handbooks and checklists. An evaluation is not an appropriate forum to discover again what is already known. All such available evidence should have been incorporated into its specification. The human–machine interface for the evaluation should be optimized as far as possible beforehand for the tasks to be performed.

Wherever practicable, the capacity limits should be set by the requirements of the evaluation and not by the equipment used. If the equipment that has to be employed for the evaluation has a poorer specification than the operational equipment will eventually have, this can impose such serious limitations on what any evaluation will be able to achieve that it may be prudent to abandon it as a technique. Yet human factors evaluations must consider future developments that could benefit air traffic control, and explore their human factors implications (David, 1991). Trying to over-control the specification of appropriate traffic scenarios and samples in the interests of generalizability carries the risk of them becoming too predictable and familiar. A human factors role is to prevent such problems. Participants in evaluations normally do their best to follow the instructions and briefings given. These may differ significantly from their normal practices, either because what is evaluated is novel or because of inadequacies in the evaluation tools. The initial briefing and training for an evaluation can have more influence than is realized. Attitudes can form very quickly and depend significantly on how easy the evaluation requirements are to understand. An optimum balanced programme of briefing, instruction, demonstration, familiarization, and training is required to ensure that the participants in the evaluation have been adequately prepared, and that they know and sympathize with its objectives, accept its unavoidable limitations, and are fully aware of the facilities and procedures. It is important not to imply that a proposal is intended for operational adoption if it is merely exploratory or an evaluation artefact.

The training for evaluations is not often utilized enough to establish which aspects of training are successful as they are and which require further development and refinement. If evaluations are too smooth, the time absorbed in communications, coordination, liaison, delays, and difficulties in conducting dialogues or in reaching agreement is under-represented. The roles of teamwork and supervision can be unrealistic in evaluations if they have not been fully developed or if measurement constraints discourage such interactions because they constitute unwanted sources of variance. Essential human factors contributions to evaluations are to recommend how much training and experience should precede the evaluation in order to obtain worthwhile results, and to recommend the number, length, and scheduling of exercises within the evaluation that will strike the best balance between the realization of its objectives and the expenditure of resources.

The objectives of an evaluation must be clear as its structure depends on them. At one extreme an exploratory evaluation may consist of feasibility studies in which options are discarded as soon as they seem flawed or impractical for any reason. At the other extreme the evaluation may in effect be a tightly controlled experiment following a carefully balanced experimental design rigorously in order to measure interactions validly. The statistical treatment of measured data is related to the objectives and to the rigour of the experimental design. Human factors specialists suggest beforehand the cluster of measures to be employed in each evaluation and their interpretation in relation to each other. Findings that seem anomalous may have to be reconciled but their emergence may justify the multiplicity of measures. Anomalies have to be interpreted but not necessarily dismissed. Subjective assessments by controllers may be able to extrapolate beyond the measured conditions. The human factors specialist must interpret the evidence in relation to the conditions under which it has been obtained, to previous evidence already known, to the effects of constraints such as the number of participants or the duration of the evaluation, and to known disparities between the evaluation and real life. The specialist must adopt an impartial view of the evaluation, although many of the other specialists and participating controllers may have pre-committed views about it.

19.5 The identification and control of relevant variables

Many variables can affect the findings of air traffic control human factors research and development. Unless they are identified, the evidence and recommendations may be an artefact of unrecognized influences and be spurious or unreliable. Their identification does not ensure that they can all be controlled successfully since it may not be possible to control independently the effects of any that are mutually incompatible. The variables that are easiest to identify and control are those which are the subject of the evaluation. Depending on the objectives, variables may be excluded altogether, tightly controlled as constants, varied systematically according to a fixed experimental protocol and measured as independent variables, allowed to vary and measured as dependent variables, or allowed to vary but not measured because they are treated as irrelevant or are unmeasurable or are so confounded with other variables that their respective influences cannot be disentangled.

Beyond the variables which form the substance of evaluations are other identifiable variables that are applied in the interpretation of the findings. These relate, for example, to the traffic samples, the procedures and instructions, the communications, the experimental conditions, limitations of the research techniques, and the previous experience of the participating controllers. If their permanent terms of reference are to act as participants in air traffic control research they will tend to identify closely with its aims but may be out of touch with their colleagues' views. If they participate because they are already familiar with the air traffic control scenarios and region that are the subject of the evaluation, they may be over influenced by current practices and judge the merits of research proposals in terms of them. Their views therefore have to be interpreted according to their experience and knowledge.

When the research objectives are formulated and a programme of work to achieve them is defined, the variables of interest are specified and carefully controlled in known ways so that their effects and interactions are measured. Other variables incidentally but not deliberately implicated in the research are not identified with comparable thoroughness and may remain undetected. Their recognition may be aided by systematic consideration of all the possible kinds of measurement, or by an examination of variables reported in the literature. The application of different theoretical and practical orientations may help to ensure the inclusion of key factors, and research experience contributes to the identification of relevant variables in advance. Checklists of variables, comparable to those used to identify the full range of factors that may be implicated in an incident or accident, can be very helpful.

The thorough identification of all variables has two main purposes. One is to ensure that any observed variance is not ascribed to the wrong causes, leading to recommendations that prove to be ineffective. The other purpose is to try to resolve incompatibilities originating from a failure to recognize further factors that are present and causing anomalies. There is a tendency to favour variables which can readily be quantified and for which measurement tools are available, over variables which may be as pervasive and influential but are more qualitative, more difficult to specify and more resistant to control in research. The processes of research may change or exclude certain variables. For example, it may be impossible to apply traditional research methods to study boredom or the effects of professional ethos or the development of norms and standards in air traffic control, although there is now evidence in the literature that all of these factors can be influential and should be controlled in research contexts. Other variables are difficult to deal with because they expand or contract to fill but not overfill the time available, as with some communication channel occupancy times, message contents, and data discarding. Some apparently trivial variables can be far more significant than they seem. It is important to identify and measure them. For example, if a controller has time for routine updating of information and for routine tidying and discarding of information, such postponable activities may indicate as well as more orthodox measures that the workload is relatively low and the task demands are not burdensome.

19.6 *The reliability and validity of evidence*

Reliability refers to consistency and repeatability, whereas validity depends on whether a measure actually measures what it purports to measure. Reliability can often be established by studying the measure itself, whereas validity usually relies on independent external criteria which themselves are reliable and valid, a requirement that is becoming impractical in complex systems such as air traffic control (Wise *et al.*, 1993). Findings from research in laboratories or using simulations may be neither valid nor reliable in real life. A major contribution is to advise on the reliability and validity of human factors evidence, recommendations, findings and conclusions, and particularly on the applicability of other human factors literature to air traffic control. Some types of evidence, for example, on reach and viewing distances, possess high reliability and validity, whereas others, such as the effects of boredom on safety, have unknown validity and reliability.

Human factors practices in ascertaining the reliability and validity of evidence before basing recommendations on it lack uniformity. Whereas the validity of the procedures for selecting air traffic controllers must be proved before they are implemented, the findings from an evaluation may be applied without any comparable checks. Research to prove validity or reliability may be perceived as unproductive by those approached to finance it, particularly if extensive work fails to validate findings—quite a common occurrence in air traffic control. Often there are no adequate tools to predict what the attainable levels of reliability or validity are likely to be. A single-minded crusade to improve reliability can lead to much unproductive effort, especially when the limitations are set by the criteria and not by the measures. The number of factors with a small but significant correlation with predicted success as an air traffic controller is large, particularly when the successful completion of training is the criterion adopted. The validity of this criterion as a predictor of success as a controller and its internal reliability as a measure have both been questioned recently. Where the probability of completing one phase of training successfully does not correlate highly with the probability of completing other phases successfully, the phases may tap different and quite specific abilities, and there may be no dominant or influential abilities that are common to all air traffic control. If there are none, the validity of any selection procedures applied to all prospective controllers must be limited. Nevertheless the quest for higher correlations with higher validity continues. Perhaps Smith's (1994) distinction between universal predictors related to work, occupational ones related to jobs, and relational ones related to others in the work environment, is applicable to air traffic control.

The consequences for reliability and validity of a change that has already occurred have not been fully appreciated. Real-time simulations and operational evaluations used to be relatively independent of the thorough planning activities that preceded them. While they retained this independence, they could function as forms of validation of the initial system policies, planning, designs and specifications. A comparatively recent development in real-time simulation has been the construction of prototypes as a tool to establish viability or feasibility, but these are not used to conduct more formal and controlled experiments or evaluations. This is sensible, provided that it is recognized that such prototyping is actually a surrogate for the planning, thinking and formulation stages of system evolution and cannot function as an alternative form of validation. Conclusions drawn from prototyping usually remain unvalidated, although an impression of validation may be gained because the prototyping processes have superficial similarities with the more independent former evaluation practices.

Many kinds of evidence can help to establish validity. Among the most common ones are the following: recurrence of the same findings in different air traffic control systems; comparable findings in other aviation contexts or in other large human–machine systems; findings that are compatible with and explicable by theories or other broad constructs; mutually supportive objective and subjective evidence; compatibility between practical findings and independent criteria; the confirmation of research findings by subsequent operational experience; compatibility between plans and their implementation; and practical experience of real-life air traffic control which confirms cumulatively the safety and efficiency of the systems as planned and implemented and the absence of human factors problems in them. Certification procedures, now becoming a topic for human factors study, may also constitute a form of validation (Wise *et al.*, 1994).

Some difficult issues of reliability or validity must now be tackled in air traffic control, as they have been postponed for too long and more adventurous thinking about them is overdue (Westrum, 1993). One concerns the validity in real life of research findings from laboratory studies, simulations and evaluations (Locke, 1986). Another concerns the dubious validity of some of the most popular criterion measures. A further issue is the validation of systems of high complexity and integrality for which independent external validation criteria of comparable complexity and integrality become impossible to devise or apply. In the future, expedient validation processes may have to be built in and applied during the system design, because it is no longer possible to devise processes that could be applied externally or retrospectively (Wise *et al.*, 1993, 1994). Historically, it has been difficult to obtain funding for the vital reliability and validity research that is needed. An encouraging aspect of the current drive towards judging the efficacy of research in such terms as cost-effectiveness is that it may force more critical consideration of the reliability and validity of processes and of evidence.

19.7 The interpretation of evidence

The human factors specialist should know the kinds of valid evidence that various techniques and measures can yield, and the kinds of question that cannot be answered validly with the practical methods available. This professional expertise is applied to gather appropriate evidence and to interpret it, and also to judge when and how more general evidence and recommendations in such sources as standards and handbooks must be adapted to become applicable in particular circumstances. Many of these sources fail to indicate the conditions which must be satisfied for their recommendations to apply, or the nature and extent of the modifications required in particular conditions. Recommendations may have to be modified for many reasons. Even for the simple example of the legibility of alphanumerics on a display, corrections may have to be made to compensate for low brightness contrast, deficiencies in the font, the height or width or height/width ratios of characters, glare or reflections, the ambient lighting, coatings and filters, the spectrum and intensity of display emissions, the viewing distance and viewing angles, the minimum eyesight standards of the operators, the visual information codings employed, attributes of the tasks, and the forms of human error that would be detected and corrected automatically.

Another aspect of the interpretation of evidence concerns how far findings generalize and what circumstances invalidate them. An obvious example of such circumstances is differences between the laboratory and real life, but other examples include very high or low task demands, fatigue, inexperience, lack of skill, various forms of computer assistance, and the observability of human tasks and functions. The interpretation of evidence must heed the practical options available. A statistically significant difference between two options does not necessarily imply that the better one should be adopted. Both may be useless; either may meet the requirements fully; or the statistically significant difference may be operationally trivial. A compromise that combines the main strengths of the measured options and avoids their main weaknesses may be superior to both of them.

The interpretation of evidence also considers the extent to which the conditions under which evidence has been gathered represent the conditions under which it will be applied. This can be quite difficult to gauge; for example, a recent study, employing instructors as subjects in a laboratory experiment with familiar traffic scenarios, illustrated some of the issues of interpretation that arise when findings seem valid within their own limits but when the criteria and conditions for their generalizability seem uncertain (Vortac *et al.*, 1994). Several kinds of criterion may be employed to assess how representative evidence is likely to be. These include attributes of individuals such as their experience and knowledge, attributes of the tasks such as the demands and consequent workload, attributes of the facilities such as the human–machine interface design and the system configuration, attributes of the organization such as its management, attributes of the preparations such as the training and instructions, and attributes of the working conditions such as the work–rest cycles and rest breaks. The generalizability of findings is partly a matter of specialist human factors interpretation but should rely whenever possible on supporting evidence rather than expert opinion. Research is never perfect, and the effects of its known imperfections on the findings must be interpreted. The value of the interpretation of evidence by human factors specialists depends on their breadth of knowledge, because the interpretation of evidence should refer to human factors as a whole and not to any narrower focus that may have been adopted in gathering it.

19.8 The dissemination of findings

Much human factors research and development in air traffic control is described initially in reports to those who have commissioned it. Most establishments conducting such research prepare their own series of reports on it, and many of these establishments are listed on pp. 443–5. If competent work has been done under conditions that allow it, the main questions should have been answered, if they can be. It is always possible in human factors research that a question has no satisfactory answer, even though the research has been exemplary. The value of research can often be enhanced by disseminating its findings more widely. How could human factors issues be recognized as specific to air traffic control, to aviation, to large human–machine systems, or to all work environments without the widespread dissemination of findings? All researchers benefit from relating other published findings to their own, by confirming them, gaining fresh insights and alternative interpretations, and preventing duplication of effort. This places an obligation on human factors researchers to disseminate the results of their work, and particularly to report unresolved questions and failures. The latter seldom imply incompetence and often seem associated with persistence and thoroughness, but if they are not publicized others may repeat the work and again be disappointed. It is therefore efficient to review and collate findings, and to present them through conferences, journals, textbooks and the media at every opportunity.

The human factors specialist in air traffic control can occupy an anomalous position in relation to academic colleagues. The career prospects and advancement of the latter may depend on the acceptance of a body of their work for publication in reputable professional journals and books. The human factors specialist, whose career depends far less on published work, may receive invitations from journal editors to submit papers for publication but be unable to accept them because no time has been officially allocated for preparation for publication, which is sometimes actively discouraged. With tight budgetary controls, the justification of the allocation of time, money and resources to disseminate findings more widely can pose a major problem.

From broader and longer experience of human factors work, the specialist builds knowledge. Examples of such knowledge, for example in relation to real-time simulation, include the kinds of question that can be answered most successfully, the most productive measures, desirable characteristics of traffic scenarios, and the theoretical perspectives that are most enlightening. The collation of this knowledge into guidelines and its subjection to critical peer review seem essential for the progress of human factors as a discipline, for its generalizability to be tested, and for scientific advancement. A further justification for the dissemination of findings is their educative role beyond human factors. It is prudent to seize every opportunity to inform others about the roles and achievements of human factors in air traffic control, and more widespread dissemination of findings is among the most effective ways to accomplish this, particularly in the current climate which is as receptive and favourable as it has ever been towards human factors in air traffic control. Nevertheless the main products of human factors in air traffic control are not within the pages of journals, but are safe and efficient air traffic control workspaces that are functional and satisfying and promote the well-being of those who work within them. The educative process cannot therefore be confined to documents but must include demonstrations of the successful application of human factors evidence.

Others will not understand many human factors activities in relation to air traffic control, and the specialist must be prepared to explain human factors methods, conclusions and evidence in non-technical language. This problem is not specific to air traffic control but applies as much to many other disciplines. Descriptions of human factors work in air traffic control should not only appear in the human factors and air traffic control literature, but should be disseminated wherever they are of interest, in the professional literature of other disciplines and through interdisciplinary conferences and specialist conferences of relevance. A human factors specialist who claims some understanding of how humans learn, remember and acquire information should be able to apply that understanding to inform others about human factors activities and convince them of their value. If human factors as a profession is unwilling or unable to do this, others will not do it on its behalf.

19.9 International collaboration

Air traffic control is an international service. No nation can isolate its air traffic control activities. Similar human factors problems in air traffic control arise concurrently in many nations because of technical advancements, commonalities in aircraft or in their equipment, or increasing air traffic demands. As more ergonomic problems are now being tackled internationally (Nielsen, 1990; Bradley and Hendrick, 1994), evidence is accumulating of limitations in the cross-cultural applicability of many recommendations (Kaplan, 1991). Positive international collaboration in human factors work on air traffic control offers considerable benefits. Even the nations with most human factors resources for air traffic control work are discovering that their resources are not sufficient for all the work that should be done. It therefore makes good sense to identify the full range of human factors problems in air traffic control through international collaboration, to apply collective expertise to planning the research needs, to commission research from different nations in accordance with their particular expertise, and to review internationally research centres with particular expertise such as the application of artificial intelligence in air traffic control (Gosling, 1993). The research can be planned and conducted internationally as a complete programme in which the parts fit together and can be interpreted in relation to each other, instead of consisting of piecemeal findings that relate only to single systems and do not generalize.

Several events reflect an increased interest in international air traffic control human factors work. The International Civil Aviation Organization has published a digest on "Human Factors in Air Traffic Control" (ICAO, 1993). The long-term NATO interest in the topic has been sustained. Air traffic control issues are covered in the Conferences on Aviation Psychology held every two years, and in the meetings of the European Association of Aviation Psychology (formerly WEAAP). European Conferences of the Human Factors and Ergonomics Society, as well as its regular meetings, address them. European air traffic control activities, centred on Eurocontrol, cover many human factors matters, and much of the research planning is now at European rather than at national level. The US National Plan for Aviation Human Factors attempted to include a comprehensive programme of related human factors research on air traffic control, as a basis for allocating resources effectively in relation to other human factors research on air traffic control within the United States and throughout the world.

Other opportunities for international collaboration are afforded by common technologies which require international agreement on their forms and applications in air traffic control. The interchangeability of air traffic control scenarios in the form of software permits in principle comparative studies between controllers in different nations performing the same tasks, and the cross-cultural adaptation of standardized tests might provide tools for other comparative studies (Hambleton, 1991). Such advances broaden the opportunities for international collaboration in human factors research and development in air traffic control, and the coordination of resources could avoid duplication and increase the cost-effectiveness of nearly all research by extending its applicability and assisting the validation of findings (Shouksmith and Burroughs, 1988). Such developments are welcome. More international pooling of the human factors resources on air traffic control and more international collaboration in the associated research would seem to benefit everyone, especially if the research programme is planned as a whole because the whole then becomes greater than the sum of its parts.

20

Human factors implications of other functions related to air traffic control

20.1 Management

Despite their relatively brief professional history, air traffic controllers have acquired a reputation in some countries as quite a difficult workforce to manage. Whether the original reasons for this were controllers' intransigence, managerial ineptitude, or mutual failures to communicate is difficult to determine retrospectively, and profitless except to prevent their recurrence. A series of formal investigations in the United States almost a quarter of a century ago put most blame on poor communications between management and workforce, and revealed simmering discontent more than a decade before the strike there by air traffic controllers. But relationships have been unsatisfactory in several other countries also. All pat explanations are probably specious. Perhaps the very attributes for which controllers are selected tend to produce independent-minded people who are quick to query any management decisions that affect their work. Perhaps management has not succeeded in explaining why certain changes are essential to maintain an efficient service, and the cost constraints on those changes. Perhaps controllers failed to convey adequately to managers their real needs in the workspace. Perhaps management tried to communicate with controllers, but controllers were unwilling to listen. A job satisfaction survey of the managers of controllers has suggested that communications could still be improved (Myers, 1990).

In the meantime, courses and textbooks have proliferated on good management practices, and there is now a vast literature and lore about them. It requires an act of faith to apply all this good advice to air traffic control, since the most fervently advocated practices have not been validated for air traffic control applications. Nevertheless the expectation must be that many of them would be helpful. This text is not about management, and only a few management concerns with particular human factors implications are mentioned here. An introductory text for managers has outlined the main interfaces between management and psychological and human factors issues (Cooper and Makin,

1984), and the ICAO Human Factors Digests are intended primarily for management, including the one on air traffic control (ICAO, 1993). A general text (McCormick and Ilgen, 1987) also indicates the scope of human factors. A recent challenging venture for both management and controllers is the proposed privatization of air traffic control. In New Zealand, air traffic control has been privatized, and several nations are watching its outcome with interest.

In practice air traffic control managers often have far less freedom of choice than controllers think they have. To controllers, the roles of management are to ensure satisfactory working conditions which are never harmful in any way, to acknowledge their high standards of safety, efficiency and professionalism, and to provide modern, reliable, safe and efficient equipment and workspaces which meet controllers' needs and are the outcome of appropriate consultations. Controllers are often less aware that management must meet financial and political objectives; must be assured that all aspects of the system are robust, reliable, serviceable and maintainable and meet or exceed the specifications; and must comply with many international and national constraints, standards, rules and regulations. The experience of both management and controllers has made them more wary of the claimed benefits of technological advances, but not to the extent of refusing to consider those that really are helpful.

Some guidelines for the management of controllers, though obvious, are worth restating. The career expectations and prospects of controllers should be realistic from the outset. In many nations, controllers do not have glowing career prospects, as comparatively few higher-level posts are open to them and most will not progress beyond the main working grade. Over-optimistic expectations lead to disappointment and frustration and are counterproductive. Rostering practices need collective agreements, and many options are available, ranging from managerial insistence on planning all rosters in great detail to the delegation of all rostering decisions to the controllers themselves within the applicable legal requirements and other constraints. An example of the latter is that rostering agreements acceptable to both management and controllers may not gain medical approval if they could have adverse effects on occupational health or stress. For controllers and their managers, the lines of responsibility must always be clear, so that there can be no doubt where the responsibility for every action lies. Most people thrive on the delegation of responsibility to them, but in air traffic control legal limitations apply and increased responsibilities can become a lever for better pay and conditions.

20.2 Public relations

Not too long ago, controllers were wont to complain that their profession was not recognized and appreciated enough by the travelling public. They sought wider acknowledgement of their work, mainly through their professional organizations and representatives at international, national and local levels. The application of public relations expertise and techniques was favoured, to give the air traffic control profession the positive promotion and publicity that it had earned. Most managements had some sympathy with this, although it was not particularly high among their own priorities. Air traffic control as a profession in the limelight might raise its national status and attract more capital expenditure and modernization, but any requests for shorter hours, better pay and improved working conditions would probably incur demands for increased productivity without compromising safety. To be effective, the publicity would have to be somewhat selective and not remind the public of previous dissensions that inconvenienced them. All publicity is not necessarily good publicity.

Two factors then intervened, and changed the public relations aspects of air traffic control. One was a series of strikes by controllers in the United States and elsewhere, which received wide publicity and dented the public image of air traffic control. The other was a growing inclination to treat air traffic control publicly as a scapegoat for delays to aircraft in the air or on the ground, whether air traffic control had any culpability or not. Delays are inevitable if the traffic scheduled by the airlines and agreed by the authorizing authorities exceeds the maximum traffic-handling capacity of an airport or route over a short period, and air traffic control tries to minimize these delays but lacks the power to refuse to sanction flights altogether in order to prevent delays. Because of these factors, the travelling public tended to perceive air traffic control as causing delays and disruptions, if they thought of it at all. Much of the limited publicity that air traffic control did receive averred it to be a particularly stress-inducing profession, despite much evidence to the contrary. Naive and ill-informed but well-publicized political statements about air traffic control, which controllers as public servants were not permitted to refute, contributed further to false public images of obstructive controllers and incompetent management. Air traffic control certainly needed better public relations.

There began to be real anxiety about the possible effects on recruitment. Who would want to join a profession that apparently lacked public respect but attracted widespread denigration? Controllers were accustomed to being unseen and unappreciated but unaccustomed to unfair criticism, and a few were dismayed by it. Human factors specialists, particularly those trained as psychologists, expect some antipathy or occasional hostility alongside recognition and appreciation of their work. Most accept that the burden of proof of the value of human factors is a professional responsibility of those who practise it. Some controllers could not understand why their profession, with its excellent safety record and high standards of proficiency, dedication and professionalism, should attract vituperative comments instead of public approbation. It seems anathema to have to defend merits that seem self-evident, or to apologize for the quality of service when there is nothing to apologize for. This denotes neither complacency nor a denial of improvements, but a belief among controllers that they generally achieve all that is possible with the resources available to them. The influence of factors and constraints beyond the immediate system was beginning to be appreciated (Wiener, 1980).

In recent years, human factors specialists have found that requests from controllers for advice on how to publicize their profession more widely have almost ceased. Slowly the public relations problems of air traffic control as a profession are easing, although much work is still needed to burnish its public image, to close the gap between that image and reality, and to place the blame for delays where it belongs so that air traffic control has no more than its fair share of it. Among the practical objectives is the attraction of a continuous supply of qualified and motivated candidates to become air traffic controllers, who have realistic and well-informed expectations about the jobs for which they are applying.

Air traffic control as a profession needs to press harder for adequate media time to respond to criticisms of it, using professional representatives who have received media training in how to present their case to best effect. Every opportunity to educate and inform the general public about air traffic control should be followed. Practical methods include more good introductory texts and articles on air traffic control, well-presented and informative talks and lectures, more publicity in schools and colleges, more incisive and better trained professional lobbyists, and more sponsorship of films, videos and television programmes as publicity and educational material. Air traffic control can be intrinsically interesting to many people who know little or nothing about it, which is a major public relations advantage. Perhaps it should become more closely associated with the larger aviation community. The objective is not to over-exalt air traffic control, but to restore its good name and match its image and its reality. After all, most controllers do enjoy being controllers.

20.3 Operational research and analysis

The techniques of operational research and analysis are applied extensively in air traffic control. There is always a requirement to predict and plan for future air traffic demands and capacities, and to ascertain whether increases in capacity, efficiency and safety would justify proposed technological or navigational advances. A further requirement is to identify the limiting factors on capacity so that resources can be concentrated on them. Operational research and analysis provide tools to trace the impact on traffic handling and on system capacities of diverse changes, such as data links, satellite-derived information, different route structures, additional runways, and revised regulations on vortex separations. As a discipline its primary concern is with system measures and capacities, and it relates to human factors only when human characteristics or variability directly affect such measures or affect them indirectly through equipment, procedures, instructions or training. Although operational research and analysis should not be equated with fast-time simulation, some of the interfaces of both with human factors are similar.

The objectives of human factors and of operational research and analysis often seem compatible, parallel or even identical but joint interdisciplinary approaches to problems have usually foundered because they share few common methods, techniques, measures or variables. Operational research and analysis questions and human factors questions are not posed in similar terms. Few common measures are accepted as valid and reliable by both disciplines. Attempts by either discipline to move towards the other, such as attempts in human factors to build mathematical models of human behaviour, still do not utilize concepts already familiar in the other discipline. Joint studies that satisfy the requirements of both disciplines seem to offer improved validation of their findings, compared with studies confined to either discipline.

Operational research and analysis can assist the identification of impending human factors problems by aiding the formulation of system objectives and by revealing future human interventions. The demands predicted on the basis of operational research and analysis show where and to what extent existing capacities must be increased, and can be coupled with human factors evidence of the conformity between theoretical and actual traffic-handling capacities. This combination of required future capacities plus existing disparities between principle and practice points towards human factors issues on which efforts should be concentrated. Disparities are not always unfavourable. Sometimes human factors studies reveal that the performance actually achieved exceeds what is achievable according to operational research and analysis measures, because it has been influenced by human factors variables which operational research and analysis do not take account of. In borderline circumstances, involving a transition between sets of rules or ambiguities over which rules are applicable, definitive answers might require both human factors and operational research and analysis contributions, since the crucial concerns in particular circumstances could be of either kind.

The lack of common approaches may exemplify more general difficulties of resolving problems in highly complex systems by integrating the contributions of different disciplines whose traditional methods, techniques and measures are incompatible simply because they have evolved independently. These different methods served their purposes well while all questions clearly belonged to one discipline or another, but their weaknesses are revealed by interdisciplinary questions. These can only become more common as systems become more integrated and complex and as fewer questions can be answered within the techniques of any one discipline. Radical rethinking may be required if the best solutions do not depend on any enterprising combination of existing techniques, but on wholly novel techniques and measures that cross traditional interdisciplinary divides.

20.4 The sensing and processing of data

In air traffic control, the data available ordain the human roles. If essential information for a task cannot be sensed and presented, the task cannot be done. Any new form of information, from satellites, data links or computations for example, has to be matched with tasks and understood if it is to be used correctly. In the past, there was often no option but to invest trust in suspect information of poor quality. Nowadays the commoner problem is insufficient trust in highly accurate and reliable data still portrayed in the same forms in which data of much inferior quality used to appear. New and unfamiliar forms of information may fail to indicate to the user flaws or failures in the data-gathering process, or the user may be unable to recognize their signs.

Primary radar displays showed when information was updated, and this had a strong influence on tasks. The equivalent data in modern systems are subject to extensive computation and smoothing. Perceptible updating may depend on technical factors such as the spacing between pixels on the display, rather than on human factors principles of just noticeable differences or on system factors such as the minimum magnitude of operationally significant changes. Frequently sensed and updated digital data with small incremental steps may look like continuously changing analogue data to the user, and thus convey a misleading impression of their nature. The minimum perceptual change on a display should be in broad accord with its just noticeable differ-ence as it is pointless to depict a change that the user cannot see. It is equally pointless to wait until a gross change has accumulated before presenting it to the user, since this degrades trend information and can mislead the user about the quality, accuracy and trustworthiness of the data. Discernible increments should depend on actual updatings or minimum perceivable differences for high-quality data, and on the minimum changes that are valid for low-quality data.

The data gathered greatly exceed the data that can be displayed. They have to be processed, condensed or discarded until the controller can be presented with a usable summary characterized by accuracy, precision, frequency of updating and discernible changes appropriate for the tasks. Data processing has two distinct functions—the presentation of information to the controller, and the basis for computations. The results of some computations may be presented to the controller as computer assistance or collated information, but many calculations are for other functions such as data smoothing and integ-rity checks and are not for presentation. Computations will employ much more accurate, frequently updated and voluminous data than the human con-troller can cope with on displays. However, the disparate requirements of computation and human use must not induce anomalies between the different forms of data. Errors and confusion in human–computer interactions result from failure to maintain compatibility among internal data sources.

Any tendency towards data gathering as an end in itself rather than a means to an end has to be resisted. In human factors terms, the sensed and processed data possess the characteristics of their presentation, and not any other characteristics they may originally have had. How the data were gathered is relevant only if it could render their presentation misleading. This should not happen but can become difficult to avoid completely when the data have to be condensed and simplified for human use. The presentation of more data leads to more problems in their satisfactory structuring and portrayal, increases the risk of clutter, and adds to the difficulties of finding and retrieving required information. The need to obtain easily all the data for every task applies to controllers, and to all other functions in air traffic control environments including monitoring, testing, calibration and assessment.

Much data processing relies on software instructions supported by tests of the integrity of the software. Principles of user-centred automation and user friendliness that have emerged comparatively recently carry implicit criticisms of earlier forms of software that ignored basic principles of human learning, understanding and application. Instructions calling for the results of complex data processing and computation to be presented to controllers are nugatory if the information is irrelevant, invalid, unacceptable, untrustworthy or misunderstood, for it will not then be used as intended. At best it will be ignored, and at worst it will mislead. The stages in data processing should be as simple and transparent as possible, to facilitate training and the user's understanding of the nature of the data and judgements of its applicability.

It is tempting to present new data such as satellite-derived information in familiar forms such as radar-like plan views. Electronic flight progress strips may retain many visual attributes of paper flight progress strips despite gross functional differences. Data equivalent to spoken messages may be synthesized for presentation as artificial speech, which conveys misleading impressions of their nature. There are other examples. Sometimes it may be essential to convince the controller that the data really are different because of new technology, new sources or new processing, and the simplest way is to employ totally new forms of presentation which make the changes apparent at once so that the user is less likely to treat the new data as equivalent to any previous form of data. Against this is the wish to continue to use existing controller skills, knowledge and experience, but this practice should be challenged more often, on the grounds that it condones the perpetuation of inappropriate habits and procedures, and falsely imputes to the new data the strengths and weaknesses of the former data. Even at the cost of extra training, it may be a prudent principle to ensure that new forms of data always look new.

20.5 Quality assurance

A long-standing problem for human factors as a discipline has been to demonstrate its cost-effectiveness by proving that the costs of the human factors resources deployed are more than recovered in the benefits achieved. The benefits of human factors can be spectacular, but there is a lack of appropriate techniques to quantify what would have happened without them. Therefore the benefits of human factors often have to be expressed as achievements rather than improvements, because the successful application of human factors removes the baseline for comparisons. This human factors problem in relation to air traffic control is shared by some other disciplines.

Quality assurance is an independent process that seeks to quantify the adequacy and acceptability of achievements by applying fixed protocols to test their compliance with specifications. It is a general technique, and is as quantitative as possible in dealing with quality. Although the notion of a human factors audit is becoming more practical as a feature of quality assurance techniques, and although some human factors standards also seem suitable, the application of quality assurance to the human factors aspects of air traffic control is still at an early stage. Independent certification of the value, competence and quality of the human factors contributions would be universally beneficial, to justify the expenditure on them, to provide criteria for such justification, and to express human factors achievements in other than human factors terms. While quality assurance has the advantage of being widely applicable, it has the disadvantage of being unproven for human factors work, and the potential problems and biases associated with certification procedures would have to be circumvented with quality assurance.

A useful start could be exploration of the common ground between quality assurance and human factors, and the extent to which the techniques of the disciplines could be compatible. Human factors specialists who are convinced that their contributions are worthwhile in terms of efficiency and safety and represent good value for money would welcome better techniques to demonstrate this. Some quality assurance measures such as the achievement of stated objectives on time to the required standard seem directly applicable to human factors work, and others could need only minor modifications, to demonstrate, for example, highly efficient levels of achieved performance or reductions in the frequency or seriousness of human error resulting from human factors recommendations. However, for such human factors objectives as the alleviation of boredom or high acceptability for innovations, quality assurance may lack valid techniques. It may be possible to show a reduction in accidents or incidents within a category following human factors work, but not to demonstrate the prevention of any particular accident or incident. There is thus a need for better links between human factors and quality assurance in air traffic control, and for better tools to quantify the efficacy and the benefits of human factors contributions.

20.6 Maintenance

For many years the professional human factors contributions to the maintenance procedures within air traffic control systems were negligible. When human factors specialists were invited to improve the workspaces of air traffic controllers, they might be asked to look at other workspaces giving human factors problems within the same centre, but no formal organization provided human factors support for the workspaces, equipment and procedures in maintenance. Recently this neglect has been acknowledged and a start made to redress it (Hopkin, 1993c). The US National Plan for Aviation Human Factors (FAA, 1990) formally acknowledged the need to include maintenance tasks.

Much standard human factors evidence, particularly on the specific requirements of workspace design for effective human use, can be applied directly to maintenance procedures and equipment, subject to some verification of the applicability of standard human factors recommendations in maintenance workspaces. Because of past neglect, the application of the extensive existing human factors knowledge to maintenance procedures and environments could probably achieve rapid improvements in terms of efficiency, reliability and satisfaction. Many maintenance personnel believe that their needs have been neglected for too long, yet air traffic control relies on the competence, correctness and professionalism of maintenance workers, and on the correct design of maintenance procedures and instructions. Sadly, it has sometimes taken a failure or fault in maintenance to force recognition of its human factors requirements.

Human factors contributions to maintenance are now evolving through a series of meetings and publications (Garland and Wise, 1993). Obvious contributions include the definition of tasks, their matching with human capabilities and limitations, the derivation and proving of appropriate selection and training procedures, the identification of safety-critical elements, and human and machine collaboration to achieve safe, efficient and cost-effective maintenance. There are human factors problems in attracting, keeping and satisfying high calibre maintenance staff, and in training them to understand their procedures fully and acquire the necessary skills. They need to be able to recognize and guard against possible human errors. Ways in which maintenance teams can work best together, and effective forms of supervision require consideration. Sources of job satisfaction and career development prospects are among further human factors issues.

The whole of human factors applies to maintenance. Many human factors research requirements in regard to maintenance have been defined (Johnson, 1993; Shepherd, 1993). The most pressing current need is to allocate sufficient human factors resources in order to build human factors experience in maintenance and its problems. At the present time the direct application of existing human factors knowledge to maintenance procedures probably has a higher priority than further research, but in the longer term a comprehensive research programme to optimize the applications of human factors to maintenance in air traffic control will be required.

21

The future

The prospects for human factors in air traffic control seem as bright now as they have ever been, but could easily be dimmed. Outright hostility to human factors as a discipline is less common than indifference or apathy. Ignorance of human factors remains widespread among those who plan, procure, commission or manage air traffic control systems. Few human factors specialists are entitled to be proud of their efforts to promote their own discipline and educate others about it. They have been too diffident over publicizing human factors achievements, or have deluded themselves that the merits of human factors as a discipline are so self-evident that it sells itself. But the world does not owe human factors a living. Given suitable opportunities, human factors can more than justify its application to air traffic control, but such opportunities have to be earned and do not present themselves without preparation and promulgation. In many nations and internationally, the roles and benefits of human factors are becoming more widely advocated, leading to a gradual expansion of human factors work in relation to air traffic control. As a discipline, human factors has been slow to develop valid techniques for costing its benefits, but these are becoming essential in an era that uses financial criteria to judge worth and to allocate resources, and that requires evidence of responsible expenditure.

Although human factors was originally conceived as an aid for system designers, its initial contributions to air traffic control often awaited the belated recognition of human factors problems in a system that was already operational or soon to become so, when the human factors specialist was asked to solve them. Usually, earlier design decisions precluded the best human factors solutions, and their retrospective introduction would have incurred unacceptable delays and costs. The less satisfactory solutions that remained practical were sometimes little better than panaceas because there were so many constraints, yet they were beneficial enough to justify human factors involvement, and the human factors specialist was able to formulate the more effective solutions that could have been recommended if the human factors contribution had been sought earlier. Effective mechanisms are still lacking for routinely introducing human factors contributions early enough in the system evolution to maximize their effectiveness, and incidentally their cost-effectiveness. Improved future mechanisms for this would represent real progress.

Human factors has not trumpeted its roles, activities and achievements in air traffic control, but it must become more proactive in educating others so that they are able to realize when they have a human factors problem and need specialist help. There is a permanent need for supportive human factors research in any context such as air traffic control where expanding demands and technical innovations generate new human factors problems. However, the most significant and urgent constraint on current progress is the deplorable under-application in air traffic control of so much human factors evidence that is already available (Heller, 1991). Human factors will not achieve its full impact on air traffic control until more human factors specialists are applying their professional knowledge directly to air traffic control systems. Human factors, as an applied discipline, is not being applied nearly enough.

As fuller acceptance is won for human factors, it has to shoulder the burden of delivering its promises on budget and on time, while maintaining high professional standards. Its true capabilities will be tested, and it must not be found wanting as it cannot afford lost opportunities. Others may find difficulty in accepting that some human factors problems have no satisfactory solution, but humans cannot do everything and some human limitations cannot be overcome or circumvented. When there is no possible solution, the human factors specialist must say so clearly and never leave the misleading impression that there is a solution to be found, for this can easily build a reputation for promising but failing to deliver. Nor is there a long-term future for deliverables from the human factors specialist that do not meet the needs and expectations of the customer who has paid for the work. The present circumstances, in which most human factors specialists in air traffic control have sufficient opportunities to satisfy their customers, depend on the continuing delivery of work of high quality that is perceived to be relevant.

In this context, human factors as a profession should perhaps give more consideration to defining and assessing the professional standards and qualifications of its members, and to providing an independent guarantee of their knowledge and competence to practise unsupervised. More familiarity with business and financial practices would often be helpful. There is no current crisis in this regard, but such guarantees could be a hallmark of human factors as a profession, and voluntary internal codes of conduct seem far better than involuntary externally imposed ones. Such a guarantee would also reassure the many who purchase human factors work but do not know enough about it to judge for themselves the quality and value of the deliverables they receive, as distinct from their practicality which they are well able to judge.

A broad range of knowledge is relevant in applications of human factors to other disciplines, and points to a real practical difficulty. In most of the topics mentioned, there are significant developments all the time. Human factors specialists must spend time in appraising the relevance of these developments to their work, in keeping their own professional knowledge up-to-date through the literature, in noting parallel developments and human factors issues in other contexts, and in learning of new theories, methods, measures and constructs. Pressures of work tend to relegate such vital but postponable activities, especially if the time spent on them cannot be assigned to contracts or costed as overheads. The pioneers of human factors recognized the strengthening support of sound theoretical and methodological foundations, but these are currently so out of favour that they may be deemed unproductive or superfluous. Human factors as a discipline will not endure without foundations, but the current difficulties in funding the time and resources to build them are formidable. A welcome future development would restore the underpinning of human factors evidence and recommendations by theories, and renew attempts to employ theories for validation purposes. The development of more productive links between basic and applied research seems opportune (Schonpflug, 1993).

When a proportion of the development costs of systems or of the training of system users must be spent on human factors work, this is a sign of progress and recognition but not a guarantee of quality. Every chance should be seized to express the benefits of human factors quantitatively in terms that are comparable with those applied to other disciplines. Statements or claims about notional but intangible human factors benefits do not lead far. The human factors contributions to air traffic control systems are generally large and durable, and can be expressed as improvements of safety, efficiency, capacity, or error rates, as fewer incidents and accidents, as greater productivity, as improved equipment serviceability, as reduced training costs, as lower attrition rates or absenteeism, as increased job satisfaction or morale, as improved working conditions and health, as increased customer satisfaction, or in many other quantifiable ways. The difficulty is not lack of evidence, but proving the links between the benefits and the human factors contributions, and then expressing them in financial terms. The most compelling forms of evidence result directly from the practical application of existing human factors knowledge to air traffic control, because this can bring immediate quantifiable benefits at low cost.

The kinds of human factors contribution to air traffic control, already numerous, are continuing to increase. Human factors specialists will need wider horizons (Westrum, 1991). Recent developments with pending human factors implications for air traffic control include virtual reality, computer graphics, adaptive computer assistance, validation as an intrinsic attribute of processes, and cultural ergonomics (Kaplan, 1991). The anticipation of future problems by deducing them from policies and plans is more productive than the retrospective solution of problems recognized years too late. If a human factors problem in air traffic control can be expressed initially in air traffic control terms rather than human factors ones, the full range of its human factors implications is more likely to be identified, as its initial human factors labelling can consign it prematurely to a particular human factors category, as an information display problem or communications problem for example, which can result in incomplete solutions with unrecognized implications.

Air traffic control will continue to generate new human factors problems, however it evolves. The more that new technical developments, automation and computer assistance are applied to air traffic control, the more new human factors problems and interesting human factors challenges there will be, some of which originate in interactions between cognitive and organizational factors (Clegg, 1994). Those who talk to controllers are often surprised by the enjoyment and satisfaction they gain from controlling air traffic. The fact that most human factors specialists in air traffic control gain comparable enjoyment and satisfaction from their own work helps them to empathize with controllers.

Computer assistance in the preparation of this text has drastically reduced routine editorial chores and retyping, but has not been applied to decisions on its content, structure or style. Authors resist computer recommendations that seem to override their professional judgement, skill and autonomy, especially if they remain fully responsible for the product and culpable for its defects. Similarly, controllers can be sceptical about the merits of computer assistance that seems to downgrade their professional expertise, knowledge, experience or skills, or to whittle away their responsibilities or autonomy. What would be the effects on the challenge, interest, satisfaction and enjoyment of the work? Human factors as a discipline must grapple with these issues in the future in relation to air traffic control and to itself, for it cannot remain aloof from the future application of computer assistance to solve human factors problems and recommend human factors decisions. Perhaps the most important future objective of human factors in air traffic control will be the definition and retention of the attributes of the controller's work that make it satisfying and enjoyable, so that there will always be able people who aspire to become air traffic controllers.

List of sources

Many reports and papers describing specific trials or research on human factors in air traffic control have been produced at establishments where there has been a continuous programme of human factors research, development or evaluation for a number of years, sometimes coupled with work by contractors under the supervision of these establishments. Few of these reports are cited in the text, as most are now of limited interest and often difficult or impossible to obtain. Appended below is a list of addresses of most of these main sources of human factors work on air traffic control, together with three further sources of relevant published papers.

DRA Centre for Human Sciences, F 131
DRA Farnborough
Hants GU14 6TD, UK
(formerly the Royal Air Force Institute of Aviation Medicine)

Scientific Group, Air Traffic Control Evaluation Unit
Bournemouth International Airport
Christchurch
Dorset BH23 6DF, UK

AD4 Division
DRA Malvern
St Andrews Road
Malvern
Worcester WR14 3PS, UK

Federal Systems Division
IBM
9201 Corporate Blvd
Rockville
MD 20850, USA

Department of Aviation
Ohio State University
Box 3022
Columbus
OH 43210–0022, USA

Center for Aviation/Aerospace Research
Embry–Riddle Aeronautical University
600 S Clyde Morris Blvd
Daytona Beach
FL 32114–3900, USA

NASA Langley Research Center
Hampton
VA 23665, USA

NASA–Ames Research Center
Moffett Field
CA 94035, USA

CTA Incorporated
6116 Executive Boulevard
Rockville
MD 20852 USA

The Mitre Corporation
1820 Dolley Madison Boulevard
McLean
VA 22102, USA

FAA Civil Aeromedical Institute
Mike Monroney Aeronautical Center
PO Box 25082
Oklahoma City
OK 73125, USA
(issues cumulative indexes of reports (Collins and Wayda 1994))

FAA Technical Center
Atlantic City Airport
NJ 08405, USA

Department of Psychology
York University
4700 Keele Street
Toronto
Ontario
Canada, M3J 1P3

Institutsdirektor
Institut fur Flugführung
DFVLR, Postfach 3267
3300 Braunschweig
Germany

CENA
Orly-Sud No. 005
94542 Orly Aérogare Cedex
Paris
France

Eurocontrol Experimental Centre
BP15
91222 Bretigny-sur-Orge Cedex
France

Further sources of publications

Human Factors and Ergonomics Society
PO Box 1369
Santa Monica
CA 90406–1369, USA

European Association for Aviation Psychology
Flugplatz
D–31673 Buckeburg
Germany
(formerly the Western European Association for Aviation Psychology
(WEAAP))

Flight Safety and Human Factors
Study Group Secretary
International Civil Aviation Organization
1000 Sherbrooke Street West
Montreal
Quebec
Canada H3A 2R2

References

Adair, D., 1985, *Air Traffic Control*, Wellingborough: Patrick Stephens.

Alexander, J. R., Alley, V. L., Ammerman, H. L., Hostetler, C. M. and Jones, G. W., 1988, *F.A.A. Air Traffic Control Operations Concepts*, Vol. III, Washington, DC: Federal Aviation Administration Report No. DOT/FAA/AP-87/01.

Altink, W. M. M., 1991, Construction and validation of a biodata selection instrument, *European Work and Organizational Psychologist*, **1**, 245–270.

Amaldi, P., 1993, Radar controller's problem solving and decision making skills, in Wise, J. A., Hopkin, V. D. and Stager, P. (Eds.), *Verification and Validation of Complex Systems: Additional Human Factors Issues*, pp. 33–57, Daytona Beach, FL: Embry-Riddle Aeronautical University Press.

Ammerman, H. L., Becker, E. S., Bergen, L. J., Davies, D. K., Inman, E. E. and Jones, G., 1986, *Operations Concept for the TCCC* [Tower Control Computer Complex] *Man–Machine Interface*, Washington, DC: Federal Aviation Administration Report No. DOT/FAA/AP-86/2.

Ammerman, H. L., Bergen, L. J., Davies, D. K., Hostetler, C. M., Inman, E. E. and Jones, G. W., 1988, *F.A.A. Air Traffic Control Operations Concepts*, Vol. VI, Washington, DC: Federal Aviation Administration, Report No. DOT/FAA/AP/87-01.

Arvey, R. D., Carter, G. W. and Buerkley, D. K., 1991, Job satisfaction: dispositional and situational influences, in Cooper, C. L. and Robertson, I. T. (Eds.), *International al Review of Industrial and Organizational Psychology*, **6**, 359–383.

Baddeley, A., 1990, *Human Memory: Theory and Practice*, Boston: Allyn and Bacon.

Bainbridge, L., 1987, Ironies of automation, in Rasmussen, J., Duncan, K. and Leplat, J. (Eds.), *New Technology and Human Error*, pp. 271–283, Chichester: Wiley.

Bainbridge, L., 1989, Development of skill, reduction in workload, in Bainbridge, L. and Quintanilla, S. A. R. (Eds.), *Developing Skills with Information Technology*, pp. 87–118, Chichester: Wiley.

Baker, C. H., 1962, *Man and Radar Displays*, Paris: NATO AGARDograph No. 60.

Baker, S., 1993, The role of incident investigation in system validation, in Wise, J. A., Hopkin, V. D. and Stager, P. (Eds.), *Verification and Validation of Complex Systems: Human Factors Issues*, pp. 239–250, Berlin: Springer-Verlag, NATO ASI Series Vol. F110.

Baldwin, R., 1991, Training requirements for automated ATC, in Wise, J. A., Hopkin, V. D. and Smith, M. L. (Eds.), *Automation and Systems Issues in Air Traffic Control*, pp. 469–479, Berlin: Springer-Verlag, NATO ASI Series Vol. F73.

Baldwin, R., 1993, Interaction of stages in validating and verifying ATC training, in Wise, J. A., Hopkin, V. D. and Stager, P. (Eds.), *Verification and Validation of Complex Systems: Human Factors Issues*, pp. 651–657, Berlin: Springer-Verlag, NATO ASI Series Vol. F110.

Ball, R. G. and Ord, G., 1983, Interactive conflict resolution in air traffic control, in Sime, M. E. and Coombs, M. J. (Eds.), *Designing for Human–Computer Interaction*, pp. 261–283, New York: Academic Press.

Bangen, H.-J., 1993, Validation problems in air traffic control systems, in Wise, J. A., Hopkin, V. D. and Stager, P. (Eds.), *Verification and Validation of Complex Systems: Human Factors Issues*, pp. 497–520, Berlin: Springer-Verlag, NATO ASI Series Vol. F110.

Barnes, R. M., 1958, *Motion and Time Study*, New York: John Wiley.

Bartram, D., 1994, Computer based assessment, in Cooper, C. L. and Robertson, I. T. (Eds.), *International Review of Industrial and Organizational Psychology*, **9**, 31–69.

Bashinski, H., Krois, P., Snyder, C. and Tobey, W., 1990, Lighting ergonomics evaluation issues for air traffic control facilities, in Das, B. (Ed.), *Advances in Industrial Ergonomics and Safety II*, pp. 947–954, London: Taylor & Francis.

Bauer, D., 1987, Use of slow phosphors to eliminate flicker in VDUs with bright background, *Displays*, **8**(1), 29–32.

Beevis, D. and St. Denis, G., 1992, Rapid prototyping and the human factors engineering process, *Applied Ergonomics*, **23**, 155–160.

Begault, D. R., 1993, Head-up auditory displays for traffic collision avoidance system advisories: a preliminary investigation, *Human Factors*, **35**(4), 707–717.

Benoit, A. (Ed.), 1973, *Air Traffic Control Systems*, Paris: NATO AGARD Conference Proceedings No. 105.

Benoit, A., (Ed.), 1975, *A Survey of Modern Air Traffic Control*, Paris: NATO AGARD-ograph No. 209 (2 volumes).

Benoit, A. (Ed.), 1980, *Air Traffic Management. Civil/Military Systems and Technologies*, Paris: NATO AGARD Conference Proceedings No. 273.

Benoit, A. (Ed.), 1983, *Air Traffic Control in Face of Users' Demand and Economy Constraints*, Paris: NATO AGARD Conference Proceedings No. 340.

Benoit, A. (Ed.), 1986, *Efficient Conduct of Individual Flights and Air Traffic* or *Optimum Utilization of Modern Technology for the Overall Benefit of Civil and Military Airspace Users*, Paris: NATO AGARD Conference Proceedings No. 410.

Benoit, A., 1991, Close ground/air cooperation in dynamic air traffic management, in Wise, J. A., Hopkin, V. D. and Smith, M. L. (Eds.), *Automation and Systems Issues in Air Traffic Control*, pp. 153–159, Berlin: Springer-Verlag, NATO ASI Series Vol. F73.

Benoit, A. and Israel, D. R. (Eds.), 1976, *Plans and Developments for Air Traffic Systems*, Paris: NATO AGARD Conference Proceedings No. 188.

Billings, C. E., 1991, *Human-Centered Aircraft Automation: A Concept and Guidelines*, Moffett Field, CA: NASA–Ames Research Center, NASA TM 103885.

Bisseret, A., 1971, Analysis of mental processes involved in air traffic control, *Ergonomics*, **14**(5), 565–570.

Bisseret, A., 1981, Application of signal detection theory to decision making in supervisory control: the effect of the operators' experience, *Ergonomics*, **24**(2), 81–94.

Boff, K. R. and Lincoln, J. E., 1988, *Engineering Data Compendium: Human Perception and Performance*, Wright Patterson Air Force Base, Ohio: Armstrong Aerospace Medical Research Laboratory.

Bohle, P. and Tilley, A. J., 1989, The impact of night work on psychological well-being, *Ergonomics*, **32**(9), 1089–1099.

Booher, H. R. (Ed.), 1990, *MANPRINT: An Approach to Systems Integration*, New York, NY: Van Nostrand Reinhold.

Boone, J. O., 1983, Radar training facility initial validation, Washington, DC: Federal Aviation Administration Report No. DOT/FAA/AM-83/9.

Bowers, J. M. and Benford, S. D. (Eds.), 1991; *Studies in Computer Supported Cooperative Work: Theory, Practice and Design*, Amsterdam: North-Holland.

Bradbury, J. N., 1991a, ICAO and civil/military coordination, in Wise, J. A., Hopkin,

V. D. and Smith, M. L. (Eds.), *Automation and Systems Issues in Air Traffic Control*, pp. 301–317, Berlin: Springer-Verlag, NATO ASI Series Vol. F73.

Bradbury, J. N., 1991b, ICAO and future air navigation systems, in Wise, J. A., Hopkin, V. D. and Smith, M. L. (Eds.), *Automation and Systems Issues in Air Traffic Control*, pp. 79–99, Berlin: Springer-Verlag, NATO ASI Series Vol. F73.

Bradley, G. E. and Hendrick, H. W. (Eds.), 1994, *Human Factors in Organizational Design and Management: IV*, Amsterdam: North-Holland.

Brenlove, M. S., 1987, *The Air Traffic System*, Ames, Iowa: Iowa State University Press.

Broach, D. and Brecht-Clark, J., 1994, Validation of the Federal Aviation Administration air traffic control specialist pre-training screen, Washington, DC: Federal Aviation Administration, Report No. DOT/FAA/AM-94/4.

Broach, D. and Manning, C. A., 1994, Validity of the air traffic control specialist non-radar screen as a predictor of performance in radar-based air traffic control training, Washington, DC: Federal Aviation Administration, Report No. DOT/FAA/AM-94/9.

Broadbent, D. E., Reason, J. T. and Baddeley, A., 1990, *Human Factors in Hazardous Situations*, Oxford: Clarendon Press.

Buck, R. O., 1984, *Aviation: International Air Traffic Control*, New York: Macmillan.

Buckley, E. P., DeBaryshe, B. D., Hitchner, N. and Kohn, P., 1983, *Methods and Measurements in Real-Time Air Traffic Control System Simulation*, Atlantic City, NJ: Federal Aviation Administration Report No. DOT/FAA/CT-83/26.

Buckley, E. P., O'Connor, W. F. and Beebe, T., 1969, *A Comparative Analysis of Individual and System Performance Indices for the Air Traffic Control System*, Atlantic City, NJ: Federal Aviation Administration, Report No. NA-69-40.

Carayon, P., 1993, Effect of electronic performance monitoring on job design and worker stress: review of the literature and conceptual model, *Human Factors*, **35**(3), 385–395.

Cardosi, K., 1993, Time required for transmission of time-critical ATC messages in an en-route environment, *International Journal of Aviation Psychology*, **3**(4), 303–313.

Cardosi, K. and Boole, P., 1991, *Analysis of Pilot Response Time to Time-Critical Air Traffic Control Calls*, Washington, DC: Federal Aviation Administration, Report No. DOT/FAA/RD-91/20.

Cardosi, K., Burki-Cohen, J., Boole, P., Mengert, P. and Disario, R., 1992, *Controller Response to Conflict Resolution Advisory*, Washington, DC: Federal Aviation Administration, Report No. DOT/FAA/NA-92/1.

Carlson, L. S. and Schultheis, S., 1993, Augmenting air traffic controller performance with integrated use of flexible automated tools, in Garland, D. J. and Wise, J. A. (Eds.), *Human Factors and Advanced Aviation Technologies*, pp. 83–90, Daytona Beach, FL: Embry-Riddle Aeronautical University Press.

Carroll, J. B., 1993, *Human Cognitive Abilities*, Cambridge: Cambridge University Press.

Carroll, J. M., 1993, Creating a design science of human–computer interaction, *Interacting with Computers*, **5**(1), 3–12.

Chapanis, A., Garner, W. R. and Morgan, C. T., 1949, *Applied Experimental Psychology*, New York: Wiley.

Chobot, R. B. and Chobot, M. C., 1991, Critical issues for decision makers in providing operator and maintainer training for advanced air traffic control systems, in Wise, J. A., Hopkin, V. D. and Smith, M. L., (Eds.), *Automation and Systems Issues in Air Traffic Control*, pp. 513–525, Berlin: Springer-Verlag, NATO ASI Series Vol. F73.

Clare, J., 1993, Requirements analysis for human systems information exchange, in Wise, J. A., Hopkin, V. D. and Stager, P. (Eds.), *Verification and Validation of Complex Systems: Human Factors Issues*, pp. 333–340, Berlin: Springer-Verlag NATO ASI Series Vol. F110.

Clegg, C., 1994, Psychology and information technology: the study of cognition in organizations, *British Journal of Psychology*, **85**, 449–477.

Cobb, B. B., Mathews, J. J. and Nelson, P. L., 1972, Attrition–retention rates of air traffic controller trainees recruited during 1960–1963 and 1968–1970, Washington, DC: Federal Aviation Administration Report No. FAA-AM-72-33.

Coeterier, J. F., 1971, Individual strategies in ATC freedom and choice, *Ergonomics*, **14**(5), 579–584.

Collins, W. E., Boone, J. O. and VanDeventer, A. D. (Eds.), 1980, *The Selection of Air Traffic Control Specialists: 1. History and Review of Contributions by the Civil Aeromedical Institute*, Washington, DC: Federal Aviation Administration Report No. DOT/FAA/AM-80/7.

Collins, W. E. and Wayda, M. E., 1994, *Index of FAA Office of Aviation Medicine Reports: 1961 through 1993*, Washington, DC: Federal Aviation Administration Report No. DOT/FAA/AM-94/1.

Connolly, D. W., 1979, *Voice Data Entry in Air Traffic Control*, Atlantic City, NJ: Federal Aviation Administration Report No. FAA-NA-79-20.

Convey, J. J., 1984, Personality assessment of ATC applicants, in Sells, S. B., Dailey, J. T. and Pickrel, E. W. (Eds.), *Selection of Air Traffic Controllers*, pp. 323–352, Washington, DC: Federal Aviation Administration, Report No. FAA-AM-84-2.

Cooper, C. L. and Makin, P. (Eds.), 1984, *Psychology for Managers*, London: Macmillan and the British Psychological Society.

Cooper, C. L. and Payne, R., 1991, *Personality and Stress: Individual Differences in the Stress Process*, Chichester: Wiley.

Costa, G., 1991, Shiftwork and circadian variations of vigilance and performance, in Wise, J. A., Hopkin, V. D. and Smith, M. L. (Eds.), *Automation and Systems Issues in Air Traffic Control*, pp. 267–280, Berlin: Springer-Verlag, NATO ASI Series Vol. F73.

Costa, G., 1993, Evaluation of workload in air traffic controllers, *Ergonomics*, **36**(9), 1111–1120.

Cox, M., 1994, *Task Analysis of Selected Operating Positions within UK Air Traffic Control*, Vol. 1, Main Report and Vol. 2 Appendices, Farnborough, Hants.: Royal Air Force Institute of Aviation Medicine Report No. 749.

Craik, F. I. M. and Lockhart, R. S., 1972, Levels of processing: a framework for memory research, *Journal of Verbal Learning and Verbal Behaviour*, **11**, 671–684.

Craik, K. J. W., 1947, Theory of the human operator in control systems. 1. The operator as an engineering system, *British Journal of Psychology*, **38**, 56–61.

Crawley, R., Spurgeon, P. and Whitfield, D., 1980, *Air Traffic Controller Reactions to Computer Assistance: A methodology for investigating controllers' motivations and satisfactions in the present system as a basis for system design*, University of Aston, Birmingham: Applied Psychology Department Report 94 (3 vols).

Crump, J. H., 1979, Review of stress in air traffic control: its measurement and effects, *Aviation, Space and Environmental Medicine*, **50**(3), 243–248.

Dailey, J. T. and Pickrell, E. W., 1984, Development of the air traffic controller occupational knowledge test, in Sells, S. B., Dailey, J. T. and Pickrel, E. W. (Eds.), *Selection of Air Traffic Controllers*, pp. 299–322, Washington, DC: Federal Aviation Administration Report No. FAA-AM-84-2.

Damos, D. L. (Ed.), 1991, *Multiple Task Performance*, London: Taylor & Francis.

Danaher, J. W., 1980, Human error in ATC system operations, *Human Factors*, **22**(5), 535–545.

David, H., 1991, Artificial intelligence and human factors in ATC: current activity at Eurocontrol Experimental Centre, in Wise, J. A., Hopkin, V. D. and Smith, M. L. (Eds.), *Automation and Systems Issues in Air Traffic Control*, pp. 174–179, Berlin: Springer-Verlag, NATO ASI Series, Vol. F73.

Davies, D. R., Matthews, G. and Wong, C. S. K., 1991, Ageing and work, in Cooper, C. L.

and Robertson, I. T. (Eds.), *International Review of Industrial and Organizational Psychology*, **6**, 149–211.

Day, P. O., 1991, Human factors in system design, in Wise, J. A., Hopkin, V. D. and Smith, M. L. (Eds.), *Automation and Systems Issues in Air Traffic Control*, pp. 201–208, Berlin: Springer-Verlag, NATO ASI Series Vol. F73.

Degani, A, and Wiener, E. L., 1993, Cockpit checklists: concepts, design and use, *Human Factors*, **35**(2), 345–359.

Della Rocco, P., Manning, C. A. and Wing, H., 1991, Selection of air traffic controllers for automated systems: applications from today's research, in Wise, J. A., Hopkin, V. D. and Smith, M. L. (Eds.), *Automation and Systems Issues in Air Traffic Control*, pp. 429–451, Berlin: Springer-Verlag, NATO ASI Series Vol. F73.

Denson, R. W., 1981, Team training: literature review and annotated bibliography, USAF Systems Command, Brooks AFB, TX: Human Resources Laboratory Report AFHRL-TR-80-40.

Diaper, D. and Addison, M., 1992, Task analysis and systems analysis for software development, *Interacting with Computers*, **4**, 124–139.

Diaper, D. and Sanger, C. (Eds.), 1993, *CSCW in Practice: An Introduction and Case Studies*, London: Springer-Verlag.

Dipboye, R. L., 1990, Laboratory vs. field research in industrial and organizational psychology, in Cooper, C. L. and Robertson, I. T. (Eds.), *International Review of Industrial and Organizational Psychology*, **5**, 1–34.

Dubois, M. and Gaussin, J., 1993, How to fit the man–machine interface and mental models of the operators, in Wise, J. A., Hopkin, V. D. and Stager, P. (Eds.), *Verification and Validation of Complex Systems: Human Factors Issues*, pp. 381–397, Berlin: Springer-Verlag, NATO ASI Series Vol. F110.

Dujardin, P., 1993, The inclusion of future users in the design and evaluation process, in Wise, J. A., Hopkin, V. D. and Stager, P. (Eds.), *Verification and Validation of Complex Systems: Human Factors Issues*, pp. 435–441, Berlin, Springer-Verlag NATO ASI Series Vol. F110.

Duke, G., 1986, *Air Traffic Control*, Shepperton: Ian Allan.

Dul, J. and Weerdmeester, B. A., 1993, *Ergonomics for Beginners: A Quick Reference Guide*, London: Taylor & Francis.

Durrett, H. J. (Ed.), 1987, *Color and the Computer*, Boston: Academic Press.

Duytschaever, D., 1993, The development and implementation of the EUROCONTROL central air traffic control management unit, *Journal of Navigation*, **46**(3), 343–352.

Edmonds, E. (Ed.), 1992, *The Separable User Interface*, London: Academic Press.

Edmondson, D. and Johnson, P., 1990, DETAIL: an approach to task analysis, in Life, M. A., Narborough-Hall, C. S. and Hamilton, W. I. (Eds.), *Simulation and the User Interface*, pp. 147–158, London: Taylor & Francis.

Edwards, D. C. (1990), *Pilot: Mental and Physical Performance*, Ames, Iowa: Iowa State University Press.

Edwards, J. L., 1991, Intelligent dialogue in air traffic control systems, in Wise, J. A., Hopkin, V. D. and Smith, M. L. (Eds.), *Automation and Systems Issues in Air Traffic Control*, pp. 137–151, Berlin: Springer-Verlag, NATO ASI Series, Vol. F73.

Empson, J., 1987, Error auditing in air traffic control, in Wise, J. A. and Debons, A. (Eds.), *Information Systems: Failure Analysis*, pp. 191–198, Berlin: Springer-Verlag, NATO Science Series Vol. F32.

Empson, J., 1991, Cognitive failures in military air traffic control, in Wise, J. A., Hopkin, V. D. and Smith, M. L. (Eds.), *Automation and Systems Issues in Air Traffic Control*, pp. 339–348, Berlin: Springer Verlag, NATO ASI Series Vol. F73.

Endsley, M. R., 1993, Situation awareness: a fundamental factor underlying the successful implementation of AI in the air traffic control system, in Garland, D. J. and

Wise, J. A. (Eds.), *Human Factors and Advanced Aviation Technologies*, pp. 117–122, Daytona Beach, FL: Embry-Riddle Aeronautical University Press.

Endsley, M. R., 1994a, Situation awareness in dynamic human decision making: theory, in Gilson, R. D., Garland, D. J. and Koonce, J. M. (Eds.), *Situational Awareness in Complex Systems*, pp. 27–58, Daytona Beach, FL: Embry-Riddle Aeronautical University Press.

Endsley, M. R., 1994b, Situation awareness in dynamic human decision making: measurement, in Gilson, R. D., Garland, D. J. and Koonce, J. M. (Eds.), *Situational Awareness in Complex Systems*, pp. 79–97, Daytona Beach, FL: Embry-Riddle Aeronautical University Press.

Endsley, M. R. and Rodgers, M. D., 1994, *Situation Awareness Information Requirements for Enroute Air Traffic Control*, Washington, DC: Federal Aviation Administration Report No. DOT/FAA/AM-94-27.

Evans, A. E., 1994, Human factors certification in the development of future ATC systems, in, Wise, J. A., Hopkin, V. D. and Garland, D. J. (Eds.), *Human Factors Certification of Advanced Aviation Technologies*, pp. 87–96, Daytona Beach, FL: Embry-Riddle Aeronautical University Press.

Evans, G. W., Johansson, G. and Carrere, S., 1994, Psychosocial factors and the physical environment: inter-relations in the workplace, in Cooper, C. L. and Robertson, I. T. (Eds.), *International Review of Industrial and Organizational Psychology*, **9**, 1–29.

FAA, 1988, Symposium on Air Traffic Control Training for Tomorrow's Technology, Oklahoma City, OK: Federal Aviation Administration.

FAA, 1990, *The National Plan for Aviation Human Factors*, Washington, DC: Federal Aviation Administration.

Fabry, J. M. and Lupinetti, A. A., 1989, Application of simulation and artificial intelligence technology for ATC training, *Proceedings of the IEEE*, **77**(11), 1762–1765.

Falzon, P. (Ed.), 1990, *Cognitive Ergonomics: Understanding, Learning and Designing Human–Computer Interaction*, New York: Harcourt Brace Jovanovich.

Farmer, E. W. (Ed.), 1994, *Aircrew Workload*, Aldershot, Hants: Avebury Aviation.

Farmer, E. W., Belyavin, A. J., Berry, A., Tattersall, A. J. and Hockey, G. R. J., 1990, *Stress in Air Traffic Control. 1. Survey of NATS Controllers*, Farnborough, Hants: Royal Air Force Institute of Aviation Medicine Report No. 689.

Farmer, E. W., Belyavin, A. J., Tattersall, A. J., Berry, A. and Hockey, G. R. J., 1991, *Stress in Air Traffic Control. 2. Effects of Increased Workload*, Farnborough, Hants: Royal Air Force Institute of Aviation Medicine Report No. 701.

Fassert, C. and Pichancourt, I, 1993, Evaluation and use of prototypes: cases in air traffic control, in Wise, J. A., Hopkin, V. D. and Stager, P. (Eds.), *Verification and Validation of Complex Systems: Additional Human Factors Issues*, pp. 69–75, Daytona Beach, FL: Embry-Riddle Aeronautical University Press.

Field, A., 1980, *The Control of Air Traffic*, Eton: Eton Publishing.

Field, A., 1985, *International Air Traffic Control: Management of the World's Airspace*, Oxford: Pergamon.

Finkelman, J. M., 1994, A large database study of the factors associated with work-induced fatigue, *Human Factors*, **36**(2), 232–243.

Finkelman, J. M., and Kirschner, C., 1980, An information-processing interpretation of air traffic control stress, *Human Factors*, **22**, 561–567.

Fisher, J., 1991, Defining the novice user, *Behaviour and Information Technology*, **10**(5), 437–441.

Fitts, P. M. (Ed.), 1951a, *Human Engineering for an Effective Air Navigation and Traffic Control System*, Washington, DC: National Research Council, Committee on Aviation Psychology.

Fitts, P. M., 1951b, Engineering psychology and equipment design, in Stevens, S. S. (Ed.), *Handbook of Experimental Psychology*, pp. 1287–1340, London: Chapman and Hall.

Fleishman, E. A. and Quaintance, M. K., 1984, *Taxonomies of Human Performance: The Description of Human Tasks*, Orlando, FL: Academic Press.

Foley, J. D., VanDam, A., Feiner, S. K. and Hughes, J. F., 1990., *Computer Graphics: Principles and Practice*, New York: Addison-Wesley.

Folkard, S. and Condon, R., 1987, Night shift paralysis in air traffic control officers, *Ergonomics*, **30**(9), 1353–1363.

Folkard, S. and Monk, T. H., 1985, *Hours of Work: Temporal Factors in Work Scheduling*, Chichester: Wiley.

Foster, H. D., 1993, Resilience theory and system evaluation, in Wise, J. A., Hopkin, V. D. and Stager P. (Eds.), *Verification and Validation of Complex Systems: Human Factors Issues*, pp. 35–60, Berlin: Springer-Verlag, NATO ASI Series Vol. F110.

Foushee, H. C. and Helmreich, R. L., 1988, Group interaction and flight crew performance, in Wiener, E. L. and Nagel, D. C. (Eds.), *Human Factors in Aviation*, pp. 189–227, San Diego: Academic Press.

Furnham, A., 1994, *Personality at Work: Individual Differences in the Workplace*, London: Routledge.

Furze, R., 1993, The organization of ATC. R. and D., in Wise, J. A., Hopkin, V. D. and Stager, P. (Eds.), *Verification and Validation of Complex Systems: Additional Human Factors Issues*, pp. 59–68, Daytona Beach, FL: Embry-Riddle Aeronautical University Press.

Gal, R. and Mangelsdorff, A. D., 1991, *Handbook of Military Psychology*, Chichester: Wiley.

Galotti, V. P., 1993, An expert air traffic control teaching machine: critical learning issues, in Wise, J. A., Hopkin, V. D. and Stager, P. (Eds.), *Verification and Validation of Complex Systems: Human Factors Issues*, pp. 635–650, Berlin: Springer-Verlag, NATO ASI Series Vol. F110.

Ganster, D. C. and Fusilier, M. R., 1989, Control in the workplace, in Cooper, C. L. and Robertson, I. T. (Eds.), *International Review of Industrial and Organizational Psychology*, **4**, 235–280.

Garcia, M., Badre, A. N. and Stasko, J. T., 1994, Development and validation of icons varying in their abstractness, *Interacting with Computers*, **6**(2), 191–211.

Garland, D. J., 1991, Automated systems: the human factor, in Wise, J. A., Hopkin, V. D. and Smith, M. L. (Eds.), *Automation and Systems Issues in Air Traffic Control*, pp. 209–215, Berlin: Springer-Verlag, NATO ASI Series Vol. F73.

Garland, D. J. and Hopkin, V. D., 1994, Controlling automation in future air traffic control: the impact on situational awareness, in Gilson, R. D., Garland, D. J. and Koonce, J. M. (Eds.), *Situational Awareness in Complex Systems*, pp. 179–197, Daytona Beach, FL: Embry-Riddle Aeronautical University Press.

Garland, D. J., Stein, E. S., Wise, J. A. and Blanchard, J. W., 1993, Situational awareness in air traffic control: a critical yet neglected phenomenon, in Garland, D. J. and Wise, J. A. (Eds.), *Human Factors and Advanced Aviation Technologies*, pp. 123–145, Daytona Beach, FL: Embry-Riddle Aeronautical Univeristy Press.

Garland, D. J. and Wise, J. A. (Eds.), 1993, *Human Factors and Advanced Aviation Technologies*, Daytona Beach, FL: Embry-Riddle Aeronautical University Press.

Gibson, R. S. 1993, Verification and validation of the training components of highly complex systems, in Wise, J. A., Hopkin, V. D. and Stager, P. (Eds.), *Verification and Validation of Complex Systems: Human Factors Issues*, pp. 627–633, Berlin: Springer-Verlag, NATO ASI Series Vol. F110.

Gilbert, G., 1973, *Air Traffic Control: The Uncrowded Sky*, Washington, DC: Smithsonian Institution Press Publication No. 4873.

Gilbreth, F. B., 1919, *Applied Motion Study*, New York: Macmillan.

Gillespie, R., 1993, *Manufacturing Knowledge: A History of the Hawthorne Experiments*, Cambridge: Cambridge University Press.

Gilliland, K. and Schlegel, R. E., 1993, *Readiness to Perform Testing: A Critical*

Analysis of the Concept and Current Practices, Washington, DC: Federal Aviation Administration Report. No. DOT/FAA/AM-93/13.

Gilson, R. D., Garland, D. J. and Koonce, J. M. (Eds.), 1994, *Situational Awareness in Complex Systems*, Daytona Beach, FL: Embry-Riddle Aeronautical University Press.

Golembiewski, R. T. (Ed.), 1993, *Handbook of Organizational Behaviour*, New York: Marcel Dekker.

Gopher, D., Weil, M. and Bareket, T., 1994, Transfer of skill from a computer game trainer to flight, *Human Factors*, 36(3), 387–405.

Gosling, G. D., 1993, Artificial intelligence in air traffic control, in Garland, D. J. and Wise, J. A. (Eds.), *Human Factors and Advanced Aviation Technologies*, pp. 31–43, Daytona Beach, FL: Embry-Riddle Aeronautical University Press.

Grandjean, E. P., 1988, *Fitting the Task to the Man: An Ergonomic Approach*, London: Taylor & Francis.

Grandjean, E. P., Wotzka, G., Schaad, R. and Gilgen, A., 1971, Fatigue and stress in air traffic controllers, *Ergonomics*, 14, 159–165.

Graves, D., 1992, *A Layman's Guide to United Kingdom Air Traffic Control*, Shrewsbury: Airlife Publishing.

Greenstein, J. S. and Arnaut, L. Y., 1987, Human factors aspects of manual computer input devices, in Salvendy, G. (Ed.), *Handbook of Human Factors*, pp. 1450–1489, Chichester: Wiley.

Gregory, R. L. (Ed.), 1987, *The Oxford Companion to the Mind*, Oxford: Oxford University Press.

Haglund, R., 1994, Presentation of a Swedish study program concerning recruitment, selection and training of student air traffic controllers: the MRU project, phase 1, in Wise, J. A., Hopkin, V. D. and Garland, D. J. (Eds.), *Human Factors Certification of Advanced Aviation Technologies*, pp. 143–161, Daytona Beach, FL: Embry-Riddle Aeronautical University Press.

Hambleton, R. K. (Ed.), 1991, Test translations for cross-cultural studies, *Bulletin of the International Test Commission*, 18, 1–2.

Hancock, P. A. 1991, The aims of human factors and their application to issues in automation and air traffic control, in Wise, J. A., Hopkin, V. D. and Smith, M. L. (Eds.), *Automation and Systems Issues in Air Traffic Control*, pp. 187–199, Berlin: Springer-Verlag, NATO ASI Series Vol. F73.

Hancock, P. A., 1993a, On the future of hybrid human–machine systems, in Wise, J. A., Hopkin, V. D. and Stager, P. (Eds.), *Verification and Validation of Complex Systems: Human Factors Issues*, pp. 61–85, Berlin: Springer-Verlag, NATO ASI Series Vol. F110.

Hancock, P. A., 1993b, On the practicalities and politics of task allocation in hybrid human–machine systems, in Garland, D. J. and Wise, J. A. (Eds.), *Human Factors and Advanced Aviation Technologies*, pp. 111–115, Daytona Beach, FL: Embry-Riddle Aeronautical University Press.

Hancock, P. A. and Chignell, M. H. (Eds.), 1989, *Intelligent Interfaces: Theory, Research and Design*, New York: North-Holland.

Hancock, P. A. and Meshkati, N. (Eds.), 1988, *Human Mental Workload*, Amsterdam: North-Holland.

Hartley, J., 1985, *Designing Instructional Text*, London: Kogan Page.

Hartley, J. F., 1992, The psychology of industrial relations, in Cooper, C. L. and Robertson, I. T. (Eds.), *International Review of Industrial and Organizational Psychology*, 7, 201–243.

Hartley, J. F. and Stephenson, G. M. (Eds.), 1992, *Employment Relations: The Psychology of Influence and Control at Work*, Oxford: Blackwell Publishers.

Harwood, K., 1993, Defining human-centered system issues for verifying and validating air traffic control systems, in Wise, J. A., Hopkin, V. D. and Stager, P. (Eds.),

Verification and Validation of Complex Systems: Human Factors Issues, pp. 115–129, Berlin: Springer-Verlag, NATO ASI Series Vol. F110.

Harwood, K. and Sanford, B., 1994, Evaluation in context: ATC automation in the field, in Wise, J. A., Hopkin, V. D. and Garland, D. J. (Eds.), *Human Factors Certification of Advanced Aviation Technologies*, pp. 247–262, Daytona Beach, FL: Embry-Riddle Aeronautical University Press.

Hausen, H.-L. (Ed.), 1984, *Software Validation: Inspection, Testing, Verification, Alternatives*, Amsterdam: North-Holland.

Hawkins, F. H., 1993, *Human Factors in Flight*, Aldershot, Hants: Ashgate Publishing.

Hearnshaw, L. S., 1987, *The Shaping of Modern Psychology*, London: Routledge.

Helander, M. (Ed.), 1991, *Handbook of Human–Computer Interaction*, Amsterdam: North-Holland.

Heller, F., 1991, The underutilization of applied psychology, *The European Work and Organizational Psychologist*, **1**(1), 9–25.

Hellier, E. J., Edworthy, J. and Dennis, I, 1993, Improving auditory warning design: quantifying and predicting the effects of different warning parameters on perceived urgency, *Human Factors*, **35**(4), 693–706.

Hendy, K. C., Hamilton, K. M. and Landry, L. S., 1993, Measuring subjective workload: when is one scale better than many? *Human Factors*, **35**(4), 579–601.

Henry, J. H., Kamrass, M. E., Orlansky, J., Rowan, T. C., String, J. and Reichenbach, R. E., 1975, *Training of US Air Traffic Controllers*, Arlington, VA: Institute for Defense Analysis Report for FAA Office of Personnel and Training.

Herschler, D. A., 1991, Resource management training for air traffic controllers, in Wise, J. A., Hopkin, V. D. and Smith, M. L., (Eds.), *Automation and Systems Issues in Air Traffic Control*, pp. 497–503, Berlin: Springer-Verlag, NATO ASI Series Vol. F73.

Herzberg, F., 1966, *Work and the Nature of Man*, New York: World Publishing Co.

Hilton Systems, 1994, *Age 60 Study* (in 4 Parts), Washington, DC: Federal Aviation Administration Report DOT/FAA/AM-94-20,21,22.

Hlibowicki, A. and Bowen, D., 1993, Electronic flight strips as an input device for automatic conflict prediction and resolution, in Garland, D. J. and Wise, A. J. (Eds.), 1993, *Human Factors and Advanced Aviation Technologies*, pp. 45–65, Daytona Beach, FL: Embry-Riddle Aeronautical University Press.

HMSO, 1989, *Human Factors For Designers of Equipment*, London: HMSO Defence Standard 00–25.

Hockaday, S. L. M., 1993, Applicaion of artificial intelligence to air traffic control, in Garland, D. J. and Wise, J. A. (Eds.), *Human Factors and Advanced Aviation Technologies*, pp. 9–17, Daytona Beach, FL: Embry-Riddle Aeronautical University Press.

Hockey, R. (Ed.), 1983, *Stress and Fatigue in Human Performance*, New York: Wiley.

Hollnagel, E., 1993, The reliability of interactive systems: simulation based assessment, in Wise, J. A., Hopkin, V. D. and Stager, P. (Eds.), *Verification and Validation of Complex Systems: Human Factors Issues*, pp. 205–221, Berlin: Springer-Verlag, NATO ASI Series Vol. F110.

Hopkin, V. D., 1970, *Human Factors in the Ground Control of Aircraft*, Paris: NATO AGARDograph No. 142.

Hopkin, V. D., 1978, An appraisal of real-time simulation in air traffic control, *Journal of Education Technology Systems*, **7**(1), 91–102.

Hopkin, V. D., 1979, Mental workload measurement in air traffic control, in Moray, N. (Ed.), *Mental Workload: its Theory and Measurement*, pp. 381–386, London: Plenum Press.

Hopkin, V. D., 1980a, Boredom, *The Controller*, **19**(1), 6–10.

Hopkin, V. D., 1980b, The measurement of the air traffic controller, *Human Factors*, **22**(5), 547–560.

Hopkin, V. D., 1982a, *Human Factors in Air Traffic Control*, Paris: NATO AGARDograph No. 275.

Hopkin, V. D., 1982b, Sleep and work–rest cycles, *The Controller*, 21(4), 44–46.

Hopkin, V. D., 1982c, *Subjective Assessment Techniques in Air Traffic Control Evaluations*, Farnborough, Hants: Royal Air Force Institute of Aviation Medicine Report No. 622.

Hopkin, V. D., 1985, Fitting machines to people in air traffic control automation, in: *Proceedings of Seminar on Informatics in Air Traffic Control*, pp. 147–163, Capri, Italy: EEC Commission Project on Transport Research.

Hopkin, V. D., 1988, Training implications of technological advances in air traffic control, in: *Proceedings of Symposium on Air Traffic Control Training for Tomorrow's Technology*, pp. 6–26, Oklahoma City, OK: Federal Aviation Administration.

Hopkin, V. D., 1989a, Man–machine interface problems in designing air traffic control systems, *Proceedings of the IEEE*, 77(11), 1634–1642.

Hopkin, V. D., 1989b, Implications of automation on air traffic control, in Jensen, R. S. (Ed.), *Aviation Psychology*, pp. 96–108, Aldershot, Hants: Gower Technical.

Hopkin, V. D., 1990, Operational evaluation, in Life, M. A., Narborough-Hall, C. S. and Hamilton, I. (Eds.), *Simulation and the User Interface*, pp. 73–83, London: Taylor & Francis.

Hopkin, V. D., 1991a, Closing remarks, in Wise, J. A., Hopkin, V. D. and Smith, M. L. (Eds.), *Automation and Systems Issues in Air Traffic Control*, pp. 553–559, Berlin: Springer-Verlag, NATO ASI Series Vol. F73.

Hopkin, V. D., 1991b, Issues in colour application, in Widdel, H. Post, D. L., Grossman, J. D. and Walraven, J. (Eds.), *Colour in Electronic Displays*, pp. 191–207, London: Plenum Press.

Hopkin, V. D., 1991c, Automated flight strip usage: lessons from the functions of paper flight strips, in *Proceedings of AIAA/NASA/FAA/HFS Symposium on Challenges in Aviation Human Factors: The National Plan*, pp. 15–17.

Hopkin, V. D., 1991d, The impact of automation on air traffic control systems, in Wise, J. A., Hopkin, V. D. and Smith, M. L. (Eds.), *Automation and Systems Issues in Air Traffic Control*, pp. 3–19, Berlin: Springer-Verlag, NATO ASI Series Vol. F73.

Hopkin, V. D., 1993a, Human factors implications of air traffic control automation, in Smith, M. J. and Salvendy, G. (Eds.), *Human–Computer Interaction: Applications and Case Studies*, pp. 145–150, Amsterdam: Elsevier.

Hopkin, V. D., 1993b, Verification and validation: concepts, issues and applications, in Wise, J. A., Hopkin, V. D. and Stager, P. (Eds.), *Verification and Validation of Complex Systems: Human Factors Issues*, pp. 9–33, Berlin: Springer-Verlag, NATO ASI Series Vol. F110.

Hopkin, V. D., 1993c, Human factors issues in air traffic control and aircraft maintenance, in Garland, D. J. and Wise, J. A. (Eds.), *Human Factors and Advanced Aviation Technologies*, pp. 19–28, Daytona Beach, FL: Embry-Riddle Aeronautical University Press.

Hopkin, V. D., 1994a, Human performance implications of air traffic control automation, in Mouloua, M. and Parasuraman, R. (Eds.), *Human Performance in Automated Systems: Current Research and Trends*, pp. 314–319, Hillsdale, NJ: Laurence Erlbaum.

Hopkin, V. D., 1994b, Situational awareness in air traffic control, in Gilson, R. D., Garland, D. J. and Koonce, J. M. (Eds.), *Situational Awareness in Complex Systems*, pp. 171–178, Daytona Beach, FL: Embry-Riddle Aeronautical University Press.

Hopkin, V. D., 1994c, Organizational and team aspects of air traffic control training, in Bradley, G. E. and Hendrick, H. W. (Eds.), *Human Factors in Organizational Design and Management: IV*, pp. 309–314, Amsterdam: North-Holland.

Hopkin, V. D., 1994d, Automation tests controller training, *Air Traffic Management Yearbook 1994/5*, pp. 60–63, London: Camrus Publishers.

Hopkin, V. D., 1994e, Colour on air traffic control displays, *Information Display*, **10**(1), 14–18.

Hopkin, V. D. and McClumpha, A., 1980, Real-time simulation: an indispensable but over-used evaluation technique, in *Modelling and Simulation of Avionics Systems and Command, Control and Communications Systems*, pp. 12, 1–6, Paris: NATO Conference Proceedings No. 268.

Hopkin, V. D. and Taylor, R. M., 1979, *Human Factors in the Design and Evaluation of Aviation Maps*, Paris: NATO AGARDograph No. 225.

Horne, J. A. and Ostberg, O., 1976, A self-assessment questionnaire to determine morningness–eveningness in human circadian rhythms, *International Journal of Chronobiology*, **4**, 97–110.

Hoshstrasser, B. D. and Small, R. L., 1993, The impact of associate systems technology on air traffic control, in Garland, D. J. and Wise, J. A., *Human Factors and Advanced Aviation Technologies*, pp. 99–106, Daytona Beach, FL: Embry-Riddle Aeronautical University Press.

Hoshstrasser, B. D. and Small, R. L., 1994, The impact of associate systems technology on an air traffic controller's situational awareness, in Gilson, R. D., Garland, D. J. and Koonce, J. M. (Eds.), *Situational Awareness in Complex Systems*, pp. 227–236, Daytona Beach, FL: Embry-Riddle Aeronautical University Press.

Huey, B. M. and Wickens, C. D. (Eds.), 1993, *Workload Transition: Implications for Individual and Team Performance*, Washington, DC: National Academy Press.

Hughes, J. A., Randall, D. and Shapiro, D., 1993a, Faltering from ethnography to design, in Wise, J. A., Hopkin, V. D. and Stager, P. (Eds.), *Verification and Validation of Complex Systems: Additional Human Factors Issues*, pp. 77–90, Daytona Beach, FL: Embry Riddle Aeronautical University Press.

Hughes, J. A., Somerville, I. Bentley, R. and Randall, D., 1993b, Designing with ethnography: making work visible, *Interacting with Computers*, **5**(2), 239–253.

Humphrey, D. G. and Kramer, A. F., 1994, Towards a psychophysical assessment of dynamic changes in mental workload, *Human Factors*, **36**(1), 3–26.

Hunt, R. W. G., 1991, *Measuring Colour*, Hemel Hempstead: Ellis Horwood.

Hunt, V. R. and Zellweger, A., 1987, The FAA's Advanced Automation System: strategies for future air traffic control systems, *Computer*, **20**, 19–32.

ICAO, 1993, *Human Factors Digest No. 8: Human Factors in Air Traffic Control*, Montreal, Canada: International Civil Aviation Organization, Circular 241-AN/145.

Ilgen, D. R. and Schneider, J., 1991, Performance measurement: a multidiscipline view, in Cooper, C. L. and Robertson, I. T. (Eds.), *International Review of Industrial and Organizational Psychology*, **6**, 71–108.

Illman, P. E., 1993, *The Pilot's Air Traffic Control Handbook*, Blue Ridge Summit, PA: TAB Books.

Isaac, A. R. and Marks, D. F. 1994, Individual differences in mental imagery experience: developmental changes and specialization, *British Journal of Psychology*, **85**, 479–500.

Ivergard, T., 1989, *Handbook of Control Room Design and Ergonomics*, London: Taylor & Francis.

Jackson, R., Macdonald, L. and Freeman, K. 1994, *Computer Generated Colour: A Practical Guide to Presentation and Display*, Chichester: Wiley.

Jensen, R. S. (Ed.), 1989, *Aviation Psychology*, Aldershot: Gower Technical.

Johnson, W. B., 1993, Enhancing human performance in maintenance with automated information and training, in Garland, D. J. and Wise, J. A. (Eds.), *Human Factors and Advanced Aviation Technologies*, pp. 179–189, Daytona Beach, FL: Embry-Riddle Aeronautical University Press.

Johnson-Laird, P. N., 1983, *Mental Models*, Cambridge: Cambridge University Press.

Jordan, N., 1968, *Themes in Speculative Psychology*, London: Tavistock Publications.

Jorna, P. G. A. M., 1991, Operator workload as a limiting factor in complex systems, in Wise, J. A., Hopkin, V. D. and Smith, M. L. (Eds.), *Automation and Systems Issues in Air Traffic Control*, pp. 281–292, Berlin: Springer-Verlag, NATO ASI Series Vol. F73.

Jorna, P. G. A. M., 1993, The human component of system validation, in Wise, J. A., Hopkin, V. D. and Stager, P., *Verification and Validation of Complex Systems: Human Factors Issues*, pp. 281–304, Berlin: Springer-Verlag, NATO ASI Series Vol. F110.

Julesz, B., 1981, Textons, the elements of texture perception, and their interactions, *Nature*, **290**, 91–97.

Kanfer, R., 1992, Work motivation: new directions in theory and research, in Cooper, C. L. and Robertson, I. T. (Eds.), *International Review of Industrial and Organizational Psychology*, **7**, 1–53.

Kanki, B. G., Folk, V. G. and Irwin, C. M., 1991, Communication variations and aircrew performance, *International Journal of Aviation Psychology*, **1**(2), 149–162.

Kaplan, M, 1991, Issues in cultural ergonomics, in Wise, J. A., Hopkin, V. D. and Smith, M. L. (Eds.), *Automation and Systems Issues in Air Traffic Control*, pp. 381–393, Berlin: Springer-Verlag. NATO ASI Series Vol. F73.

Karson, S. and O'Dell, J. W., 1974, Personality makeup of the American air traffic controller, *Aerospace Medicine*, **45**, 1001–1007.

Kavanagh, M. J., Hurst, M. W. and Rose, R., 1981, The relationship between job satisfaction and psychiatric health symptoms for air traffic controllers, *Personnel Psychology*, **34**, 691–707.

Keenan, T., 1989, Selection interviewing, in Cooper, C. L. and Robertson, I. T. (Eds), *International Review of Industrial and Organizational Psychology*, **4**, 1–23.

Kendal, B. 1990, Air Navigation Systems. The Beginnings of Directional Radio Techniques for Air Navigation, 1910–1940, *Journal of Navigation*, **43**(3), 313–330.

Kerns, K., 1992, Data-link communications between controllers and pilots: a review and synthesis of the simulation literature, *International Journal of Aviation Psychology*, **1**(3), 181–204.

Kinney, G. C. and Culhane, L. G., 1978, *Colour in Air Traffic Control Displays: Review of the Literature and Design Considerations*, McLean, VA: Mitre Corporation Report No. 7728.

Kirlik, A, 1993, Modeling strategic behaviour in human–automation interaction: why an "aid" can (and should) go unused, *Human Factors*, **35**, 221–242.

Kirwan, B. I., 1994, *A Guide to Practical Human Reliability Assessment*, London: Taylor & Francis.

Kirwan, B. and Ainsworth, L. M. (Eds.), 1992, *A Guide to Task Analysis*, London: Taylor & Francis.

Klein, G. A., Orasanu, J. Calderwood, R. and Zsambok, C. E. (Eds.), 1993, *Decision Making in Action: Models and Methods*, Norwood, NJ: Ablex.

Kleinbeck, U., Quast, H.–H., Thiery, H. and Hacker, H. (Eds.), 1990, *Work Motivation*, New York: Lawrence Erlbaum.

Knepell, P. L. and Arangno, D. C., 1993, *Simulation Validation: A Confidence Assessment Methodology*, Los Alamitos, CA: IEEE Computer Society Press.

Lane, S. E. J., 1991, The implementation and impact of automatic data processing on UK military ATC operations, in Wise, J. A., Hopkin, V. D. and Smith, M. L. (Eds.), *Automation and Systems Issues in Air Traffic Control*, pp. 47–53, Berlin: Springer-Verlag NATO ASI Series Vol. F73.

Langan-Fox, C. P. and Empson, J. A. C., 1985, "Actions not as planned" in military air traffic control, *Ergonomics*, **28**, 1509–1521.

Layton, C., Smith, P. J. and McCoy, C. E., 1994, Design of a co-operative problem-solving system for en-route flight planning: an empirical evaluation, *Human Factors*, **36**(1), 94–119.

Leroux, M., 1993, The role of verification and validation in the design process of knowledge based components of air traffic control systems, in Wise, J. A., Hopkin, V. D. and Stager, P. (Eds.), *Verification and Validation of Complex Systems: Human Factors Issues*, pp. 357–373, Berlin: Springer-Verlag, NATO ASI Series Vol. F110.

Levesley, J., 1991, From under the headset: the role of the air traffic controllers' professional association in present and future air traffic control systems development, in Wise, J. A., Hopkin, V. D. and Smith, M. L. (Eds.), *Automation and Systems Issues in Air Traffic Control*, pp. 55–59, Berlin: Springer-Verlag, NATO ASI Series Vol. F73.

Life, M. A., Narborough-Hall, C. S. and Hamilton, W. I. (Eds.), 1990, *Simulation and the User Interface*, London: Taylor & Francis.

Lind, M., 1991, Modelling control tasks in complex systems, in Wise, J. A., Hopkin, V. D. and Smith, M. L. (Eds.), *Automation and Systems Issues in Air Traffic Control*, pp. 218–233, Berlin: Springer-Verlag NATO ASI Series F73.

Lindgren, R. N., 1991, Social, political and regulatory issues concerning harmonization of interacting air traffic systems in Western Europe, in Wise, J. A., Hopkin, V. D. and Smith, M. L., (Eds.), *Automation and Systems Issues in Air Traffic Control*, pp. 101–106, Berlin: Springer-Verlag NATO ASI Series Vol. F73.

Locke, E. A., 1986, *Generalizing from Laboratory to Field Settings*, Lexington, MA: Lexington Books.

Logie, R. H., 1993, Working memory and human–machine systems, in Wise, J. A., Hopkin, V. D. and Stager, P. (Eds.), *Verification and Validation of Complex Systems: Human Factors Issues*, pp. 341–353, Berlin: Springer-Verlag, NATO ASI Series Vol. F110.

Logie, R. H, and Denis, M. (Eds.), 1991, *Mental Images in Human Cognition*, New York: Elsevier Science Publishers.

Lohse, G., Walker, N., Biolsi, K. and Rueter, H., 1991, Classifying graphical information, *Behaviour and Information Technology*, **10**(5), 419–436.

Lord, R. G. and Maher, K. J., 1989, Cognitive processes in industrial and organizational psychology, in Cooper, C. L. and Robertson, I. T. (Eds.), *International Review of Industrial and Organizational Psychology*, **4**, 49–91.

Luffsey, W. S., 1990, *Air Traffic Control: How to Become an FAA Air Traffic Controller*, New York: Random House.

Lysacht, R. J., Hill, S. G., Dick, A. O., Plamondon, V. D., Linton, P. M., Wierwille, W. W., Zaklad, A. L., Bittner, A. C. and Wherry, R. J., 1989, *Operator Workload: Comprehensive Review and Evaluation of Operator Workload Methodologies*, Fort Bliss, TX: US Army Research Institute Report 851.

Mackworth, N. H., 1950, *Researches on the Measurement of Human Performance*, London: HMSO, Medical Research Council Special Report Series 268.

Manning, C. A. and Aul, J. C., 1992, *Evaluation of an Alternative Method for Hiring Air Traffic Control Specialists with Prior Military Experience*, Washington, DC: Federal Aviation Administration Report No. DOT/FAA/AM-92/5.

Manning, C. A. and Broach, D., 1992, *Identifying Ability Requirements for Operators of Future Automated Air Traffic Control Systems*, Washington, DC: Federal Aviation Administration Report No. DOT/FAA/AM-92/26.

Manning, C. A., Kegg, P. S. and Collins, W. E., 1989, Selection and screening programs for air traffic control, in Jensen, R. S. (Ed.), *Aviation Psychology*, pp. 321–341, Aldershot, Hants: Gower Technical.

Marr, D., 1982, *Vision*, San Francisco: W. H. Freeman.

Marten, D., 1993, European ATC harmonization and integration programme, *Journal of Navigation*, **46**(3), 326–335.

Maslow, A. H., 1976, *The Farther Reaches of Human Nature*, Harmondsworth: Penguin Books.

McAlindon, P. J., 1991, Aircraft traffic forecast and communications requirements in the year 2000, in Wise, J. A., Hopkin, V. D. and Smith, M. L. (Eds.), *Automation and Systems Issues in Air Traffic Control*, pp. 107–120, Berlin: Springer-Verlag, NATO ASI Series Vol. F73.

McCormick, E. J., 1957, *Human Engineering*, New York: McGraw-Hill.

McCormick, E. J. and Ilgen, D. R., 1987, *Industrial and Organizational Psychology*, London: Routledge.

McGuigan, F. J., 1989, *Experimental Psychology: Methods of Research*, Hemel Hempstead: Prentice Hall.

Megaw, E. D., 1991, Ergonomics: trends and influences, in Cooper, C. L. and Robertson, I. T. (Eds.), *International Review of Industrial and Organizational Psychology*, **6**, 109–148.

Meister, D., 1985, *Behavioural Analysis and Measurement Methods*, New York: Wiley.

Meister, D. and Rabideau, G. F., 1965, *Human Factors Evaluation in System Development*, New York: Wiley.

Melton, C. E., 1982, *Physiological Stress in Air Traffic Controllers: A Review*, Washington, DC: Federal Aviation Administration Report No. DOT/FAA/AM-82/17.

Melton, C. E. and Bartanowicz, R. S., 1986, *Biological Rhythms and Rotating Shift Work: Some Considerations for Air Traffic Controllers and Managers*, Washington, DC: Federal Aviation Administration Report No. DOT/FAA/AM-86/2.

Mies, J. M., Colmen, J. G. and Domenech, O., 1977, *Predicting Success of Applicants for Positions as Air Traffic Control Specialists in the Air Traffic Service*, Washington, DC: Education and Public Affairs Inc. for Federal Aviation Administration Contract DOT-FA75WA-3646.

Millar, R., Crute, V. and Hargie, O., 1992, *Professional Interviewing*, London: Routledge.

Mital, A. and Kumar, S. (Eds.), 1994, Fatigue, *Human Factors*, **36**(2), 195–349 (Special Issue).

Mogford, R., 1991, Mental models in air traffic control, in Wise, J. A., Hopkin, V. D. and Smith, M. L. (Eds.), *Automation and Systems Issues in Air Traffic Control*, pp. 235–242, Berlin: Springer-Verlag NATO ASI Series Vol. F73.

Mogford, R., 1994, Mental models and situation awareness in air traffic control, in Gilson, R. D., Garland, D. J. and Koonce, J. M. (Eds.), *Situational Awareness in Complex Systems*, pp. 199–207, Daytona Beach, FL: Embry-Riddle Aeronautical University Press.

Mohler, S., 1983, The human element in air traffic control: aeromedical aspects, problems and prescriptions, *Aviation, Space and Environmental Medicine*, **54**(6), 511–516.

Moray, N. (Ed.), 1979, *Mental Workload: Its Theory and Measurement*, New York: Plenum.

Moser, H. M., 1959, *The Evolution and Rationale of the ICAO Word Spelling Alphabet*, Bedford, MA: USAF Research and Development Command Report No. AFCRC-TN-59-54.

Muckler, F. A. and Seven, S. A., 1992, Selecting performance measures: "objective" versus "subjective" measurement, *Human Factors*, **34**, 441–456.

Muir, B. M., 1987, Trust between humans and machines, and the design of decision aids, *International Journal of Man–Machine System*, **27**, 527–539.

Mundra, A. D., 1989, *Ghosting: Potential Applications of a New Controller Automation Aid*, McLean, VA: Mitre Corporation Report MW-89W00030.

Murrell, K. F. H., 1965, *Ergonomics: Man in his Working Environment*, London: Chapman and Hall.

Myers, C. S. (Ed.), 1929, *Industrial Psychology*, London: Oxford University Press.

Myers, C. and Davids, K., 1993, Tacit skill and performance at work, *Applied Psychology: An International Review*, **42**(2), 117–137.

Myers, J. G., 1990, *Management Assessment: Implications for Development and Training*, Washington, DC: Federal Aviation Administration Report No. DOT/FAA/AM-90/2.

Myers, J. G. (Ed.), 1992, *A Longitudinal Examination of Applicants to the Air Traffic Control Supervisory Identification and Development Program*, Washington, DC: Federal Aviation Administration Report No. DOT/FAA/AM-92/16.

Nadler, E.D., DiSario, R., Mengert, P. and Sussman, E. D., 1990, *A Simulation Study of the Effects of Communication Delay on Air Traffic Control*, Washington, DC: Federal Aviation Administration Report No. DOT/FAA/CT-90/6.

Narborough-Hall, C. S., 1985, Recommendations for applying colour coding to air traffic control displays, *Displays*, **6**(3), 131–137.

Narborough-Hall, C. S., 1987, Automation implications for knowledge retention as a function of operator control responsibility, in Diaper, D. and Winder, R. (Eds.), *People and Computers 11*, pp. 269–282, Cambridge: Cambridge University Press.

Narborough-Hall, C. S. and Hopkin, V. D., 1988, Human factors contributions to air traffic control evaluations, in Megaw, E. D. (Ed.), *Contemporary Ergonomics 1988*, pp. 142–147, London: Taylor & Francis.

NASA, 1989, *Man–System Integration Standards*, Houston, TX: National Aeronautics and Space Administration NASA-STD-3000 (2 Vols.).

Nealey, S. M., Thornton, G. C., Maynard, W. S. and Lindell, M. K., 1975, *Defining Research Needs to Insure Continued Job Motivation of Air Traffic Controllers in Future Air Traffic Control Systems*, Seattle, WA: Battelle Memorial Institute for Federal Aviation Administration, Contract DOT/FA/74WAI-499.

Nielsen, J. (Ed.), 1990, *Designing User Interfaces for International Use*, Amsterdam: North-Holland.

Nolan, M. S., 1990, *Fundamentals of Air Traffic Control*, Belmont, CA: Wadsworth.

Norman, D. and Bobrow, D., 1975, Data limited and resource limited processing, *Cognitive Psychology*, **7**, 44–60.

Norman, D. A. and Draper, S. W., 1986, *User-Centered System Design: New Perspectives on Human Computer Interaction*, Hillsdale, NJ: Lawrence Erlbaum.

Norman, K. L., 1991, *The Psychology of Menu Selection: Designing Cognitive Control at the Human–Computer Interface*, Norwood, NJ: Ablex.

Nye, L. G. and Collins, W. E., 1991, *Some Personality Characteristics of Air Traffic Control Specialist Trainees: Interactions of Personality and Aptitude Test Scores with FAA Academy Success and Career Expectations*, Washington, DC: Federal Aviation Administration Report No. DOT/FAA/AM-91/8.

Nye, L. G., Schroeder, D. J. and Dollar, C. S., 1993, *Relationships of Type A Behavior with Biographical Characteristics and Training Performance of Air Traffic Controllers*, Washington, DC: Federal Aviation Administration Report No. DOT/FAA/AM-94/13.

Nyfield, G., 1991, Automation issues for the selection of controllers, in Wise, J. A., Hopkin, V. D. and Smith, M. L. (Eds.), *Automation and Systems Issues in Air Traffic Control*, pp. 453–459, Berlin: Springer-Verlag NATO ASI Series Vol. F73.

Nygren, T. E., 1991, Psychometric properties of subjective workload measurement techniques: implications for their use in the assessment of perceived mental workload, *Human Factors*, **33**(1), 17–34.

O'Hanlon, J. F., 1981, Boredom: practical consequences and a theory, *Acta Psychologica*, **49**, 53–82.

Older, H. J. and Cameron, B. J., 1972, *Human Factors Aspects of Air Traffic Control*, Washington, DC: National Aeronautics and Space Administration, NASA Report CR-1957.

Orr, N. W. and Hopkin, V. D., 1968, The role of the touch display in air traffic control, *The Controller*, **7**(4), 7–9.

Paap, K. P. and Roske-Hofstrand, R. J., 1988, Design of menus, in Helander, I. M. (Ed.), *Handbook of Human–Computer Interaction*, pp. 205–235, New York: Elsevier/North-Holland.

Parasuraman, R., Molloy, R. and Singh, I. L., 1993, Performance consequences of automation-induced "complacency", *International Journal of Aviation Psychology*, **3**, 1–24.

Parsons, H. M., 1972, *Man–Machine System Experiments*, Baltimore: Johns Hopkins Press.

Parsons, H. M., 1985, Automation and the individual: comprehensive and comparative views, *Human Factors*, **27**, 99–112.

Perrow, C., 1984, *Normal Accidents*, New York: Basic Books.

Perry, T. S. and Adam, J. A., 1991, Special Report: air traffic control, *IEEE Spectrum*, **28**(2), 22–36.

Pheasant, S., 1986, *Bodyspace: Anthropometry, Ergonomics and Design*, London: Taylor & Francis.

Phillips, M. D., Bashinski, H. S., Ammerman, H. L. and Fligg, C. M., 1991, A task-analytic approach to dialogue design, in Helander, I. M. (Ed.), *Handbook of Human–Computer Interaction*, pp. 835–857, Amsterdam: Elsevier/North-Holland.

Phillips, M. D. and Tisher, K., 1985, Operations concept formulation for next generation air traffic control systems, in Shackel, B. (Ed.), *Human–Computer Interaction: INTERACT '84*, pp. 895–900, New York: Elsevier/North-Holland.

Pickrel, E. W., 1984, Research contributions at the Office of Aviation Medicine (OAM), in Sells, S. B., Dailey, J. T. and Pickrel, E. W. (Eds.), *Selection of Air Traffic Controllers*, pp. 113–118, Washington, DC: Federal Aviation Administration Report No. FAA-AM-84-2.

Pitts, J., Kayten, P. and Zalenchak, J., 1993, The National Plan for Aviation Human Factors, in Wise, J. A., Hopkin, V. D. and Stager, P. (Eds.), *Verification and Validation of Complex Systems: Human Factors Issues*, pp. 529–540, Berlin: Springer-Verlag NATO ASI Series Vol. F110.

Pozesky, M. T. (Ed.), 1989, Special issue on air traffic control, *Proceedings of the IEEE*, **77**(11), 1603–1775.

Preece, J., Rogers, Y., Sharp, H., Benyon, D., Holland, S. and Carey, T., 1994, *Human Computer Interaction*, Wokingham, Berks: Addison-Wesley.

Prinzo, O. V. and Britton, T. W., 1993, *ATC/Pilot Voice Communications: A Survey of the Literature*, Washington, DC: Federal Aviation Administration Report No. DOT/FAA/AM-93/20.

Quintanilla, S. A. R. (Ed.), 1991, Work centrality and related work meanings, *European Work and Organizational Psychologist*, **1**, 2/3.

Rajecki, D. W., 1990, *Attitudes*, Oxford: W. H. Freeman.

Rantanen, E., 1994, The role of dynamic memory in air traffic controllers' situational awareness, in Gilson, R. D., Garland, D. J. and Koonce, J. M. (Eds.), *Situational Awareness in Complex Systems*, pp. 209–215, Daytona Beach, FL: Embry-Riddle Aeronautical University Press.

Rasmussen, J., 1986, *Information Processing and Human–Machine Interaction*, Amsterdam: North-Holland.

Ratcliffe, S., 1975, Principles of air traffic control, in Benoit, A. (Ed.), *A Survey of Modern Air Traffic Control* (Vol. 1), pp. 5–19, Paris: NATO AGARDograph No. 209.

Ratcliffe, S., 1991, Safe vertical separation of aircraft, *Journal of Navigation*, **44**(3), 386–391.

Ratcliffe, S., 1994, The air traffic capacity of two-dimensional airspace, *Journal of Navigation*, **47**(1), 33–40.

Raylor, A., 1993, *Air Traffic Control: Today and Tomorrow*, Shepperton: Ian Allan.

Reason, J. T., 1990, *Human Error*, Cambridge: Cambridge University Press.

Reason, J. T. 1993, The identification of latent organizational failures in complex systems, in Wise, J. A., Hopkin, V. D. and Stager, P. (Eds.), *Verification and Validation of Complex Systems: Human Factors Issues*, pp. 223–237, Berlin: Springer-Verlag, NATO ASI Series Vol. F110.

Reason, J. T. and Zapf, D. (Eds.), 1994, Errors, error detection and error recovery, *Applied Psychology: An International Review*, **43**(4), 427–584 (Special Issue).

Reber, A. S., 1985, *The Penguin Dictionary of Psychology*, Harmondsworth: Penguin Books.

Reynolds, L., 1994, Colour for air traffic control displays, *Displays*, **15**(4), 215–225.

Rinalducci, E. J., Smither, J. A.-A. and Bowers, C., 1993, The effects of age on vehicular control and other technological applications, in Wise, J. A., Hopkin, V. D. and Stager, P. (Eds.), *Verification and Validation of Complex Systems: Additional Human Factors Issues*, pp. 149–166, Daytona Beach, FL: Embry-Riddle Aeronautical University Press.

Robertson, A., Grossberg, M. and Richards, J., 1979, *Validation of Air Traffic Control Workload Models*, Cambridge, MA: Federal Aviation Administration Report No. DOT/FAA/RD-79-83.

Rock, D. B., Dailey, J. T., Ozur, H., Boone, J. O. and Pickrel, E. W., 1981, *Selection of Applicants for the Air Traffic Controller Occupation*, Washington, DC: Federal Aviation Administration Report No. DOT-FAA-AM-82-11.

Rodgers, M. D., 1993, *An Examination of the Operational Error Database for Air Route Traffic Control Centers*, Washington, DC: Federal Aviation Administration Report No. DOT/FAA/AM-93/22.

Rodgers, M. D. and Blanchard, R. E., 1993, *Accident Proneness: A Research Review*, Washington, DC: Federal Aviation Administration Report No. DOT/FAA/AM-93/9.

Rodgers, M. D. and Drechsler, G. K., 1993, *Conversion of the CTA, Inc. en route Operations Concepts Database into a Formal Sentence Outline Job Task Taxonomy*, Washington, DC: Federal Aviation Administration, Report No. DOT/FAA/AM-93/1.

Rodgers, M. D. and Duke, D. A., 1993, *SATORI: Situation Assessment Through the Re-creation of Incidents*, Washington, DC: Federal Aviation Administration Report No. DOT/FAA/AM-93/12.

Rodgers, M. D. and Duke, D. A., 1994, SATORI: situation assessment through the re-creation of incidents, in Gilson, R. D., Garland, D. J. and Koonce, J. M. (Eds.), *Situational Awareness in Complex Systems*, pp. 217–225, Daytona Beach, FL: Embry-Riddle Aeronautical University Press.

Rose, R. M., Jenkins, C. D. and Hurst, M. W., 1978, *Air Traffic Controller Health Change Study*, Washington, DC: Federal Aviation Administration Report No. FAA-AM-78-39.

Rosekind, M. R., Gander, P. H., Miller, D. L., Gregory, K. B., Smith, R. M., Weldon, K. J., Co, E. L., McNally, K. L. and Lebacqz, J. V., 1994, Fatigue in operational settings: examples from the aviation environment, *Human Factors*, **36**(2), 327–338.

Rouse, W. B., 1991, *Design for Success: A Human-Centered Approach to Designing Successful Products and Systems*, Chichester: Wiley.

Ryberg, P., 1993, Adaptive air traffic management, in Garland, D. J. and Wise, J. A. (Eds.), *Human Factors and Advanced Aviation Technologies*, pp. 67–72, Daytona Beach, FL: Embry-Riddle Aeronautical University Press.

Sackett, P. R., Burns, L. R. and Ryan, A. M., 1989, Coaching and practice effects in personnel selection, in Cooper, C. L. and Robertson, I. T. (Eds.), *International Review of Industrial and Organizational Psychology*, **4**, 145–183.

Saint-Exupéry, A. de, 1939, *Wind, Sand and Stars*, London: Heinemann.

Salas, E., Stout, R. J. and Cannon-Bowers, J. A., 1994, The role of shared mental models in developing shared situational awareness, in Gilson, R. D., Garland, D. J. and Koonce, J. M. (Eds.), *Situational Awareness in Complex Systems*, pp. 297–304, Daytona Beach, FL: Embry-Riddle Aeronautical University Press.

Salvendy, G. (Ed.), 1987, *Handbook of Human Factors*, New York: Wiley.

Sanders, A. F. and Roelofsma, P. H. M. P., 1993, Performance evaluation of human–

machine systems, in Wise, J. A., Hopkin, V. D. and Stager, P. (Eds.), *Verification and Validation of Complex Systems: Human Factors Issues*, pp. 315–332, Berlin: Springer-Verlag NATO ASI Series Vol. F110.

Sanders, M. S. and McCormick, E. J. (Eds.), 1993, *Human Factors in Engineering and Design*, New York: McGraw-Hill.

Sandiford, W. K., 1991, Meeting the ATC challenge through simulation, in Wise, J. A., Hopkin, V. D. and Smith, M. L. (Eds.), *Automation and Systems Issues in Air Traffic Control*, pp. 181–184, Berlin: Springer-Verlag NATO ASI Series Vol. F73.

Sarter, N. B. and Woods, D. D., 1991, Situation awareness: a critical but ill-defined phenomenon, *International Journal of Aviation Psychology*, **1**, 45–57.

Schonpflug, W., 1993, Applied psychology: newcomer with a long tradition, *Applied Psychology: An International Review*, **42**(1), 5–30.

Schroeder, D. J., Broach, D. and Young, W. C., 1993, *Contributions of Personality Measures to Predicting Success of Trainees in the Air Traffic Control Specialist Nonradar Screen Program*, Washington, DC: Federal Aviation Administration Report No. DOT/FAA/AM-93/4.

Schroeder, D. J., Touchstone, R. M., Stern, J. A., Stoliarov, N. and Thackray, R., 1994, *Maintaining Vigilance on a Simulated ATC Monitoring Task Across Repeated Sessions*, Washington, DC: Federal Aviation Administration Report No. DOT/FAA/AM-94/6.

Sears, A., 1991, Improving touchscreen keyboards: design issues and a comparison with other devices, *Interacting with Computers*, **3**(3), 253–269.

Sears, A., Plaisant, C. and Shneiderman, B., 1993, A new era for high precision touchscreens, in Hartson, R. and Hix, D. (Eds.), *Advances in Human–Computer Interaction*, pp. 1–33, Norwood, NJ: Ablex.

Sells, S. B., Dailey, J. T. and Pickrel, E. W., 1984, *Selection of Air Traffic Controllers*, Washington, DC: Federal Aviation Administration, Report No. DOT/FAA/AM-84/2.

Sen, T. and Boe, W. J., 1991, Confidence and accuracy in judgements using computer displayed information, *Behaviour and Information Technology*, **10**(1), 53–64.

Senders, J. W. and Moray, N. P., 1991, *Human Errors: Their Causes, Prediction, and Reduction*, Hillsdale, NJ: Lawrence Erlbaum.

Shepherd, W. T., 1993, Issues in aviation maintenance human factors, in Garland, D. J. and Wise, J. A. (Eds.), *Human Factors and Advanced Aviation Technologies*, pp. 207–212, Daytona Beach, FL: Embry-Riddle Aeronautical University Press.

Sherr, S. (Ed.), 1988, *Input Devices*, New York, Academic Press.

Shirom, A., 1989, Burnout in work organizations, in Cooper, C. L. and Robertson, I. T. (Eds.), *International Review of Industrial and Organizational Psychology*, **4**, 25–48.

Shneiderman, B, 1992, *Designing the User Interface*, New York: Addison-Wesley.

Shouksmith, G. and Burroughs, S., 1988, Job stress factors for New Zealand and Canadian air traffic controllers, *Applied Psychology: An International Review*, **37**(3), 263–270.

Simolunas, A. S. and Bashinski, H. S., 1991, Computerization and automation: upgrading the American air traffic control system, in Wise, J. A., Hopkin, V. D. and Smith, M. L. (Eds.), *Automation and Systems Issues in Air Traffic Control*, pp. 31–38, Berlin: Springer-Verlag, NATO ASI Series Vol. F73.

Simpson, C. A., McCauley, M. A., Roland, E. F., Ruth, J. C. and Williges, B. H., 1987, Speech controls and displays, in Salvendy, G. (Ed.), *Handbook of Human Factors*, pp. 1490–1525, Chichester: Wiley.

Sinaiko, H. W., (Ed.), 1961, *Selected Papers on Human Factors in the Design and Use of Control Systems*, New York: Dover.

Sinaiko, H. W. and Buckley, E. P., 1957, *Human Factors in the Design of Systems*, Washington, DC: US Naval Research Laboratory NRL Report No. 4996.

Smith, K. and Hancock, P. A., 1994, Situation awareness in adaptive, externally directed consciousness, in Gilson, R. D., Garland, D. J. and Koonce, J. M. (Eds.), *Situational Awareness in Complex Systems*, pp. 59–68, Daytona Beach, FL: Embry-Riddle Aeronautical University Press.

Smith, M., 1994, A theory of the validity of predictors in selection, *Journal of Occupational and Organizational Psychology*, **67**, 13–31.

Smith, M. and George, D., 1992, Selection methods, in Cooper, C. L. and Robertson, I. T. (Eds.), *International Review of Industrial and Organizational Psychology*, **7**, 55–97.

Smith, M. L., 1991, Adaptive training to accommodate automation in the air traffic control system, in Wise, J. A., Hopkin, V. D. and Smith M. L. (Eds.), *Automation and Systems Issues in Air Traffic Control*, pp. 481–495, Berlin: Springer-Verlag, NATO ASI Series Vol. F73.

Smith, P. A. and Wilson, J. R., 1993, Hypertext and expert systems: the possibilities for integration, *Interacting with Computers*, **5**(4), 371–384.

Smith, R. C., 1973, Comparison of the job attitudes of personnel in three air traffic control specialties, *Aerospace Medicine*, **44**(8), 918–927.

Smith, R. C., 1980, *Stress, Anxiety and the Air Traffic Control Specialist: Some Conclusions from a Decade of Research*, Washington, DC: Federal Aviation Administration Report No. DOT/FAA/AM-80/14.

Smolensky, M. W., 1993, Automation in the workspace: at what price productivity?, in Garland, D. J. and Wise, J. A. (Eds.), *Human Factors and Advanced Aviation Technologies*, pp. 165–174, Daytona Beach, FL: Embry-Riddle Aeronautical University Press.

Smolensky, M. W. and Hitchcock, L., 1993, When task demand is variable: verifying and validating mental workload in complex "real world" systems, in, Wise, J. A., Hopkin, V. D. and Stager, P. (Eds.), *Verification and Validation of Complex Systems: Human Factors Issues*, pp. 305–313, Berlin: Springer-Verlag NATO ASI Series Vol. F73.

Soede, M., Coeterier, J. F. and Stassen, H. G., 1971, Time analyses of the tasks of approach controllers in ATC, *Ergonomics*, **14**(5), 591–601.

Solso, R. L., 1991, *Cognitive Psychology*, Hemel Hempstead: Allyn and Bacon.

Sonnentag, S., Brodbeck, F. C., Heinbokel, T. and Stolte, W., 1994, Stressor–burnout relationship in software development teams, *Journal of Occupational and Organizational Psychology*, **67**, 327–341.

Spector, P. E., Brannick, M. T. and Coovert, M. D., 1989, Job analysis, in Cooper, C. L. and Robertson, I. T., *International Review of Industrial and Organizational Psychology*, **4**, 281–328.

Sperandio, J. C., 1971, Variation of operator's strategies and regulating effects on workload, *Ergonomics*, **14**(5), 571–577.

Sperandio, J. C., 1978, The regulation of working methods as a function of workload among air traffic controllers, *Ergonomics*, **21**, 193–202.

Stager, P., 1991, Error models for operating irregularities: implications for automation, in Wise, J. A., Hopkin, V. D. and Smith, M. L. (Eds.), *Automation and Systems Issues in Air Traffic Control*, pp. 321–338, Berlin: Springer-Verlag NATO ASI Series Vol. F73.

Stager, P., 1993, Validation in complex systems: behavioural issues, in Wise, J. A., Hopkin, V. D. and Stager, P. (Eds.), *Verification and Validation of Complex Systems: Human Factors Issues*, pp. 99–114, Berlin: Springer-Verlag NATO ASI Series Vol. F110.

Stager, P. and Hameluck, D., 1990, Ergonomics in air traffic control, *Ergonomics*, **33**(4), 493–499.

Stager, P. and Paine, T. G., 1980. Separation discrimination in a simulated air traffic control display, *Human Factors*, **22**(5), 631–636.

Stammers, R. C. and Bird, J. M., 1980, Controller evaluation of a touch input air traffic data system: an indelicate experiment, *Human Factors*, **22**, 582–589.

Stein, E. S., 1985, *Air Traffic Controller Workload: An Examination of Workload Probe*, Atlantic City, NJ: Federal Aviation Administration Report No. DOT/FAA/CT-TN 84/24.

Stein, E. S., 1988, *Air Traffic Controller Scanning and Eye Movements: A Literature Review*, Atlantic City, NJ: Federal Aviation Administration Report No. DOT/FAA/CT-TN 88/24.

Stein, E. S., 1992, *Air Traffic Control Visual Scanning*, Atlantic City, NJ: Federal Aviation Administration Report No. DOT/FAA/CT-TN 92/16.

Stein, E. S. and Garland, D. J., 1993a, *Air Traffic Control Working Memory: Considerations in Air Traffic Control Tactical Operations*, Washington, DC: Federal Aviation Administration Report No. DOT/FAA/CT-TN 93/37.

Stein, E. S. and Garland, D. J., 1993b, A practical guide for research planning, in Garland, D. J. and Wise, J. A. (Eds.), *Human Factors and Advanced Aviation Technologies*, pp. 233–243, Daytona Beach, FL: Embry-Riddle Aeronautical University Press.

Stern, J. A., Boyer, D. and Schroeder, D., 1994, Blink rate: a possible measure of fatigue, *Human Factors*, **36**(2), 285–297.

Stewart, T. F. M., 1991, Editorial, *Behaviour and Information Technology*, **10**(3), i–ii.

Still, A. and Costall, A. (Eds.), 1991, *Against Cognitivism: Alternative Foundations for Cognitive Psychology*, Hemel Hempstead: Harvester Wheatsheaf.

Stokes, A. F. and Kite, K., 1994, *Flight Stress: Stress, Fatigue and Performance in Aviation*, Aldershot, Hants: Avebury Technical.

Stokes, A. F., Wickens, C. D. and Kite, K. 1990, *Display Technology: Human Factors Concepts*, Warrendale, PA: Society of Automotive Engineers.

Stonor, T., 1991, Air traffic control today and tomorrow, *Journal of Navigation*, **44**(2), 143–151.

Streufert, S. and Nogami, G. Y., 1989, Cognitive style and complexity: implications for I/O psychology, in Cooper, C. L. and Robertson, I. T., (Eds.), *International Review of Industrial and Organizational Psychology*, **4**, 93–143.

Strube, M. J. (Ed.), 1991, *Type A Behaviour*, London: Sage.

Stuart-Hamilton, I., 1994, *The Psychology of Ageing: An Introduction*, London: Jessica Kingsley.

Stubler, W. F., Roth, E. M. and Mumaw, R. J., 1993, Integrating verification and validation with the design of complex man–machine systems, in Wise, J. A., Hopkin, V. D, and Stager, P. (Eds.), *Verification and Validation of Complex Systems: Human Factors Issues*, pp. 159–172, Berlin: Springer-Verlag, NATO ASI Series Vol. F110.

Talaga, J. and Beehr, T. A., 1989, Retirement: a psychological perspective, in Cooper, C. L. and Robertson, I. T. (Eds.), *International Review of Industrial and Organizational Psychology*, **4**, 185–211.

Tattersall, A. J., Farmer, E. W. and Belyavin, A. J., 1991, Stress and workload management in air traffic control, in Wise, J. A., Hopkin, V. D. and Smith, M. L. (Eds.), *Automation and Systems Issues in Air Traffic Control*, pp. 255–266, Berlin: Springer-Verlag NATO ASI Series Vol. F73.

Taylor, F. V., 1957, Psychology and the design of machines, *American Psychologist*, **12**(5), 249–258.

Taylor, F. V. and Garvey, W. D., 1959, The limitation of a "Procrustean" approach to the optimization of man–machine systems, *Ergonomics*, **2**(2), 187–194.

Taylor, F. W. 1911, *The Principles of Scientific Management*, New York: Harper and Row.

Thackray, R. I., 1980, *Boredom and Monotony as a Consequence of Automation: A Consideration of the Evidence Relating Boredom and Monotony to Stress*, Washington, DC: Federal Aviation Administration Report No. DOT/FAA/AM-80/1.

Thackray, R. I. and Touchstone, R. M., 1989, *A Comparison of Detection Efficiency on* an *Air Traffic Control Monitoring Task With and Without Computer Aiding*, Washington, DC: Federal Aviation Administration Report No. DOT/FAA/AM-89/1.

Travis, D., 1991, *Effective Colour Displays: Theory and Practice*, London: Academic Press.

Trites, D. K. and Cobb, B. B., 1964, Problems in air traffic management, III. Implications of training-entry age for training and job performance of air traffic control specialists, *Aerospace Medicine*, **35**, 336–340.

Tsang, P. S. and Vidulich, M. A., 1994, The roles of immediacy and redundancy in relative subjective workload assessment, *Human Factors*, **36**(3), 503–513.

Tufts College, 1949, *Handbook of Human Engineering Data for Design Engineers*, Medford, MA: Tufts College Institute for Applied Experimental Psychology, US Navy Special Devices Center, Tech. Dept. S.D.C. -199-1-1.

Turing, A. M., 1950, Computing machinery and intelligence, *Mind*, **59**, 433–460.

Turner, J. E., 1990, *Air Traffic Controller*, New York: ARCO.

VanDeventer, A. D., Taylor, D. K., Collins, W. E. and Boone, J. O., 1983, *Three Studies of Biographical Factors Associated with Success in Air Traffic Control Specialist Screening/Training at the FAA Academy*, Washington, DC: Federal Aviation Administration Report No. DOT/FAA/AM-83/6.

Van Laar, D. and Flavell, R., 1993, *Human Factors in Colour Displays: Principles for Effective Design*, Hemel Hempstead: Ellis Horwood.

Vidulich, M., Dominguez, C., Vogel, E. and McMillan, G., 1994, *Situation Awareness: Papers and Annotated Bibliography*, Wright Patterson AFB., OH: Air Force Material Command Report AL/CF-TR-1994-0085.

Volckers, U., 1991, Application of planning aids for air traffic control: design principles, solutions, results, in Wise, J. A., Hopkin, V. D. and Smith, M. L. (Eds.), *Automation and Systems Issues in Air Traffic Control*, pp. 169–172, Berlin: Springer-Verlag NATO ASI Series Vol. F73.

Vortac, O. U., Edwards, M. B., Fuller, D. K. and Manning, C. A., 1994, *Automation and Cognition in Air Traffic Control: an Empirical Investigation*, Washington, DC: Federal Aviation Administration Report No. DOT/FAA/AM-94/3

Vortac, O. U., Edwards, M. B., Jones, J. P., Manning, C. A. and Rotter, A. J., 1993, En Route Air Traffic Controllers' Use of Flight Progress Strips: A Graph Theoretic Analysis, *International Journal of Aviation Psychology*, **3**(4), 327–343.

Walker, W. J. V., 1993, UK airspace planning: the new ICAO airspace classification system, *Journal of Navigation*, **46**(3), 336–342.

Wallace, M. D. and Anderson, T. J., 1993, Approaches to interface design, *Interacting with Computers*, **5**(3), 259–278.

Walraven, J., 1985, The colours are not on the display: a survey of non-veridical perceptions that may turn up on a colour display, *Displays*, **6**, 35–42.

Warr, P., 1993, In what circumstances does job performance vary with age?, *European Work and Organizational Psychologist*, **3**(3), 237–249.

Westin, A. F., 1992, Two key factors that belong in a macroergonomics analysis of electronic monitoring: employee perceptions of fairness and the climate of organizational trust or distrust, *Applied Ergonomics*, **23**, 35–42.

Westrum, R., 1991, Automation, information and consciousness in air traffic control, in Wise, J. A., Hopkin, V. D. and Smith, M. L. (Eds.), *Automation and Systems Issues in Air Traffic Control*, pp. 367–380, Berlin: Springer-Verlag NATO ASI Series Vol. F73.

Westrum, R., 1993, Cultures with requisite imagination, in Wise, J. A., Hopkin, V. D. and Stager, P. (Eds.), *Verification and Validation of Complex Systems: Human Factors Issues*, pp. 401–416, Berlin: Springer-Verlag NATO ASI Series Vol. 110.

Westrum, R., 1994, Is there a role for a "test controller" in the development of new air traffic control equipment?, in Wise, J. A., Hopkin, V. D. and Garland, D. J. (Eds.),

Human Factors Certification of Advanced Aviation Technologies, pp. 221–228, Daytona Beach, FL: Embry-Riddle Aeronautical University Press.

Whitefield, A. and Hill, B., 1994, Comparative analysis of task analysis products, *Interacting With Computers*, **6**(3), 289–309.

Whitfield, D. and Jackson, A., 1982, The air traffic controller's "picture" as an example of a mental model, in Johannsen, G. and Rijnsdorp, J. E. (Eds.), *Analysis, Design and Evaluation of Man–Machine Systems*, pp. 45–52, New York: Pergamon.

Whitfield, D. and Stammers, R. B., 1978, The air traffic controller, in Singleton, W. T. (Ed.), *The Analysis of Practical Skills, Vol. 1: The Study of Real Skills*, pp. 209–232, Lancaster, UK: MTP Press.

Wickens, C., 1992, *Engineering Psychology and Human Performance*, New York: HarperCollins.

Widdel, H., Post, D. L., Grossman, J. D. and Walraven, J. (Eds.), 1991, *Colour in Electronic Displays*, London: Plenum Press.

Wiener, E. L., 1980, Midair collisions: the accidents, the systems and the realpolitik, *Human Factors*, **22**(5), 521–533.

Wiener, E. L., Kanki, B. G. and Helmreich, R. L. (Eds.), 1993, *Cockpit Resource Management*, San Diego, CA: Academic Press.

Wiener, E. L. and Nagel, D. C. (Eds.), 1988, *Human Factors in Aviation*, San Diego, CA: Academic Press.

Wieringa, P. A. and Stassen, H. G., 1993, Assessment of complexity, in Wise, J. A., Hopkin, V. D. and Stager, P. (Eds.), *Verification and Validation of Complex Systems: Human Factors Issues*, pp. 173–180, Berlin: Springer-Verlag NATO ASI Series Vol. F110.

Wierwille, W. W. and Eggemeier, F. T., 1993, Recommendations for mental workload measurement in a test and evaluation environment, *Human Factors*, **35**(2), 263–281.

Williams, J. E. D., 1990, Air navigation systems. Heading references 1909–1959, *Journal of Navigation*, **43**(1), 58–87.

Williams, K. W. (Ed.), 1994, *Summary Proceedings of the Joint Industry-FAA Conference on Development and Use of PC-based Aviation Training Devices*, Washington, DC: Federal Aviation Administration Report No. DOT/FAA/AM-94/25.

Wilpert, B. and Qvale, T., 1993, *Reliability and Safety in Hazardous Work Systems*, (Eds), Hove, East Sussex: Lawrence Erlbaum.

Wilson, J. R. and Corlett, E. N. (Eds.), 1994, *Evaluation of Human Work: A Practical Ergonomics Methodology*, London: Taylor & Francis.

Wilson, J. R. and Rutherford, A., 1990, Mental models: theory and application in human factors, *Human Factors*, **31**, 617–634.

Wing, H., 1991, Selecting for air traffic control: the state of the art, in Wise, J. A., Hopkin, V. D. and Smith, M. L. (Eds.), *Automation and Systems Issues in Air Traffic Control*, pp. 409–427, Berlin: Springer-Verlag NATO ASI Series Vol. F73.

Wing, J. and Manning, C. A. (Eds.), 1991, *Selection of Air Traffic Controllers: Complexity, Requirements, and Public Interest*, Washington, DC: Federal Aviation Administration Report No. DOT/FAA/AM-91/9.

Wise, J. A. and Debons, A. (Eds.), 1987, *Information Systems: Failure Analysis*, Berlin: Springer-Verlag, NATO Science Series Vol. F32.

Wise, J. A., Hopkin, V. D. and Garland, D. J. (Eds.), 1994, *Human Factors Certification of Advanced Aviation Technologies*, Daytona Beach, FL: Embry-Riddle Aeronautical University Press.

Wise, J. A., Hopkin, V. D. and Smith, M. L. (Eds.), 1991, *Automation and Systems Issues in Air Traffic Control*, Berlin: Springer-Verlag, NATO ASI Series Vol. F73.

Wise, J. A., Hopkin, V. D. and Stager, P. (Eds.), 1993, *Verification and Validation of Complex Systems: Human Factors Issues*, Berlin: Springer-Verlag NATO ASI Series Vol. F110.

Wise, J. A., Hopkin, V. D. and Stager, P. (Eds.), 1994, Verification and Validation of Complex Systems: Additional Human Factors Issues, Daytona Beach, FL: Embry-Riddle Aeronautical University Press.

Witt, L. A., 1994, *Perceptions of Organizational Support and Affectivity as Predictors of Job Satisfaction*, Washington, DC: Federal Aviation Administration Report No. DOT/FAA/AM-94/2.

Woods, D. D., 1991, The cognitive engineering of problem representations, in Weir, G. R. S. and Alty, J. L. (Eds.), *Human–Computer Interaction in Complex Systems*, pp. 169–188, New York: Academic Press.

Woods, D. D. and Sarter, N. B., 1993, Evaluating the impact of new technology on human–machine cooperation, in Wise, J. A., Hopkin, V. D. and Stager, P. (Eds.), *Verification and Validation of Complex Systems: Human Factors Issues*, pp. 133–158, Berlin: Springer-Verlag, NATO ASI Series Vol. F110.

Woodson, W. E., 1954, *Human Engineering Guide for Equipment Designers*, Berkeley, CA: University of California Press.

Wrightson, A. M., 1993, Re-usable models of complex requirements, in Wise, J. A., Hopkin, V. D. and Stager, P. (Eds.), *Verification and Validation Of Complex Systems: Additional Human Factors Issues*, pp. 25–30, Daytona Beach, FL: Embry-Riddle Aeronautical University Press.

Yoeli, M. (Ed.), 1991, *Formal Verification of Hardware Design*, Los Alamitos, CA: IEEE Computer Society Press.

Index